Springer Series in Language and Communication 17

Editor: W.J.M. Levelt

Springer Series in Language and Communication

Editor: W.J.M. Levelt

(continued after Index)

Language Awareness and Learning to Read

Edited by
John Downing and Renate Valtin

Springer-Verlag New York Berlin Heidelberg Tokyo

John Downing

Department of Psychological Foundations in Education
University of Victoria
Victoria, British Columbia, V8W 2Y2
Canada

Renate Valtin

Central Institute for Education Science and Curriculum Development
Freie Universität
1000 Berlin 33
Federal Republic of Germany

Series Editor

Professor Dr. Willem J. Levelt

Max-Planck-Projektgruppe für Psycholinguistik
Berg en Dalseweg 79, Nijmegen, The Netherlands

Library of Congress Cataloging in Publication Data
Main entry under title:
Language awareness and learning to read.
   (Springer series in language and communication ; 17)
   Includes bibliographies and index.
   I. Reading—Addresses, essays, lectures. 2. Children—Language—Addresses,
essays, lectures. I. Downing,
John A. II. Valtin, Renate. III. Series: Springer
series in language and communication ; v. 17.
LB1050.22.L36 1984   372.4   83-12464

With 13 Figures

Typeset by Publishers Service, Bozeman, Montana.
Printed and bound by R. R. Donnelley & Sons, Harrisonburg, Virginia.
Printed in the United States of America.

9 8 7 6 5 4 3 2 1

ISBN 0-387-90890-0   Springer-Verlag   New York   Berlin   Heidelberg   Tokyo
ISBN 3-540-90890-0   Springer-Verlag   Berlin   Heidelberg   New York   Tokyo

Preface

During the 1970s there was a rapid increase in interest in metacognition and metalinguistics. The impetus came from linguistics, psychology, and psycholinguistics. But with rather unusual rapidity the work from these scientific disciplines was taken over in education. This new direction in these various areas of academic study was taken simultaneously by several different investigators. Although they had varying emphases, their work sometimes appears to be overlapping; despite this, it has been rather difficult to find a consensus. This is reflected in the varying terminology used by these independent investigators—"linguistic awareness," "metacognition," "metalinguistic ability," "task awareness," "lexical awareness," and so on.

For educators these developments presented a glittering array of new ideas that promised to throw light on children's thinking processes in learning how to read. Many reading researchers and graduate students have perceived this as a new frontier for the development of theory and research. However, the variety of independent theoretical approaches and their accompanying terminologies has been somewhat confusing.

The aim of this book is to bring together in an integrative manner some of the chief authors in this field. We have attempted to get these authors to react to one another as well as to present their own theoretical position. Our book is designed especially for university and college teachers and their graduate students in language development, especially reading. We had in mind that the book should have special value for graduate-level seminars where the aim is to sharpen critical thinking on theory. Our stance is multidisciplinary and this book should be a useful basis for discussion in psychology and linguistics, as well as in education. Hopefully, this integration of these varying theoretical positions on the central issue of language awareness and its relation to learning to read will lead to a clarification of the problem and release new energy for its investigation.

Victoria, Canada   JOHN DOWNING
Berlin, FRG   RENATE VALTIN

Contents

Contributors

Ioanna Berthoud-Papandropoulou, University of Geneva, F.P.S.E., Geneva, Switzerland

John Downing, Department of Psychological Foundations in Education, University of Victoria, Victoria, B.C., V8W 2Y2, Canada

Linnea C. Ehri, Department of Education, University of California at Davis, Davis, CA, 95616, U.S.A.

Jerry L. Johns, College of Education, Northern Illinois University, De Kalb, IL, 60115, U.S.A.

George W. Ledger, Department of Psychology, Hollins College, Hollins, VA, 24020, U.S.A.

Ignatius G. Mattingly, Department of Linguistics, University of Connecticut; and the Haskins Laboratories, New Haven, CN, 06510, U.S.A.

Douglas Pidgeon, 80 Dorling Drive, Ewell, Surrey KT 17 3BH, England

Ellen Bouchard Ryan, Department of Psychiatry, McMaster University, Hamilton, Ontario L8S 4KJ, Canada

Anne Sinclair, University of Geneva, F.P.S.E., Geneva, Switzerland

Harry Singer, Department of Education, University of California at Riverside, Riverside, CA, 92502, U.S.A.

Renate Valtin, Department of Education, Freie Universität, Berlin, Federal Republic of Germany

Alan J. Watson, Sidney College of Advanced Education, St. George Institute of Education, Oatley 2223, Australia

CHAPTER 1

Introduction

John Downing and Renate Valtin

Language awareness, linguistic awareness, lexical awareness, and *metalinguistics* are terms that have appeared with increasing frequency in reading theory literature and research in the 1970s and 1980s. Our book is the latest in this area of development which borders on the disciplines of linguistics, psycholinguistics, experimental and child psychology, and education. The varied terminology reflects the differing focuses of these several disciplines, but one can detect two consistent threads running through these varied works. One thread has a linguistic or psycholinguistic orientation with Chomsky (1957, 1965) and Chomsky and Halle (1968) as the references, and the other thread is psychological or psycho-educational.

The psycholinguistic thread is best represented by an earlier paper by Mattingly (1972) in the volume *Language by Ear and by Eye* (Kavanagh & Mattingly, 1972) and Mattingly's chapter in this text. Underpinning this work is the active support of the (American) National Institute of Child Health and Human Development for basic and applied research into communicating by language, which has led to such publications as *Communicating by Language: The Reading Process* (Kavanagh, 1968), *Speech and Language in the Laboratory, School and Clinic* (Kavanagh & Strange, 1978), and *Orthography, Reading, and Dyslexia* (Kavanagh & Venezky, 1980).

The psychological or psycho-educational thread itself encompasses several fairly distinct threads of work. Possibly recognition for the earliest steps in this direction should be accorded to the Russian psychologists and educators who studied this problem. Vygotsky's (1934) *Thought and Language* stimulated psychologists in his own country, Russia, which resulted in Russia's early entry into this field. One can also discern the influence of Vygotsky's thinking (though later) on language awareness research and theory in other countries. The Russian work is represented by such authors as Egorov (1953), Elkonin (1973a, 1973b), and Luria (1946). Another strand of psycho-educational research may be noted in the works of Clay (1972), Downing (1970, 1971-1972), and Reid (1966). A German study by Bosch (1937) about the objectivation of language (Vergegenständlichung) as a prerequisite for learning to read also deserves to be mentioned in this strand, though it has not been referred to in the Anglo-American literature.

Yet another strand is that represented by such works as the outcome of the Nijmegen Max-Planck project on the child's conception of language (Sinclair, Jarvella & Levelt, 1978), the book by Ferreiro and Teberosky (1979), and such articles as that of Papandropoulou and Sinclair (1974). In this last-mentioned strand the influence of Piaget seems to be considerable.

The research of A. M. Liberman and I. Y. Liberman and their associates at Haskins Laboratories might have been included in our psycholinguistic thread (especially since Mattingly is one of those associates), but the definitive linguistic and acoustic research of the Liberman group is drawn upon by all threads and most strands referred to above.

In June 1979 an attempt was made to bring these various threads and strands together at the International Research Seminar on Linguistic Awareness and Learning to Read at the University of Victoria in Canada. This book is *not* the proceedings of that seminar. The seminar did not consist of paper presentations. Instead, all of the papers were printed and circulated in advance so that the seminar could be devoted to discussion. In this way, all of the positions of the participants could be further developed by the seminar. This is what we present in our book. Some of the seminar participants were invited to contribute a chapter to our book. Their chapters are their positions *after* the experience of the seminar. In addition, some other authors who have made prominent contributions to this area of research have been invited to join our team.

Three main criteria were used in choosing the authors and the topics of this book. First, the topic should represent a broad range of aspects of language awareness, relating both to featural and functional aspects; second, a link should be drawn between theoretical insights and practical consequences for methods of learning to read. Last, but not least, the authors represent a wide range of continents and countries, thus allowing a review of research studies from different countries. While most articles refer to reading, some others also present ideas about spelling, since both processes have many subskills in common.

Mattingly, in his chapter "Reading, Linguistic Awareness and Language Acquisition," presents a slightly altered version of his paper that served as the keynote address at the above-mentioned International Seminar in Victoria, Canada. He refines and elaborates his earlier concept of linguistic awareness, basing his linguistic and psycholinguistic assumptions largely on Chomsky's transformational grammar. In clarifying the relationship of reading to speaking and listening, he discusses two types of evidence. First, he describes practical orthographies and their relationship to linguistic structure. He suggests that practical orthographies are not phonetic but that they convey the lexical content of a sentence by transcribing words morphemically. Mattingly proposes that different orthographies require different degrees of linguistic awareness. Second, he presents some experimental evidence suggesting that orthographic information in reading is phonetically recoded and stored in short-term memory used for spoken language, and that phonetic recoding has no connection with the character of the orthography since it occurs with logographies and alphabetic orthographies. Finally, he offers some hypotheses concerning the parallels between reading and language acquisition and proposes that differences in children's readiness and ability to learn to read are related to different patterns of language acquisition.

Downing, in his chapter "Task Awareness in the Development of Reading Skill," reviews evidence that reading belongs to the class of behavior that psychologists call "skill." He suggests that since reading is a member of the skill category, psychological research findings on skill learning in general can be applied to learning how to read in particular. One such finding is that the initial step in learning a skill is to conjure up a conception of the task to be accomplished in performing the skill. This is task awareness, also referred to as the "cognitive phase" in skill development. Downing proposes a "cognitive clarity theory" of learning to read. Briefly, this theory states that the development of reading skill progresses through the learner's increasing awareness and comprehension of the functions and linguistic features of speech and writing. Downing outlines a variety of metacognitive and metalinguistic abilities that are involved in task awareness in reading. He differentiates two types of linguistic concepts: functional concepts regarding the purposes of reading and writing and featural concepts regarding those features of spoken language that are to be represented by written symbols. He presents a comprehensive review of research studies in this domain and shows that there is an important relationship between metacognitive awareness of the reading task and success or failure in learning to read. In the next step, results of a new test, the Linguistic Awareness in Reading Readiness (LARR) test, and its predictive value for reading achievement are presented. Finally, Downing discusses the nature of the relationship between metacognitive development and learning to read and argues in favor of an interaction hypothesis.

Downing points out in his chapter that there is a lack of studies concerned with the awareness of functions of both oral and written language, while there is an abundance of studies related to the awareness of features of language. Johns, in his chapter "Students' Perceptions of Reading: Insights From Research and Pedagogical Implications," addresses the question of how students view the task of reading. After a short overview of the present research on this topic, he presents data of his own large-scale investigation, interviewing pupils in the United States about their knowledge of what reading is, and what one has to do and to learn when reading. He discusses studies about the relationship between students' perception of reading and their reading ability and provides practical suggestions for helping students develop a perception of reading with an emphasis on meaning. Research and instructional activities that have implications for disabled readers are also presented.

The next four chapters discuss children's developing awareness of features of language. In their chapter "Children's Thinking About Language and Their Acquisition of Literacy," Sinclair and Berthoud-Papandropoulou examine different aspects of children's knowledge about language: about language as a general object of thought, about its spoken and written form, and about properties and characteristics of the written system. The authors argue that—since our alphabetic writing system represents aspects of the sounds—learners have to construct new conceptions about the relationship between words and meaning in written texts. Based on research studies mainly from French- and Spanish-speaking countries, they show that the ability to grasp the alphabetic principle is preceded by a slow development of various conceptual substages. These substages may be characterized as to the amount of dissociation different formal properties of oral and written language have from their meaning. The authors present detailed data concerning parallels in the children's thinking about oral and written language that refer to different aspects: the relationship between signifier and signified, linguistic units, formal properties of verbal signifiers, and links between sounds and letters. While they generally suggest a link between this type of knowledge and cognitive development, they also hypothesize that the characteristics of the written text—with its material permanence and discrete properties—may in general stimulate and guide children's thinking about language.

The interrelationship between language awareness, learning to read, and cognitive development has been further explored by Watson in his chapter "Cognitive Development and Units of Print in Early Reading." His concern is the understanding of units of print (letters, words, and sentences). He argues that the appreciation of these units is a demanding cognitive task that requires a high mental flexibility. In order to explain the relationship between the knowledge of this aspect of language awareness and learning to read, the author discusses three hypotheses: a language acquisition hypothesis (as suggested by Mattingly), a print experience view developed on the basis of information processing theory, and a conceptual reasoning perspective derived from cognitive developmental theory. Watson prefers this latter hypothesis and presents an empirical study for its validation. He used a longitudinal method to assess the effects of operativity, oral language, and reading progress on a measure of linguistic awareness—the units of print test. His results, mainly derived from path analysis, are consistent with a with a cognitive developmental theory of print awareness which suggests that gaining knowledge of the units of print is the result of a productive match of the child's developmental status and the instructional methods. Watson points to further studies that indicate the relevance of cognitive developmental factors for the explanation of language awareness and learning to read. Finally, some current issues in research and theory which are explained by a cognitive developmental view are considered, as are practical implications of the findings.

The possible explanation mentioned by Sinclair and Berthoud-Papandropoulou that characteristics of the written text may guide children's thinking about language is further investigated by Ehri. In her chapter "How Orthography Alters Spoken Language Competencies in Children Learning to Read and Spell," Ehri presents ample theoretical and empirical evidence to show how experiences with printed language influence children's competencies with speech. She claims that when children learn to read and spell, they acquire a visual representational system which enables them to see what they hear and say. Printed spellings provide a view of words in speech, and letters provide a view of phonetic units within speech. The process of learning to read and spell teaches the child about the lexical composition of sentences and the phonetic composition of words. Ehri presents some of her studies demonstrating that the presence of spelling in memory influences how speakers perceive words in spoken language tasks, how well they remember words, and how spelling may shape pronunciation. A comparison of novice and mature spellers revealed that beginners do not detect and represent as many sound segments as mature spellers. Besides this impact on the conceptualization of the sound level of speech, experience with print may also influence children's lexical awareness, and Ehri presents some data to substantiate this view. She further argues that having a visual model of speech may facilitate metalinguistic processes. Having a concrete picture of language in one's mind makes it easier to treat language as an object, inspect its form, and manipulate structural units.

While Ehri mainly refers to lexical and phonological awareness, Ryan and Ledger in their chapter are concerned with "Learning to Attend to Sentence Structure: Links Between Metalinguistic Development and Reading." They describe young children's developing ability to focus attention on sentence structure and the role of sentence-processing subskills in learning to read effectively. First, the data concerning children's evaluations of sentence acceptability and other structure-based sentence tasks are reviewed. During the years from age five to eight, children show major performance changes on tasks requiring them to override their natural tendency to deal with sentences in terms of meaning and to attend to the structural features of language. Then the authors discuss the value of training studies for selecting among alternative cognitive interpretations of metalinguistic development. They refer to investigations of the links between these sentence-processing subskills and learning to read. Although moderate relationships have been clearly established, the direction of causation remains a matter of dispute. Finally, the authors suggest a model of metalinguistic development which is based on two underlying dimensions: analyzed linguistic knowledge and control (ability to direct attention deliberately to information relevant to problem solution). This model provides a framework for understanding the cognitive foundations of both reading and grammatical awareness as well as a basis for classifying the diverse array of tasks used to assess grammatical awareness.

The next two chapters are more directly related to issues of linguistic awareness and their implication to reading instruction. Pidgeon, in his chapter "Theory and Practice in Learning to Read," is concerned with theoretical aspects of the learning-to-read process and classroom practices, dealing mainly with the early stages of reading and the closely related prereading activities. First, he presents a brief account of learning to read methods mainly in the Anglo-American countries and describes their underlying theoretical framework. He outlines how various conceptualizations of the difference between oral and written language influence the theories of learning to read, and shows in what way the concept of linguistic awareness had an impact on notions of theories and methods of learning to read. Pidgeon then presents results of a study that was carried out in British infant schools. The aim was to observe the practices employed in teaching beginners to read and to find out about the knowledge that teachers possess or—very often—lack to justify their classroom activities. He finally proposes a framework for a learning-to-read theory and outlines the pedagogical consequences. This program is built on principles of the concept of mastery learning and emphasizes pupils' linguistic awareness, conceived of as the understanding of how language sounds are structured to form words and how this structure can be represented on paper.

While Pidgeon mainly refers to the early stages of reading, Singer is concerned with both beginning and skilled readers in his chapter "Learning to Read and Skilled Reading: Multiple Systems Interacting Within and Between the Reader and the Text." Based on the substrata theory of reading that views reading as a multidimensional process—including perceptual, linguistic, cognitive, motivational, affective, and physiological components—Singer argues that any theory of reading must be inadequate if one of these aspects is overemphasized. He criticizes Mattingly's concept of reading because it draws solely upon linguistics. Singer then provides an overview on the multiple subskills the child has to master in learning to read and describes conditions for learning to read and for teaching reading. Referring to this framework, he then analyzes Mattingly's concept of linguistic awareness and investigates two of his claims. The first claim refers to the role of phonological processes in beginning and mature readers. Singer argues that phonological processes might indeed be bypassed when deaf subjects learn to read, but that there is no evidence that beginning readers of an alphabetic script bypass phonological processes. Furthermore, there seems to be no evidence for the other claim inherent in Mattingly's position that there is a critical age for learning to read. Finally, Singer discusses the nature of the relationship between linguistic awareness and its pedagogical implications.

Since the authors in this book connect different meanings with the term "language awareness," Valtin, in her two chapters with which this book is brought to a close, tries to analyze this rather vague and global concept into its conceptual components. Based on newer German research, she outlines developmental stages of language awareness and presents as a theoretical framework Leontev's (1973, 1975) model of speech that allows a differentiation between various forms of awareness as well as a suggestion as to their developmental order and their relation to the learning of written language. This model claims that the child acquires the linguistic notions of word and phoneme mainly through instruction in reading and writing (spelling) in school. Some research studies are presented that show how the preschool child's everyday concept of a word undergoes a transformation under the influence of reading instruction in school, in the sense that a multiple-core-concept of a word is narrowed to a concept that is mainly oriented to visual strategies. After a review of Anglo-American studies on phonemic segmentation, some newer European studies are presented. They provide evidence that phonemic synthesis and analysis develop as a function of various factors: general cognitive level, preschool reading ability, performance level at the beginning of school, and length of reading instruction. Though no direct experimental evidence is available, the results are consistent with the view that phonemic segmentation develops largely as a consequence of learning to read and that the two interact with each other. Some open theoretical questions regarding the roots of the child's knowledge of how to segment speech and how to classify speech sounds are then discussed.

After considering featural concepts of language, Valtin goes on to discuss the child's understanding of functions of oral language. Evidence is presented that preschool children are implicitly aware of the communicative function of language. Moreover, as one study of metacommunicative behavior shows, preschool children are able to make judgments and verbal statements about inappropriate communication behavior. Finally, some consequences regarding methods of assessing aspects of language awareness are presented, and pedagogical consequences regarding different stages of metalinguistic knowledge are outlined. An attempt is made throughout the chapter to synthesize some ideas of the preceding chapters and to relate them to newer theoretical and empirical studies primarily from European countries.

CHAPTER 2

Reading, Linguistic Awareness, and Language Acquisition

Ignatius G. Mattingly

Introduction

Most of us who speculate about the reading process begin by considering the nature of the relationship between listening to speech and reading text. Are these two mental processes essentially the same, apart from a difference in input modality? Or are they essentially quite different, despite their shared linguistic character? My view is that reading, though closely related to listening, is different from it in some very crucial respects.

In an earlier paper (Mattingly, 1972), I attempted to characterize the difference in terms of "primary" and "secondary" linguistic activity. I suggested that, while the primary linguistic activities of speaking and listening are natural in all normal human beings, secondary linguistic activities, such as versification and reading, are parasitic on these primary activities, and require "linguistic awareness," a specially cultivated metalinguistic consciousness of certain aspects of primary linguistic activity. I still believe this distinction to be a valid one, but I now think that linguistic awareness is not a matter of consciousness, but of access. This access is probably largely unconscious, but the degree of consciousness is not very relevant. Moreover, what the linguistically aware person has access to is not his linguistic activity—the processes by which he actually produces and understands sentences—but rather his knowledge of the grammatical structure of sentences. Finally, I would not now wish to imply that secondary activity is less "natural" than primary linguistic activity. I will argue, in fact, that reading involves not only the mechanisms of speech understanding but also those of language acquisition, and that it is just as natural, and in a sense more "linguistic," than listening to speech.

It may perhaps disarm criticism to some degree if, before proceeding further, I distinguish two modes of mental activity that might conceivably be regarded as reading. In the first mode, which might be called "analytic" reading, the reader identifies written words in a sentence as corresponding to specific items in his mental lexicon and makes a grammatical analysis, as a result of which he may be said to understand the sentence. In the second mode, which might be called "impressionistic" reading, the reader tries to guess the meaning of the text just by looking at the words, without making specific lexical identifications and without making a grammatical analysis. This mode of reading relies on the fact that a written word, just because it is a familiar orthographic pattern, and not because it corresponds to a lexical item, is capable of evoking a rich network of semantic associations.

In the following, I am concerned almost entirely with analytic reading, justifiably, I feel. It may well be that, relying on the semantic associations of orthographic patterns and on *a priori* knowledge, a reasonably intelligent impressionistic reader can get the general sense of a text. Analytic reading may be slower and more laborious than impressionistic reading; it would not be surprising if the evocation of semantic associations by a familiar written word were shown to occur much more rapidly than the identification of a word as a specific lexical item. Analytic reading may even be a relatively rare act on the part of a skilled reader; depending on the nature of the text and his motivation in reading it, he may be reading impressionistically most of the time. Yet I believe that, useful as it may be to be able to read impressionistically, a person is not a reader if he cannot read a sentence anaytically when it is really essential for his understanding of a text to do so.

Some Linguistic and Psycholinguistic Assumptions

According to the generative linguists, the ideal speaker-hearer's knowledge about the structure of his language (his linguistic competence) is mentally represented in a grammar. The grammar may be viewed as a device for specifying the linguistically relevant aspects of any and all sentences in the language. It consists of syntactic, phonological, and semantic components, each of which is a set of ordered rules; and a lexical component, each entry in which specifies the peculiar syntactic, phonological, and semantic properties of a word in the language. The rules of the syntactic component "generate," i.e., derive, the phrase marker that represents the syntactic structure of a sentence. Rules of lexical insertion (also part of the syntactic component) relate the words of the sentence to lexical entries. The rules of the phonological component generate the phonetic representation—the intended or perceived pronunciation of the sentence—given the phrase marker and the lexically specified phonological properties of each word. Analogously, the rules of the semantic component generate the semantic representation—the meaning of the sentence—given the phrase marker and the lexically specified semantic properties of each word. The phrase marker, the lexical content of the sentence, and the phonetic and semantic representations constitute the "surface-structure" description of the sentence (Chomsky, 1975, 1977).

An actual speaker-hearer's "grammatical knowledge" is no doubt very imperfect. It is also tacit knowledge: The speaker-hearer knows the grammar of his language, but need not "know that he knows" it, or be able to formulate it coherently. Yet grammatical knowledge is accessible, in the sense that the speaker-hearer has intuitions about grammaticality. He is able to say whether a certain phonetic contrast is distinctive in his language, whether a certain syntactic pattern is acceptable, or whether a certain sentence is meaningful. The validity of these intuitions is corroborated by the success of linguists in reconstructing descriptively adequate grammars. But there are limitations on the scope of grammatical knowledge. The speaker-hearer has very limited intuitions, for example, about the acoustic properties of the speech signal that can be shown to determine his phonetic perceptions (Liberman, Cooper, Shankweiler & Studdert-Kennedy, 1967). Accordingly, the grammar has nothing to say about the complex relationships between the phonetic representation of a sentence and its acoustic realization.

A child acquiring the grammar of his native language is rather in the position of a linguist (Chomsky, 1965). Given a theory of language specifying the structural properties that all grammars share, and data as to correspondences between sound and meaning, he proceeds to construct the lexicon and the grammatical rules. The child's position is different from that of the linguist mainly in that his general theory of language is innately given, and superior to any general theory so far explicitly formulated by linguists. But having a task similar to the linguist's, he must have psychological mechanisms for doing what linguists do: making hypotheses about rules and about the content of lexical entries, constructing hypothetical phonetic and semantic representations, and comparing them with the available data.

The relationship of grammatical knowledge to actual speaking and listening in real time appears to be rather indirect. The task of the speaker is to determine (and realize articulatorily) the phonetic representation of a sentence, starting with information (his knowledge, motives, and intentions) that most directly constrains the semantic representation; the task of the listener is to determine the semantic representation, starting with information (the auditory properties of the acoustic signal) that most directly constrains the phonetic representation. It is rather unlikely that either speaker or listener obtains the required representation by generating it, or that either obtains intermediate representations by applying generative rules in reverse order. Instead, various analytic mechanisms— mechanisms for speech perception, phonological analysis, lexical search, and semantic analysis—seem to be in operation. (See the discussion of these questions in Fodor, Bever & Garrett, 1974, Chap. VI.) These performance mechanisms are heuristic in character, apparently using pragmatic, obviously fallible strategies. For example, a possible parsing strategy for English appears to be: Assume that the elements of any sequence of the form NP V(NP) are the subject, verb, and object of the same clause (Fodor et al., 1974, p. 345). A possible strategy for lexical search, if it is assumed that for each lexical entry all possible phonetic transcriptions are listed, would be to compare the input phonetic string with each stored transcription until a match is found. What the strategies actually are is, of course, one of the questions that psycholinguists are trying to answer.

While it is useful to postulate the existence of specific analytic mechanisms with particular functions, it is probably incorrect to envision these mechanisms as a series of modules, the output of each being the input to the next. To account for our perceptions, such a model would require provisions for feedback of information from later mechanisms to earlier ones, and the more feedback is assumed, the more arbitrary the functional separation of the different mechanisms appears. It seems more prudent to make the weaker assumption: that the analytic mechanisms somehow collaborate in concurrently reconstructing the various parts of the surface-structure description, given an input that is itself of mixed character. Thus, the reconstruction of the phonetic representation in listening is based not only on acoustic information but also on semantic presuppositions and hypotheses; and the semantic, syntactic, and lexical mechanisms play a part as well as the speech perception mechanism. In the process, other parts of the surface-structure description are partially determined as well.

The sets of strategies used by the various mechanisms might be referred to as "performance knowledge." Performance knowledge, unlike grammatical knowledge, seems to be rather inaccessible. The speaker-hearer has relatively few intuitions about performance. What is known about performance mechanisms—the speech perception mechanism or the parsing mechanism, for example—has therefore been learned, in the main, by experimental inference rather than by linguistic analysis.

Though performance knowledge is not the same thing as grammatical knowledge, the two must somehow be related. Which strategies it is appropriate to try in what order obviously depends upon the grammar of the language, even though the performance mechanisms are not doing grammatical derivations. It has been suggested that the capacity to acquire language includes the ability to compute the optimal set of analytic strategies for a given grammar (Fodor et al., 1974, p. 372). Thus, as grammatical knowledge develops during language acquisition, performance knowledge would increase as well.

How much must an actual speaker-hearer know about the grammar of his language to ensure a degree of performance knowledge sufficient for ordinary speaking and understanding? Perhaps relatively little, in comparison with what the ideal speaker-hearer of linguistic theory knows. It is quite believable, for example, that a person might have parsing strategies that could cope, much of the time, with passive constructions, without having complete grammatical knowledge of the rules for generating passive sentences.

To put the matter somewhat differently, the grammatical knowledge a language learner is potentially capable of acquiring far exceeds the functional requirements of performance. But if this is so, we should not find it surprising that some speaker-hearers continue acquiring the grammar of their language indefinitely, while others essentially abandon language acquisition once the performance mechanisms are adequately equipped for the purposes of ordinary communication. To suggest the existence of individual differences in language acquisition, in grammatical knowledge, and in linguistic performance is in no way to deny Chomsky's claim that all human beings have a specific innate linguistic capacity. Obviously, the actual linguistic development of any individual may be determined not only by this capacity but also by many nonlinguistic factors.

If one takes seriously the conception of the infant language learner as a linguist, one might suppose that, during the period of active language acquisition, grammatical knowledge would be highly accessible (access, to repeat, does not imply consciousness). But if grammatical knowledge is not directly used in linguistic performance, it is plausible that after language acquisition has ceased to be a major preoccupation, grammatical knowledge should tend to become less accessible. (How conscious the child is of what is accessible to him is, of course, a further question.)

Orthography and Reading

To clarify the relationship of reading to speaking and listening, I will discuss two kinds of evidence. The first kind of evidence derives from consideration of practical orthographies and their relationship to linguistic structure. The second kind is experimental: Performance in tasks that are similar to reading can be compared with performance in tasks involving production or perception of spoken language.

What aspects of an utterance do practical orthographies transcribe? The apparent heterogeneity of orthographies might suggest that there is no one answer to this question; the traditional classification of orthographies into logographic, syllabary, and alphabetic types seems to imply that each type transcribes sentences in a different way. Yet if there is not some clear sense in which all orthographic types are alike, the unappealing possibility that a separate account of the reading process must be given for each type will have to be entertained.

I begin with the assertion that a practical orthography is always a linguistic transcription. If this point seems overly obvious, consider that nonlinguistic orthographies are quite conceivable. One might imagine, for example, an orthography based on some physical description of speech. But though it is, of course, quite possible to display the acoustic waveform of an utterance, or its spectral composition, or its pattern of articulatory movement, no such physical orthographies exist. The reason for their absence is not simply that other orthographic modes were already established by the time such displays became technically feasible, but rather, I suggest, that the information in such displays is useless to the reader, because the physical attributes of an utterance encode its linguistic structure (Liberman et al., 1967) and the required perceptual decoding mechanisms are available not to the eye but only to the ear. Hence, the quite serious attempts that have been made to teach the deaf to read spectrograms have had limited success (Potter, Kopp & Green, 1947). Even though, in the course of understanding a spoken utterance, the listener converts the acoustic waveform to a spectral representation and (at least in the view of some researchers) recovers the corresponding articulatory gestures, displays of such information do not seem to be what a reader requires.

Again, one can imagine another sort of nonlinguistic orthography: one that transcribed the meanings of utterances. A truly semantic orthography would have several highly desirable properties: the reader (or writer) really could go directly to (or from) meaning; all speakers of a given language would agree on how a particular text could be read aloud in that language; yet the text could be read by literate speakers of any language. Unfortunately, no such general-purpose semantic orthography exists, and it is more than doubtful that one is feasible. The nearest approaches are the notational systems of logic, mathematics, and the sciences, where the semantic domain of the orthography is precisely and narrowly specified; or the ideography now often used in signs in public places, where the possible set of messages is arbitrarily limited. The picture writing of the American Indians, although not so clearly limited, can hardly be said to transcribe the meanings of particular utterances. But no one understands how to design a general-purpose semantic writing system that would not be hopelessly ambiguous and nonspecific.

At this point, it might be suggested that I have overlooked some apparent counterexamples. Wilkins (1668) and later Bell (1867) proposed writing systems in which the symbols depict appropriate vocal-tract configurations; are not these systems, which certainly might have been adopted had historical circumstances been different, physical orthographies? And is not Chinese writing, in which many characters or character elements suggest meanings, an instance of a semantic orthography? But of course in both cases, the symbols of the orthography actually stand for linguistic units—phonemes in the one case, morphemes in the other —and the most that can be said is that the physical or semantic aspects of the design of a symbol help the reader to learn and remember the particular linguistic unit it stands for. The symbols of Bell and Wilkins cannot capture the dynamic behavior of the vocal tract in an actual utterance, and it is necessary to know Chinese in order to read Chinese writing.

But it does not suffice to observe merely that orthographies are necessarily linguistic, for there are further limitations. The only syntactic information that orthographies supply about a sentence are the clause and sentence boundaries, indicated by capitalization and punctuation, and the sequence of lexical items. The reader must somehow infer the syntactic structure from the word order. (It is interesting that while the reader clearly needs syntactic information, no one has suggested that the beginning reader could be aided by an initial teaching orthography that provided syntactic information explicitly, in the form, say, of tree diagrams).

Moreover, practical orthographies transcribe lexical items morphemically. In the lexical component of the grammar, phonological information about the word appears in a morphophonemic representation (Chomsky & Halle, 1968). For example, in the lexicon of English, the words *heal, health, healthy* have the representations /hēl/, /hēl+θ/, /hēl+θ+y/; /hēl/, /θ/ and /y/ being morphemes. (The phonological symbols used here are to be taken as convenient abbreviations for sets of distinctive-feature values, and "+" indicates a morpheme boundary). A morpheme has semantic as well as phonological value, but of course the semantic value of a word is not necessarily predictable from the semantic values of its component morphemes.

It is obviously possible to transcribe this representation in a number of equivalent ways: using a distinct symbol for each morpheme (the logographic mode), or for each syllable (the syllabary mode) or for each morphophoneme (the alphabetic mode), or even for each distinctive-feature value (too cumbersome an approach for a practical orthography). But practical orthographies are morphemic in the sense that, with limited exceptions, a morpheme is transcribed in the same way wherever it occurs.

The morphemic character of logographic systems, such as the one used for Chinese and borrowed for Japanese, is quite obvious, since a separate character is used for each morpheme of a word. (It is perhaps worth insisting on the distinction between words and morphemes in this connection. Modern Chinese has many one-morpheme words but also a great many compounds. Thus, a reader who knows the characters for only a few thousand common morphemes is able to read many thousands of words.)

Since a morpheme is a pairing of semantic and phonological values, it is not surprising to find that the most common kind of Chinese character consists of a "radical" or semantic element, itself a character standing for a morpheme of related meaning, and a "phonetic" element, a character standing for a morpheme that is, or once was, phonologically similar (Martin, 1972). It is also clear that, in this logographic system, the characters stand for morphemes as such, and not for sequences of morphophonemes, because a character never corresponds to an *arbitrary* sequence of morphophonemes, and because homophonous morphemes are regularly assigned distinct characters. On the other hand, as I have already argued, the fact that morphemes, which are meaning-bearing elements, are so obviously the units of transcription does not compel the conclusion that Chinese and Japanese readers must "go directly to meaning." The point is rather that words of a sentence are transcribed according to the morphemic structure of their lexical entries.

The essentially morphemic character of alphabetic and syllabary systems is perhaps less obvious. In the first place, it has to be shown that an alphabetic orthography, such as that of English, is not, as sometimes assumed, a phonetic orthography.

In the grammar, the phonetic representation is generated by the application of phonological rules to the morphophonemic forms in the lexicon. Thus the rules shorten the long vowel of /hēl+θ/ ("laxing"), yielding [helθ] ; and they add a following glide ("diphthongization") and shift the quality of this same vowel in /hēl/ ("vowel shift"), yielding [hiyl] . Other rules assign varying degrees of stress to both words, depending upon their position and syntactic function in the sentence. Similarly, through the application of the relevant phonological rules, morphophonemic /tele+græf/, /tele+græf+ik/, /tele+græf+y/ yield phonetic [télǝgræf] , [tèlǝgrǽfik] , [tǝlégrǝfiy] . Thus, in the phonetic representation, an underlying morpheme is not consistently represented as it is in the morphophonemic representation (Chomsky & Halle, 1968). Clearly, as Chomsky (1965) has argued, the conventional spellings of these and most other English words correspond far more closely to the morphophonemic than to the phonetic forms. The morphophonemic character of an alphabetic orthography is, of course, more obvious in the case of a language with a relatively "deep" phonology, such as English or French. The orthography of a language with a shallow phonology will inevitably be closer to the phonetic representation, since the morphophonemic representation itself is closer to the phonetic representation. This seems to be true for the orthographies of Finnish, Vietnamese, and Serbo-Croatian, for example, which are often loosely said to be "phonetic," but are not really exceptions.

For some languages with simple syllable structure, syllabary systems are used, but it is still the case that the transcription is at the morphophonemic level. For example, it is a rule of Japanese phonology that an underlying voiceless stop becomes voiced noninitially (Martin, 1972). In the Romanized forms *kana, hiragana*, the effect of this rule is explicit: The initial /k/ of /kana/ in the one-morpheme word becomes [g] in the compound. But in the hiragana syllabary system that is one way of transcribing Japanese, the same kana character is used for the syllable /ka/ in both words.

Yet it must be admitted that some orthographies, though not phonetic, are less than perfectly morphophonemic. In Turkish, the alternations determined by the Vowel Harmony rule are transcribed, perhaps because there are numerous borrowed words not subject to this law. In Spanish, infinitives are transcribed without the phonologically deleted final /e/ of the morphophonemic representation; thus, /decire/, "to say," is written *decir* (Harris, 1969). In Sanskrit, the predictable alternations between aspirated and inaspirated stops (Grassman's Law) are transcribed. And it is too bad that, in English orthography, *fashion*, *delusion, cylinder* are not spelled **facion*, **deludion*, **cylindr*, respectively (see Klima, 1972, for discussion). Of course, English orthography makes no attempt to cope with genuine morphological irregularities: the past tense of /θink/ is written *thought*, not **thinked*.

Granted that alphabetic systems are morphophonemic (though imperfectly so), it is now argued that they are morphophonemic in order to represent morphemes consistently, rather than to represent the morphophonemes as such. If alphabetic orthographies were entirely systematic, it might not be possible to demonstrate this convincingly. But, in the case of English, the inconsistencies can be turned to account. Thus, it is quite easy to demonstrate that in English a morphophoneme may be written in a number of different ways, and that, worse still, the transcriptions of one morphophoneme overlap with those of another. For example, morphophonemic /e/ is spelled *ee* or *ie* or *ea* or *eCe* or *iCe; ie* and *iCe* can also spell /i/ and *ea* can spell /e/ and /æ/. But notice that these variations often serve to distinguish homophonous morphemes from one another, as in *sea, see; meet, meat, mete;* and apart from the cases of phonological alternations just mentioned, a particular morpheme is generally spelled in the same way in its occurrences in different words: Discrepancies like *proceed, recede* are quite exceptional.

In the case of syllabary systems the point can be made in a different way. A syllabary is preferable either to an alphabetic or to a logographic system from the standpoint of learnability and convenience. But since the alphabetic principle became well known, syllabaries have not been widely used. It seems to be a desirable, if not essential, condition for using a syllabary that not only should the syllable structure of the language be simple, but also that morpheme boundaries should coincide with syllable boundaries. If this condition is not met—and it is not in many Indo-European languages, English included—morphemes cannot be consistently transcribed by a syllabary. The limited use of syllabaries thus probably attests to the basic importance of the morpheme.

In sum, a practical orthography conveys the lexical content of a sentence by transcribing the words morphemically. Differences in orthographic type reduce to whether a morpheme is written as a single symbol or as a sequence of symbols corresponding to morphophonemes or morphophonemic syllables. But we need to ask why this should be so. Though a phonetic transcription, as we have seen, does not render lexical items invariantly, this lack of invariance is also character-istic of the phonetic representation that the listener presumably recovers. Indeed, if reading is simply a matter of turning symbols into speech and then listening to them, a phonetic transcription would seem to be the most obvious mode of orthography. If the reader is capable of recovering the syntactic phrase marker from word order, should he not also be capable of recovering lexical representa-tions from a phonetic transcription? But, in fact, reading a phonetic transcription of any length is intolerably burdensome even for an experienced phonetician. We are forced to conclude that there must be something special about lexical represen-tations that makes them the only possible basis for practical orthographies, even though by no means the only conceivable one. (I return to this question later.)

This characterization of practical orthographies suggests that in the actual process of reading, the analysis of a sentence begins with its lexical content and not with its phonetic representation. (I will also return to this point later.) It also suggests that lexical items are recognized by virtue of their morphological and (in the case of alphabets and syllabaries) their morphophonemic structure. But it would be a mistake to conclude that such structure is the only basis for word recognition. The semantic associations of an orthographic form (the basis for what I have called impressionistic reading) apparently tell the reader very quickly that the word is one he has seen before. If the word is very familiar, they may also be sufficient for lexical identification. The effect is enhanced if the orthographic form is "glyphic," i.e., compact and visually distinct (Brooks, 1977). On the other hand, if the word is one that the reader knows but has never seen in written form before, it is obvious that the morphophonemic information is usually essential to identify the word.

More interesting is the case of the word that is only fairly familiar. In this case, it is likely that semantic associations serve only to narrow down the field, quite rapidly no doubt, to a group of semantically related entries. At this point, a reader who cannot exploit the internal morphophonemic structure of the words has no alternative but to guess, and poorly trained or aphasic readers will often substitute a semantically related word for the correct one. But, for the reader who can use this internal structure, lexical search is unambiguous and self-terminating; the word is there, with a morphophonemic representation consistent with its orthographic form, or it is not.

It has been shown experimentally that the internal structure of words facilitates recognition, and continues to do so even after the words have become quite familiar (Brooks, 1977). One might suppose that the more advanced the material being read, the more often the reader would be reading low-frequency, "fairly familiar" words, and the more important the ability to exploit the morphophonemic information in the orthography would become.

It would appear, then, that it would be to the advantage of a reader to be phonologically mature, to know the phonology of the language, so that the morphophonemic representations of words in his personal lexicon match the transcriptions of the orthography. If he is phonologically mature, he has, in the course of acquiring English, mastered the Laxing, Diphthongization, and Vowel Shift rules, and he has inferred that [hĩyl] and [helθ] can both be derived from /hēl/, /θ/ being a separate morpheme. Thus, he has /hēl/ and /hēl+θ/ as morphophonemic representations in his lexicon and not /hĩyl/ and /helθ/. If he has not, in fact, gone through this process, the spellings *heal* and *health* will presumably seem to him arbitrary rather than regular.

This knowledge is, of course, a form of what has earlier been called "grammatical" knowledge, and it is of great significance that such knowledge is directly exploited in reading but not necessarily in listening. As has already been suggested, it is possible that the listener does not always reconstruct the morphophonemic representations of words. As long as his lexical search strategy has somehow paired the phonetic forms [hĩyl] and [helθ] with their lexical entries, he can analyze and understand the sentence. If his morphophonemic representations are immature, the only consequence is that the semantic information in the entry for *health* may not be as rich: He does not associate "health" with "healing."

Yet it would seem that for both beginning and experienced readers, *access* to morphophonemic representations is of even more importance than the maturity of these representations. The need for such access does not arise for the listener understanding sentences, presumably because he has innate automatic mechanisms for lexical search. If a reader has such access, that is, if he can bring his grammatical knowledge to bear on the task of reading, then the orthography will seem like a rational way of transcribing utterances in his language. Without access to grammatical knowledge, not only particular spellings, but the very idea of transcribing an utterance segmentally, will seem strange and arbitrary.

The state of having access to one's grammatical knowledge is what I meant by linguistic awareness[1] in my earlier paper (Mattingly, 1972). At that time I

---

[1] My distinction between phonological maturity and linguistic awareness is perhaps slightly artificial. Klima (1972) has suggested that the morphophonemic representation may be less accessible than certain shallower levels of derivation. If so, it would be difficult to distinguish empirically between lack of phonological maturity and lack of linguistic awareness, especi-

believed that this awareness had a metalinguistic, somewhat unnatural character. It now seems to me that such a state of awareness is eminently natural, since it is a mental state resembling that of the language learner. The language learner has access to morphophonemic representations because he is in the process of establishing them. Practical orthographies presuppose that the reader has the same sort of access to these representations.

To return to the question of what differences in the reading process are implied by differences in orthographic type, it would seem that what is primarily involved is the degree of linguistic awareness required. Logographic systems are the least demanding in this respect, since access only to morphological and not to the morphophonemic aspects of the representation is required, but the obvious price paid is that a larger set of characters must be remembered. Alphabetic systems, on the other hand, are the most demanding. For a language with appropriate morphological and phonological properties, a syllabary appears to be a happy compromise.

Some Experimental Evidence: Phonetic Recoding

The orthographic evidence, then, suggests that a reader uses his grammatical knowledge to establish the lexical content of the sentence. But such evidence suggests nothing about how, given this information, the reader is able to understand the sentence, that is, how he reconstructs the phrase marker and the semantic representation. He might conceivably make use of grammatical knowledge; he might use some analytic mechanism peculiar to reading and quite independent of spoken-language analysis; or he might use the analytic mechanisms that the listener uses.

The grammatical interpretation of earlier parts of a sentence depends generally on information in later parts. Since spoken sentences are physical events in real time, a listener must have a way of representing this early information in memory until he is prepared to analyze it. Yet to represent physical events in memory at all requires some analysis of these events. Thus, the listener is compelled to make a rapid preliminary analysis that can then be deepened and refined in light of later information.

This preliminary representation is stored in short-term memory. Analysis of errors in short-term recall suggests that the information being stored is phonetic (Wickelgren, 1965a, 1966). However, since other sorts of linguistic information must obviously be stored in short-term memory as well (Fodor et al., 1974, Chap. VI), it would be more cautious to say that the short-term representation is *at least* phonetic.

ally in the case of "nonproductive" phonological rules, like Vowel Shift. (Moskowitz [1973], however, appears to have surmounted this difficulty.) Pending clarification, we are assuming parsimoniously, that linguistic awareness means having access to the appropriate units of one's morphophonemic representations, while phonological maturity means controlling the phonological rules and having morphophonemic representations in one's lexicon approximating those of an ideal speaker-hearer of one's language.

Reading is a real-time process, just like listening. It is hardly relevant that the lexical information remains before the reader on the page. (A reader who does exploit this fact, making numerous regressions as he scans the list, is a reader in trouble). Whatever the analytic mechanisms he uses, he must make use of the results of earlier analysis in the course of current analysis, and it does not seem to occur to him to note his preliminary results in the margin. Thus, like the listener, he requires some form of temporary storage. Iconic storage, in which visual information is initially represented, is unsuitable for the purpose because of its very brief duration (Sperling, 1960). One might entertain the possibility of an "orthographic" short-term memory, analogous to "phonetic" short-term memory; or of a "semantic" short-term memory, in which words were represented by their meanings. Many individuals, however, report "inner speech" while reading, and some readers engage in actual articulatory or acoustic activity. These observations suggest that "phonetic" short-term memory itself provides temporary storage of information during reading.

There is, in fact, considerable evidence that if, in an experimental situation, orthographic material is to be temporarily remembered, "phonetic recoding" occurs (for a review, see Conrad, 1972). One experimental paradigm is considered in detail here because it provides opportunities for both semantic short-term memory and orthographic short-term memory to manifest themselves (Waugh & Norman, 1965; Kintsch & Buschke, 1969). In this paradigm a subject is asked to remember a list of words presented one by one, fairly rapidly. His recall of the list is then immediately tested by presenting a "probe" word and asking him to report the word that preceded the probe on the list. The typical finding is that words appearing early on the list and words appearing near the end of the list are better recalled than words in intermediate position. Recall of the later words is ascribed to a short-term memory representation still available at the time of the probe; in the case of intermediate words, this representation has decayed. Recall of the earlier words is ascribed to the subject's attempt to retain the list in long-term memory. When a list consists of semantically similar items, long-term but not short-term recall is reduced; when a list consists of phonetically similar items, short-term but not long-term recall is reduced. No effect is observed for words that are orthographically but not phonetically similar. These effects are the more impressive in that the phonetic and semantic similarities of the items are obvious to the subject, and he is free to use any available mnemonic strategy.

The effect of semantic similarity on long-term recall suggests that long-term memory is semantically structured, as one would expect. The effect of phonetic similarity on short-term recall suggests that the short-term memory representations are phonetically structured. This interpretation is corroborated by analysis of errors in other experiments on short-term recall of orthographic material. Such analysis reveals phonetic confusions similar to those observed in short-term recall of spoken material (Wickelgren, 1965b). The most reasonable inference is that the same short-term memory is used for both types of material. The phonetic similarity effect, considered together with the absence of other short-term effects, also suggests that there is no equally convenient alternative to phonetic short-term memory with an appropriate duration—no orthographic or semantic short-term memory, for example. Had such alternatives been available, the subjects in Wickelgren's test could have avoided the disadvantages of storing phonetically similar items in phonetic short-term memory.

It is of considerable interest that the effects described are obtained not only for alphabetic but also for logographic stimuli. If the logographic kanji characters used in writing Japanese are presented to native speakers of Japanese in a probe paradigm, results parallel to those already described for English are obtained: When a list consists of kanji with phonetically similar readings, and only then, short-term recall is adversely affected (Erickson, Mattingly & Turvey, 1972, 1977). Comparable results have been obtained for Chinese characters (Tzeng, Hung & Wang, 1977).

To the extent that inference from the recall of visually presented lists to actual reading is justified, these results suggest that orthographic information in reading is indeed "phonetically recoded" and stored in the short-term memory used for spoken language.

The validity of the inference to actual reading is strengthened by a related experiment. Liberman et al. (1977) tested the short-term recall of children considered good readers and children considered poor readers. Strings of five letters were briefly presented, and after each presentation the subjects were asked to write down the letters in their given order. It was found that the performance of the good readers was better than that of the poor ones. More interestingly, it was found that the recall of a string in which the letter names rhymed was *poorer*, relative to the recall of a control string, for good readers than for poor readers, presumably because the good readers more consistently employed phonetic short-term memory to retain the strings. Thus, there seems to be a direct relationship between reading ability and phonetic recoding.

What is the significance of phonetic recoding? It is sometimes assumed—indeed, the term itself implies as much—that phonetic recoding takes place because of the supposed "phonetic" character of orthographies. According to this view, the reader converts letters (or letter patterns) to sounds, representing this information in short-term memory. This view is appealing because it seems to lead to a tidy statement on the relationship between reading and listening: In both activities, a phonetic representation is established that then serves as the basis for subsequent analysis. But this view is unsatisfactory for several reasons. First, it has been shown that neither logographies nor alphabetic orthographies are phonetic transcriptions; yet phonetic recoding occurs with both. Second, this view ignores the lexical character of all practical orthographies. What they give the reader are word identities, and hence, the phonological, syntactic, and semantic information in lexical entries. Third, this view implies that in the analysis of a spoken sentence, the speech-perception mechanism is separate from, and does not interact with, other analytic mechanisms.

Phonetic recoding has no connection with the character of the orthography. As Kleiman (1974) has demonstrated, it does not occur simply as a consequence of word recognition. Rather, what phonetic recoding means is that reader and listener are almost certainly employing the same analytic mechanisms. If the use of phonetic short-term memory reflects the reconstruction of the phonetic representation by these mechanisms in the course of analysis, it is very likely that they are doing the rest of the job as well.

Since the input to the analytic mechanisms in reading is lexical, and the required output is semantic, it might seem strange that the reconstruction and temporary storage of the phonetic representation cannot be dispensed with, as it is in artificial schemes for understanding printed text. But the analytic mechanisms constitute an intricate special-purpose system for understanding spoken sentences by working out their surface structure. The phonetic representation, though not logically required in reading, is an integral part of the product.

This point is clear, indeed, from the probe experiments. What the subject tries to do is to remember an entire list of words. The only way he can do this is to form semantic representations in long-term memory, in other words, to treat each item on the list as a one-word sentence to be analyzed, and he succeeds in doing so for the early words on the list. The formation of the short-term phonetic representation appears to be an essential part of the process. On the other hand, subjects who do not give evidence of phonetic recoding, like the poor readers of Liberman et al. (1977), are probably not using any of their analytic mechanisms. If the formation of a phonetic representation could be readily dispensed with, Liberman et al. would have found *good* readers who gave no evidence of phonetic recoding.

It must be emphasized that the phonetic representation formed by a listener or a reader is abstract, like other mental representations, not a reenactment of speech production. Silent rehearsal, actual articulatory movement, or subvocalization—forms of behavior that are with some justice regarded as marks of a slow and inefficient reader—are not an essential concomitant of the phonetic representation. Rather they are evidence that, for some reason, analysis of the sentence is proceeding so slowly that the information in short-term memory needs to be refreshed. Obviously, a skilled reader has no reason to employ such devices, but their absence does not suggest the absence of phonetic recoding. As for the subjective phenomenon of "inner speech," I do not know whether it is to be regarded as merely the consciousness of the phonetic representation or as a form of rehearsal. Since some very skilled readers report it, I am inclined to the former conclusion. But in any case, though it is no doubt a consequence of phonetic recoding, it is not a necessary consequence.

It is also often assumed, because of a similar misunderstanding, that the phenomenon of phonetic recoding means that reading speed is constrained by the relatively low rates at which speech can be uttered or the somewhat higher rates at which speech, if the signal is specially manipulated, may be understood. There is no justification for this view. How fast phonetic representations pass through short-term memory must depend on the rates at which orthographic forms can be recognized and sentences can be analyzed, not on the rates of speech production or speech perception.

Reading and Language Acquisition:
Some Conjectures and Conclusions

From the point of view adopted in this chapter, the real mystery is that the analytic mechanisms can be used in reading *at all*. It has been stressed that these mechanisms are innate, highly specialized, and inaccessible; it would seem entirely reasonable if it were the case that only a spoken sentence could be understood. Yet in reading, the boundary between grammatical knowledge and performance is crossed. The accessible information carried by the orthography is somehow able to trigger these inaccessible processes. Moreover, this information is not at all equivalent to the auditory information that initiates the process of understanding speech; as has been emphasized, it is lexical information.

My explanation, which must be regarded as purely conjectural, is that the reader takes advantage of a language-acquisition procedure. Part of the task of a language learner, of course, is to increase his stock of lexical entries. Consider how he might go about this. Suppose that his data consist of the phonetic representation and the phrase marker of a sentence containing a word that is new to him, and that, in addition, he has gathered from the situational context a tentative notion as to the semantic representation. To explain these data, he hypothesizes the existence of a word whose lexical entry has certain semantic, syntactic, and phonological features. To test his hypothesis, he supplies the analytic mechanisms with this postulated lexical information, as well as with previously determined lexical information about other words in the sentence under consideration. If the analysis yields the observed phonetic representation and the assumed semantic representation, the correctness of the proposed lexical entry is corroborated. That the analytic mechanisms can accept lexical information as input is perhaps not too surprising. It has already been argued that understanding speech is a nonlinear process with semantic and syntactic as well as auditory input.

This quite speculative but, I feel, not totally implausible account of what must be involved in learning a new word assumes that the language learner is innately equipped to initiate the analysis of a sentence, starting with lexical information that is accessible to him. If we are willing to believe that the reader exploits this aspect of his language-learning capacity, a tentative explanation is available of the reader's otherwise surprising ability to make use of mechanisms that might be supposed to be reserved for the listener.

It was observed earlier that orthographies appealed to grammatical knowledge, i.e., to knowledge of the language in the form in which it is acquired by the speaker-hearer. If the reader is to recognize words efficiently, it is important for him not only to have such knowledge (phonological maturity) but to have access to it (linguistic awareness), just as he did when the knowledge was originally acquired. The present discussion has led us in the same direction. It has been suggested that the reader's ability to make use of the analytic mechanisms is part of his capacity for language acquisition. If this is really the case, then the resemblance between the reader and the language learner can be extended. Not only do they both have access to grammatical knowledge, but also they both activate the analytical mechanisms with lexical information.

It has been mentioned already that the course of language acquisition varies considerably. At one extreme is the individual who virtually abandons language acquisition as soon as he has developed the relatively modest body of analytic strategies needed to get by; at the other extreme, the "word child" who continues indefinitely to add to his grammatical knowledge.

I now offer the further conjecture that differences in children's readiness and ability to learn to read are related to these different patterns of language acquisition. The child who is still actively acquiring language at the time he begins to read will be relatively mature phonologically, so that the orthography will correspond to a considerable extent with his morphophonemic representations. Having access to these representations, he will be linguistically aware, and the orthography will seem to him a plausible way of representing sentences. The anaysis of a sentence on the basis of its lexical content will present no problems, since he has continued to use this analytic procedure in the course of learning new words. Moreover, he will, as a linguist, see that reading is a source of fresh data. If he does not already have the morphophonemic forms /hēl/ and /hēl+θ/ in his lexicon, and the associated rules in his phonology, the orthographic forms *heal* and *health* will prompt him to revise his grammar accordingly. Thus, the linguistic curiosity that has motivated his continuing language acquisition will motivate his learning to read as well.

On the other hand, the child who is no longer very actively acquiring language will surely find learning to read very difficult and unsatisfying. His morphophonemic representations will be less mature than they might be, so that the discrepancies between the orthography and the morphophonemic representations will be substantial. More seriously, these representations, being part of his grammatical knowledge, will have become less accessible to him; he will be lacking in linguistic awareness. As a result, the orthography will seem a mysterious and arbitrary way to represent sentences. Finally, since his capacity for language learning will not have been recently exercised, he may well have lost some of his ability to analyze a sentence on the basis of its lexical content.

Because most people continue to be capable of learning new words when they must, and even new languages when circumstances compel them to, it does not seem likely that the capacity for language acquisition atrophies completely. If not, there is certainly reason to hope that in poor readers this capacity is merely dormant, a muscle that has grown flabby from disuse. With proper instruction and appropriate environmental stimulation, it can be reawakened. But, obviously, it would be even better if language acquisition had not been allowed to falter in the first place, if there had been no awkward interval between the period of learning to talk and the period of learning to read. This observation is not to be construed as a demand for very early reading instruction, but rather as a plea for linguistic stimulation above and beyond speaking and listening during preschool years: storytelling, word games, rhymes and riddles, and the like. The value of such stimulation is certainly appreciated by most specialists in reading. But its justification is not merely that it prepares the child for the experience of learning to read, but that it helps to keep active psychological mechanisms that are indispensable in learning to read.

To summarize what must appear to be a rather discursive argument, it is my contention that written language, far more than speech, places a direct demand on the individual's acquired knowledge of language—what has been called grammatical knowledge; and that such knowledge, consequently, must be accessible to the reader, as it presumably is to the language learner. Moreover, it appears that although reading and listening use the same analytic mechanisms—hence "phonetic recoding"—analysis of a sentence is accomplished in reading, unlike listening, from an input that corresponds to the lexical content of the sentence. It is conjectured that the reader is able to do this by means of what is really a language-acquisition procedure. Reading thus has much in common with language acquisition, and the child who has continued to acquire language beyond what is required for performance is likely to learn to read more easily than the child whose language-acquisition capacity has become dormant.

Acknowledgments

Support by the International Reading Association and by Grant HD-01994 from the National Institute of Child Health and Human Development is gratefully acknowledged.

CHAPTER 3

Task Awareness in the Development of Reading Skill

John Downing

Introduction

Reading behavior and its acquisition is of interest to many different kinds of people—laymen, professional educators, scientists such as linguists and psychologists, and so on. They have different terminologies to describe and discuss reading, and this has often resulted in ambiguity and misunderstandings. This chapter is written by a psychologist who will attempt to analyze the development of reading behavior in terms of established concepts in the discipline of psychology.

Psychology is a science of behavior. Therefore, we might begin by asking how people use the term "reading" in everyday life. Then we can pose a second question—What kind of behavior do people describe as "reading"? Which leads to the crucial question—To what category of behavior does reading belong in the taxonomy of scientific psychology?

In answer to the first question, one can observe that people do not confine the application of the word "reading" to interactions with books or other forms of text. People are said to "read" maps, charts, graphs, clocks, and gauges. Fortune tellers claim to "read" the lines in people's hands or the tea leaves in cups. The farmer "reads" the sky to forecast the weather. Hunters "read" the spoor of game. The deaf "read" lips. The blind "read" the raised dots of braille by feeling them with their fingers. All of these behaviors commonly described as "reading" involve the *interpretation of signs*. Since some of these behaviors have been practiced in prehistoric times, it is reasonable to assume that there are fundamental psychological and neurological factors common to all kinds of sign interpreting behavior that would influence the more modern behavior of text reading. This observation gives the concept of "developmental dyslexia" face validity. Although it is impossible for special brain mechanisms to have evolved for the interpretation of that modern phenomenon *text*, it is quite possible that reading disability can sometimes be caused by failures or lags in the development of the neurological bases of abilities which are prerequisite for interpreting signs or learning how to interpret signs.

However, there is one very important difference between, on the one hand, ancient forms of sign-interpreting behavior such as reading animal footprints when hunting and, on the other hand, modern forms of sign-interpreting behavior such as reading text. The signs read by the hunters were not deliberately created with any communicational intent, whereas the signs in a written or printed text are arbitrary symbols deliberately created and used for the purpose of communication. It is this latter kind of behavior that is the concern of this book. Therefore, our final and crucial question stated more narrowly is—To what category of behavior does the interpretation of symbols in reading text belong in the taxonomy of scientific psychology?

Reading as a Skill

A popular term among reading specialists in education is "the reading process." But "process" is a vague term even in everyday speech. What does "the reading process" mean? Is it the things that happen in your head when you read? If we leave out the reference to "reading," we are left with "things that happen in your head." Not surprisingly, "process" has no specific technical meaning in psychology. "Process" is a term of broad generality denoting some systematic change in behavior or its underlying mechanism without specifying its form or characteristics. The shift from the generality and vagueness of "process" to more specific explanatory terminology is well brought out in the following psycholological definition of reading behavior:

> Reading is a process by which the child can, on the run, extract a sequence of cues from printed texts and relate these, one to another, so that he understands the precise message of the test. The child continues to gain in this skill throughout his entire education, interpreting statements of ever-increasing complexity (Clay, 1972, p. 8).

Thus, Clay's choice of words specifies that the undefined events that occur in the reading process belong to that category of behavior that psychologists have labeled "skill." She uses the term without any explanation simply because most psychologists would accept that reading is an example of that special type of behavior called "skill." For example, Holmes' (1970) well-known Substrata-Factor theory of reading states that "reading is an audio-visual verbal-processing skill of symbolic reasoning . . . " (p. 187). Many other psychologists have applied the term "skill" to reading (for example, Lansdown, 1974; Singer, 1966).

Skill is an old, established category in the taxonomy of psychology. Research on skill learning is one of the oldest interests of scientific psychology. Well-designed experiments on the acquisition of skill were begun towards the end of the nineteenth century, and continuous and rather noncontroversial progress in this field has been made to the present day. Borger and Seaborne (1966) provide a modern definition of this specific category of behavior: "particular, more or less complex, activities which require a period of deliberate training and practice to be performed adequately and which often have some useful function." They indicate that their concern, as psychologists, is with "the individual, with competence, with proficiency. It may manifest itself in the exercise of recognized skills like driving, in the playing of games like tennis or marbles, also in more widespread accomplishments such as riding a bicycle, the tying of shoe laces or just walking." Borger and Seaborne describe how psychologists have investigated "the conditions under which skill develops, the factors making for more or less rapid achievement of a given criterion, and to do this in a way which is only incidentally concerned with the particular type of activity involved" (pp. 127-128). In other words, psychologists have established that there is a category of behavior whose members have certain universal characteristics. This generality of features among all skills has been even more widely applied in psychology in recent years. Thus, although at one time psychologists made distinctions between types of skills such as "verbal skills" and "psychomotor skills," the differences are now considered to be merely in surface appearance. Thus, Whiting (1975) states that, although "verbal, mental, perceptual, social and motor are common adjectives in relation to skills," it would "be wrong . . . to assume that the process involved in learning any of these skill categories is essentially different from the learning of another" (p. 6). Whiting and den Brinker (1980) emphasize that all skills "are intelligently carried out" (p. 2).

What are the chief features of this psychological category "skill?" Can we be certain that reading is an example of a skill? A brief description by McDonald (1965, p. 387) brings out the essential characteristics of a skill. He writes that, "From a psychological point of view, playing football or chess or using a typewriter or the English language correctly demands complex sets of responses—some of them cognitive, some attitudinal, and some manipulative." McDonald takes as an example "skill" in playing baseball, and he points to the player's need "to perform sets of responses with ease, quickness and economy of motion." But McDonald emphasizes that it is not merely a matter of motor behavior. The player "must also *understand*" the game and must like playing the game and have appropriate attitudes about winning and sportmanship." The total performance ". . . is a complex set of processes—cognitive, attitudinal, and manipulative. This complex integration of processes is what we usually mean when we refer to 'skill' . . .".

Clearly, reading behavior fits this description. Reading involves complex integration of many behaviors—cognitive, attitudinal, and manipulative. But it may be felt that McDonald's description is too general, too global, to determine whether reading should be classified as skill. A more detailed analysis of the features of skill in general should be compared with what is known about characteristics of reading behavior in particular. Such an analysis and comparison was made by Downing and Leong (1982). They analyzed several reviews of psychological studies of skill and found 20 frequently mentioned characteristics of a skill:

(1) It involves a *highly complex* pattern of behavior.

(2) A skilled performer executes this complex pattern *smoothly*, without faltering.

(3) This unhesitating smoothness is the product of *integration* of the many different aspects of behavior involved in the complex pattern.

(4) *Timing* is a very important aspect of this integration. Sometimes *speed* is a criterion of a high level of skill, but often *flexibility* in timing is more important because timing must vary according to the changing needs of the situation in which the skill is performed.

(5) The smooth performance of a skill is derived from the performer's being *ready for a wide range of events* that may or may not occur from moment to moment.

(6) This readiness for action depends on the performer's *anticipation* of future events.

(7) The skilled performer runs through the behavior pattern *automatically* but reverts to more flexible and conscious control if an unusual situation arises.

(8) The function of this automaticity is to *free the skill performer's attention for other activities.*

(9) *Consciousness of one's own activities and their functions* is a characteristic of skill performance at some times but not at others.

(10) A skill performance involves a continuous stream of reactions to external and internal *cues.*

(11) These reactions depend on the performer's attention to relevant cues and her/his *translation* of them into appropriate action.

(12) The two principal sources of cues are: (a) *changes in the outside environment*, and (b) *changes within the performer.*

(13) *Feedback* is a constant regulator of the pattern of behavior, all the time homing in on the goal of greater accuracy in the performance of the skill.

(14) The skill performer's utilization of cues depends on *selective attention.* As skill develops, the performer's sensitivity to cues changes in several ways.

(15) One type of change in sensitivity that occurs as a skill is learned is the *shift from external cues to internal cues.*

(16) Another type of change that occurs as skill develops is that the performer utilizes *perceptual units of increasing size.*

(17) *Larger units of action* are also developed as skill grows.

(18) Yet another change that is observed as skill develops is *cue reduction and addition.*

(19) The *ability to cope with stress increases* as skill develops. Under stress less skilled performers tend to regress to an even more primitive level of performance.

(20) Skills may be broken down into smaller units of behavior called *"subskills" or "subroutines."* Earlier behaviorist theories proposed a hierarchy of subskills which implied that there was a rigid order in which the subskills must be learned in developing a skill. However, more recent work indicates that skill is *modular* in organization. Subskills, once mastered,

become modules that are available for use in a variety of known and new contexts. This modular organization may be hierarchical in terms of control without there being any necessary sequence of either learning or performance.

The above list of characteristics of skill is very much abbreviated. Downing and Leong (1982) provide substantial descriptions of each characteristic and cite sources of evidence. It should be noted also that is is recognized that each of the 20 listed features may not be distinct from the others. This overlapping arises because Downing and Leong wished to avoid distorting the original sources which they were reviewing.

Downing and Leong went on to seek an answer to the question—"How do observations of reading behavior fit the psychologist's generalized description of skill performance?" (p. 18). For each item in the list of characteristics of skill development in general, evidence was found from reading research for the existence of the same feature in reading behavior or its development. There was only one exception, item (19), regarding the effects of stress on skill performance. This feature does not appear to have been investigated with respect to reading. Apart from this item, Downing and Leong found that "the fit is very good. Therefore, we conclude that psychological research findings on skill acquisition in general can be applied with confidence to the specific skill of reading" (p. 28).

Development of Reading Skill

This application of research is attempted in Downing's and Leong's chapter, "Principles of Skill Acquisition in Reading" (pp. 29-49). They give prime consideration to what has been termed the "cognitive phase" in the development of skill.

Fitts' (1962) review of the research on skill learning led him to conclude that there are three phases in the development of any skill. These may be termed the "cognitive," "mastering," and "automaticity" phases. They occur in that order, although, of course, they are really one continuous process without any distinct boundary between them. Furthermore, it should be noted that, in a very complex skill such as reading, these three phases continually recur as the learner meets each new subskill during the many years needed to become a fully skilled reader.

The initial *cognitive phase* is when the learner, according to Cronbach (1977, p. 396), "in an unfamiliar situation must find out what to do." Thus, the beginner "is getting in mind just what is to be done" (p. 398). Therefore, in teaching a skill or subskill, it is important that the task should be clearly understandable in the initial stages. The results of research on learning to fly a plane, for example, showed that the average number of hours needed to learn to fly solo was reduced from eight to four when special attention was given to helping students to understand their tasks (Williams & Flexman, 1949; Flexman, Matheny & Brown, 1950).

The usual length of this phase in adults is comparatively brief—a few hours or days—but it may be much longer in children learning to read.

In the *mastering phase,* learners work to perfect their performance of the skill. They practice until they achieve a high level of accuracy with practically no errors. This stage may last for days, months, or even years, depending on the complexity of the skill and opportunities for practice.

But, even when the skill has been mastered, there remains a very important stage ahead. This is the *automaticity phase,* which comes about through over-learning (practice beyond the point of mastery). When this is accomplished, expert performers can run through the skill behavior effortlessly—automatically. They continue to do so, except when some unusual problem arises that makes it necessary for them to become conscious of their activities again.

As was mentioned above, these three phases of skill development recur whenever some new subskill in a complex skill has to be acquired. But it is in the initial stage of learning a complex skill that a large number of new subskills must be faced all at once. Therefore, the cognitive aspect of skill acquisition is especially significant in the child's first weeks and months of reading instruction. If children fail to comprehend their reading instruction in the beginning stage, then they cannot move on to the mastering phase. They remain trapped in the cognitive phase and may lose faith in their own ability to understand what they are supposed to do in reading lessons. From these considerations, it becomes clear that the cognitive aspect of developing the skill of reading is of utmost importance.

This conclusion was first given prominence by Vernon (1957) in her very extensive and thoroughly critical review of research on the causes of reading disability. She concluded that "the fundamental and basic characteristic of reading disability appears to be cognitive confusion . . ." (p. 71). Vernon wrote that "the fundamental trouble appears to be a failure in development of this reasoning process" (p. 48). She described the cognitively confused child as being "hopelessly uncertain and confused" as to why certain successions of printed letters should correspond to certain phonetic sounds in words. In her more recent extension of her survey of research on reading disability, Vernon (1971) develops this theme further:

> It would seem that in learning to read it is essential for the child to real-ize and understand the fundamental generalization that in alphabetic writing all words are represented by combinations of a limited number of visual symbols. Thus it is possible to present a very large vocabulary of spoken words in an economical manner which requires the memoriz-ing of a comparatively small number of printed symbols and their asso-ciated sounds. But a thorough grasp of this principle necessitates a fairly advanced stage of conceptual reasoning, since this type of organi-zation differs fundamentally from any previously encountered by children in their normal environment (p. 79).

With regard to simple associationistic views of learning the letter/sound code, Vernon asserts "The employment of reasoning is almost certainly involved in understanding the variable associations between printed and sounded letters. It might appear that certain writers suppose that these associations may be acquired through rote learning. But, even if this is possible with very simple letter-phoneme associations, the more complex associations and the correct application of the rules of spelling necessitate intelligent comprehension" (p. 82).

The insight gained from Vernon's studies of reading disability is that learning to read is essentially a problem-solving task in which the child applies reasoning abilities to understanding the communicative and linguistic relationships between speech and writing. According to Vernon, cognitive confusion is the chief symptom of reading disability. Therefore, if we generalize from her finding, we can postulate that *cognitive clarity* should be the typical characteristic of the successful reader. But, of course, we should not expect perfect cognitive clarity about the reading task from the beginning. Indeed, Vernon (1957) likened the cognitive confusion of the older reading-disabled child to the state of the beginner. Possibly, the word "confusion" may be distasteful to educators of young children. It may seem to imply a negative evaluation of a child who has as yet had little opportunity to achieve cognitive clarity about literacy tasks. Indeed, it is not quite correct to say that the young beginner in learning the skill of reading is "cognitively confused." From the adult's point of view, the child appears to be confused because he/she is unskilled and makes what the adult considers to be errors. But the young beginner usually does not feel confused. The child is quite satisfied with his/her present notion of the task and his/her progress in achieving it. As Piaget pointed out, the child, in trying to understand something, constructs or reinvents it. There is no need to feel confused just because the construction is incomplete. This gradual building up of a logical schema of a task was well described by Luria (1976):

> Every familiar school problem constitutes a complex psychological structure in which the final goal (formulated as the problem's question) is determined by specific conditions. Only by analyzing these conditions can the student establish the necessary relations between the components of the structure in question: he isolates the essential ones and disregards the inessential ones. By getting a preliminary fix on the problem's conditions, the student formulates a general strategy for its solution; in other words, he creates a general logical schema that determines the direction for further search. The schema in turn determines the reasoning tactics and the choice of operations that can lead to the making of a decision. Once this is done, the student moves on to the last stage, merging the results with the specified conditions. If the results are in agreement, he is finished; if any of the conditions remain unmet and the results disagree with the initial conditions, the search for the necessary solution continues (pp. 117-118).

Bruner (1971) emphasizes problem solving and the holistic nature of skill development when he writes: "In broad outline, skilled action requires recognizing the features of a task, its goal, and means appropriate to its attainment, a means of converting this information into appropriate action, and a means of getting feedback that compares the objective sought with present state attained" (p. 112).

These two quotations from Luria and from Bruner bring together two of the most fundamental features of skill development: (1) Integration is the key process in the performance of a skill; (2) The learner of a skill always tries to understand the task to be performed—from the very outset. In other words, the learner of a skill must attempt to approximate the skilled act as a whole from the very beginning. One learns the skill of chess by playing chess, one learns the skill of fishing by fishing, one learns to talk by talking, and one learns to read and write by reading and writing. At every step of the way progress is made by approximating the integrated whole skill. In this way, integration is learned through practice. Practice in integration is only supplied by performing the whole skill or as much as is a part of the learner's "preliminary fix." This explains Bamberger's (1976) paradox:

> Many children do not read books because they cannot read well enough. They cannot read well because they do not read books (p. 61).

Samuels (1976a) writes "Students will learn to read only by reading" (p. 325). It also explains why, in general, children seem to learn to read by almost any method. Singer (1966) states that, "Children have learned to read by means of a wide variety of methods and materials. . . . However, all the necessary elements for reading are present in the materials employed by each method so that pupils in learning to read through any of these methods could have used their capabilities for selecting their own unit of perception, their own conceptualized mediational response systems, and developed their own mental organization for attaining speed and power of reading" (pp. 116-117).

In other words, reading is a skill, and, therefore, no matter what framework of teaching methods and materials we set reading in, its essential psychological features assert themselves. The brain processes that determine the course of skill development operate constantly in learners despite the variety of methods and materials used in reading instruction. As Hoskisson (1975 a) puts it, "Perhaps one of education's greatest delusions is that we teach children to read. All that may really occur is that materials are presented to a child in one form or another and he uses them to solve the reading problem" (p. 446).

These remarks should not be misunderstood as negative cynicism about teaching methods and materials. On the contrary, teachers, their methods, and their materials can make a difference in children's success and failure in learning to read. They make a difference to the extent that they approximate the natural course of skill development. As Henderson (1980) has pointed out, what children can learn "depends upon the conceptual frame they bring to the task" (p. 2).

Task Awareness in Reading

In the preceding section of this chapter, we have shown that an essential feature of all skill learning is what Luria called getting a preliminary fix, or, in other words, conjuring up a conception of the task to be accomplished in performing the skill. There will be many such preliminary fixes as the learner makes succeeding attempts to become more skillful, each fix approximating more closely the ideal performance. This is *task awareness*. Although this awareness is usually of the task as a whole, that "whole" involves a variety of related concepts.

The term "awareness" would be regarded with suspicion by many psychologists. It is associated with psychology's great debate on "consciousness." This writer also uses the term "awareness" with some trepidation, and proposes Piaget's view of awareness as fitting the present discussion. Sinclair (1978) reviews two of Piaget's (1974a, 1974b) books which indicate the psychological nature of awareness. She writes:

> When Piaget speaks about awareness, he means the subject's gradual awareness of the how and why of his actions, and their results and of the course of his reasoning—but not of what makes his way of acting or thinking possible or necessary. Thus the research on "becoming aware" is to be interpreted as becoming aware of the how, and eventually the why, of specific actions and of the how, and eventually the why, of certain interactions between objects (p. 193). According to Piaget, in all intentional actions . . . the acting subject is aware of at least two things: the goal he wants to reach and, subsequent to his action, the result he has obtained (success, partial success or failure). From these modest beginnings, awareness proceeds in two different, but complementary directions. Especially when the action fails, but also when the subject is pleasantly surprised by success, or, at the ages when this can be done, when he is asked questions himself, he will construct a conceptual representation of at least some of the features of the actions he has performed and of some of the reactions and properties of the objects he acted upon (p. 195).

That "awareness" implies consciousness is suggested when Piaget (1976) writes that "The results of cognitive functioning are relatively conscious, but the internal mechanisms are entirely, or almost entirely, unconscious. For example, the subject knows more or less what he thinks about a problem or an object; he is relatively sure of his beliefs. But, though this is true of the results of his thinking, the subject is usually unconscious of the structures that guide his thinking" (p. 64). This passage from Piaget suggests that what we call task awareness is a matter of the learner's *beliefs* about the task to be attempted. It is also a matter of *metacognition*—what the learner knows that he/she knows about the task. In the case of a language task such as learning how to perform the reading skill, task awareness is also a matter of *metalinguistics*—what the learner knows about his/her own language behavior. The content of these beliefs, metacognition, and metalinguistics depends on the development of specific linguistic concepts.

It may be instructive to speculate about the conceptual foundations for the creative thinking that led to the inventions of writing that occurred independently in various cultures. The philologists Gelb (1952) and Jensen (1970) have traced the prehistory and history of the creation of the alternative methods of writing language, and from this it seems clear that the inventors and developers of visible symbols for language did so on the basis of two sets of ideas. First, they had ideas about the purposes of spoken language. They developed concepts of communication. They then took a leap forward in realizing that the auditory symbols of speech could be translated into visible symbols. When this idea was grasped, a second set of ideas developed. A way had to be found to analyze speech so that it could be represented by visible symbols in an economic code. Different people had different ideas about how to analyze speech. Thus, today there exist different ways to write language. Nevertheless, despite this great variety of writing systems, all visible language rests on these two basic kinds of concepts discovered by the orginators and developers of writing:

(1) *Functional concepts:* the communication purposes of writing.
(2) *Featural concepts:* the features of spoken language that were to be represented by written symbols.

These same two groups of linguistic concepts are the ones required for developing beliefs, metacognition, and metalinguistics in awareness of the task of learning how to read. In effect, the learner is an archaeologist who is faced with the problem of finding out the intentions of the people who have produced writing. Why did they make those visible symbols? The answer has two parts: (1) they intended to communicate some meaning; (2) they intended to code certain features of speech. Ferreiro and Teberosky (1979) make the point clearly and succinctly, "Reading is not deciphering; writing is not copying." The real task of literacy development is the "intelligent construction by the child" of these two skills of literacy (pp. 344-345). In other words, beginners have to rediscover those same basic functional and featural concepts that led to the invention of the writing system used in their language.

These considerations led the present author to put forward the "cognitive clarity theory" of the construction of reading skill in his book, *Reading and Reasoning* (Downing, 1979, p. 37), from which the following summary is quoted:

(1) Writing or print in any language is a visible code for those aspects of speech that were accessible to the linguistic awareness of the creators of that code or writing system; (2) this linguistic awareness of the creators of a writing system included simultaneous awareness of the communicative function of language and certain features of spoken language that are accessible to the speaker-hearer for logical analysis; (3) the learning-to-read process consists in the rediscovery of (a) the functions and (b) the coding rules of the writing system; (4) their rediscovery depends on the learner's linguistic awareness of the same features of communication and language as were accessible to the creators of the writing system; (5) children approach the tasks of reading instruction in a normal state of cognitive confusion about the purposes and technical features of language; (6) under reasonably good conditions, children work themselves out of the initial state of cognitive confusion into increasing cognitive clarity about the functions and features of language; (7) although the initial stage of literacy acquisition is the most vital one, cognitive confusion continues to arise and then, in turn, give way to cognitive clarity throughout the later stages of education as new subskills are added to the student's repertory; (8) the cognitive clarity

theory applies to all languages and writing systems. The communication aspect is universal, but the technical coding rules differ from one language to another.

The above original set of postulates, however, must be revised now in view of new data recently obtained by this writer. Research conducted in Papua New Guinea on hypotheses derived from the cognitive clarity theory discovered that children who had never been to school had higher scores on a test of linguistic awareness than children attending schools in which they received their instruction in a second language. Hence, it is not appropriate to describe the child who has not yet started school as being "cognitively confused." On the contrary, the children in Papua New Guinea schools being taught in a second language were more cognitively confused than those who lived in villages where there were no schools. Children receiving instruction in their mother tongue, however, achieved higher scores on the linguistic awareness test than the children who had never been to school (Downing, in press; Downing & Downing, in press). Hence, postulates 5, 6, and 7 of the cognitive clarity theory must be revised as follows:

(5) children approach the tasks of reading instruction with only partially developed concepts of the functions and features of speech and writing;
(6) under reasonably good conditions, children develop increasing cognitive clarity about the functions and features of language;
(7) although the initial stage of literacy acquisition is the most vital one, conceptual challenges continue to arise and thus broaden the range of cognitive clarity throughout the later stages of education as new subskills are added to the student's repertory.

Concepts of the Functions of Literacy

Piaget (1959) reports an experiment in which children of primary school age were grouped in pairs. One member of the pair was the "explainer" and the other was the "listener." The adult experimenter gave the explainer a diagram of a water tap and told him how it worked. The adult experimenter got the child to repeat back the description to make sure that he had understood it. Then the explainer was told to go and explain the tap to the other child, the listener. Next, the listener was asked to explain the tap to the adult experimenter. The words of both the explainer and the listener were recorded. The results of both communication acts—the speaking of the explainer and the listening of the listener—were remarkably poor. Another experiment in retelling a story produced similar results.

Piaget's explanation of these results is that children are slow to give up their egocentric point of view in such communication activities. The explainers explained inefficiently because they believed that the listeners knew just as much as they did themselves. Similarly, the listeners believed that their understanding was equal to that of the explainers and hence their listening was inefficient also. Neither the explainers nor the listeners put themselves at the point of view of the other partner in communication. Piaget points out that at this stage children have had little experience with communication problems. They have not yet appreciated what Piaget described as the appalling density of the human mind. Thus, Piaget showed that children of the typical school-beginning age have relatively little awareness of the communication functions of language.

More precisely on the point of awareness of literacy functions, Vygotsky (1934) found, in his study of school beginners in Russia, that they had only a vague idea of the purposes of written language. More recently, Reid (1966) conducted focused interviews with five year olds beginning school in Scotland. She found that they showed a general lack of awareness of the purposes of written language. Downing (1970) replicated and extended Reid's study and confirmed that young beginners have difficulty in understanding the purposes of literacy.

Downing conducted two studies of the influence of home background on the development of children's concepts of the purposes of literacy. Both studies used the Orientation to Literacy test developed by Evanechko, Ollila, Downing and Braun (1973), which is a paper and pencil test in which subjects are required to select and circle a picture which illustrates a specified function of literacy. In the first study (Downing, Ollila & Oliver, 1975), two groups of Canadian-Indian beginners were found to be significantly less aware of the purposes of reading and writing than were non-Indian beginners attending the kindergarten classes in the same school districts. This supported the hypothesis that Indian children's concepts of the functions of writing are less well developed because they come from a home background with no cultural tradition of literacy. In a second study (Downing, Ollila & Oliver, 1977), the same test was administered to 787 kindergarten children in a stratified random sample of schools in a large city in western Canada. It was found that children from lower socioeconomic classes scored at a significantly inferior level than children of higher socioeconomic class.

The Orientation to Literacy test used in the above-mentioned studies was later redesigned to test a wider range of concepts with a greater number of items and improved reliability. The new test is named Understanding Literacy Functions and it is part of a battery called the Linguistic Awareness in Reading Readiness (LARR) test (Downing, Ayers & Schaefer, 1983). The Understanding Literacy Functions subtest is designed to discover whether or not the child understands purposes of literacy. The subject is asked, for example, to draw circles around persons in a row of pictures who are enjoying a story, learning that there is a sale on, finding what music to listen to, telling someone a story, remembering what to buy at the store, and so on.

The two other subtests in the LARR battery are Recognizing Literacy Behavior and Technical Language of Literacy. The first of these investigates whether or not the child can recognize the kind of activities which are involved in reading and writing. For example, the child is asked to draw a circle around the part of the TV screen or cereal box which people read. The child is asked to circle each person who is reading or each person who is writing, and so on. The Technical

Language of Literacy test will be described later in this chapter in the section on featural concepts of literacy.

Each of the LARR subtests has two parallel forms, "A" and "B."

Downing et al. have conducted two major investigations of the reliability and validity of the LARR test which also provide data relevant to a consideration of children's concepts of the functions and features of language as postulated in the cognitive clarity theory. These investigations will be reported in a later section of this chapter.

For the most part, the other chapters in this volume have little or nothing to say about the child's awareness of the functions of literacy. The exception is Johns' chapter on students' perceptions of reading, which does include some further evidence on children's functional concepts.

It seems surprising that the study of children's awareness of language in speech and writing should have so little to say about concepts of their functions. The neglect is surprising in view of the emphasis placed on the influence of the reader's purpose on her/his style of performance of the skill of reading.

It is quite well established that the purpose of the reading act is closely related to its technique. Purpose in reading has been likened to the gear-shift system in an automobile. The shift system changes dynamically according to the driver's (or reader's) purpose and the level of difficulty of the road (or book).

Two classical studies indicated the key significance of purpose in reading. Both Gray (1917) and Judd and Buswell (1922) found that eye movements in reading change according to reading purpose. The eyes are literally extensions of the brain and provide an opportunity to observe changes in brain processes during reading. A considerable body of research exists which shows that changes in eye movements and hence brain processes occur when readers change their purposes for reading and when the level of difficulty of the text changes (Buswell, 1926; Tinker, 1958, 1965).

Research by other methods also has shown how the reader's purpose influences her/his style of reading. For example, Postman and Senders (1946) found in their experiment with college students that their reading comprehension was significantly influenced by the purpose that they were instructed to adopt. Rickards and August (1975), also studying college students, found that those "who were free to underline any one sentence per paragraph recalled significantly more incidental material than those explicitly instructed to underline the one sentence per paragraph that was most important to the overall meaning of each paragraph" (p. 864). This suggests that awareness of purpose is an important factor in comprehension. Russell (1970) believed that "The dominant factor in comprehension . . . is the purpose of the reader, stated or unstated" (p. 170). Different purposes require different reading techniques, and this suggests the need for experience and practice in reading for a variety of purposes if flexibility of style is to be developed as the gear-shift system in reading skill. Therefore, Malmquist (1973) has proposed that awareness of the purposes of reading should be fostered from the earliest stages of schooling.

Nevertheless, there is a paucity of research evidence of children's awareness of the functions of literacy and the way in which this is or is not related to the development of skill. However, some further evidence on this question will be presented later in this chapter when we return to the research on the LARR test.

Concepts of Features of Language

There is a much larger body of research evidence on children's linguistic awareness and their developing concepts of features of language. Indeed, the major part of this volume deals with metalinguistic development. We will leave it to the reader of this book to explore those other chapters on this aspect of task awareness. We shall confine ourselves here to some of our own studies of children's featural concepts with brief citations of some other related studies.

Luria (1946) proposed what has become known as the "glass window theory." He stated:

> The first important period in a child's development is characterized by the fact that, while actively using grammatical speech and signifying with words the appropriate objects and actions, the child is still not able to make the word and verbal relations an object of his consciousness. In this period a word may be used but not noticed by the child, and it frequently seems like a glass window through which the child looks at the surrounding world without making the word itself an object of his consciousness and without suspecting that it has its own existence, its own structural features (p. 61).

For example, a child who hears the sentence "The wolf ate Red Riding Hood's grandmother," may conjure up some image of that event, without realizing that the utterance included the word "the," the word "wolf," the word "ate," and so on.

A number of investigators (Chappell, 1968; Holden & MacGinitie, 1972; Huttenlocher, 1964; Karpova, 1955), using a variety of research methods, found that the speech segments of children's perception do not coincide with the unit "word" or "phoneme" as usually understood by adults such as teachers.

Liberman, Shankweiler, Fischer, and Carter (1974) conducted an experiment in which children were trained to tap out speech segments. No four year old was able to learn the phoneme segmentation task, but 46 percent succeeded in acquiring syllable segmentation. For five year olds the respective percentages were 17 and 48, while 70 and 90 percent of six year olds succeeded in learning to segment phonemes and syllables, respectively. Hakes (1980), using the same phoneme segmentation training task, confirmed that four and five year olds find it very difficult to learn how to segment utterances into phonemes. The ability to learn to segment by phonemes increases with age. By age 8 most children can accomplish this task.

It may seem odd that young children's listening comprehension and meaningful utterances utilize phonemic discriminations and yet they cannot distinguish the phoneme unit itself. But there is an important psychological difference between using language and making judgments about language. Gleitman and Gleitman (1979) state:

> It always turns out that giving language judgments—retrieving and making use of one's intuitions—is relatively hard, compared to talking and understanding. Thus it is not surprising that we find extensive individual and population differences in performance on the harder judgmental tasks, compared to lesser differences in talking and understanding. We believe this is because judgmental performances require a higher order of self-consciousness than do speech performances. To give a language judgment, one must take a prior cognitive process (linguistic performance) as the object of a yet higher-order cognitive process (reflection about language performance, or, as we have called it, *metalinguistic* performance) which may have properties of its own (p. 105).

Thus, as Liberman and Shankweiler (1979) pointed out, a child may be able to distinguish between pairs of spoken words like "bet" and "best" and to recognize each as a different word in her/his vocabulary, without being aware that "bet" has three phonemes while "best" has four.

But the phoneme segment is also particularly difficult to distinguish because, as Liberman, Cooper, Shankweiler, and Studdert-Kennedy (1967) demonstrated, phoneme boundaries are not clearly definable since phonetic segments are often co-articulated. For example, a consonant may be merged with a related vowel. The word "dog," for instance, has three phonemes, but studies of acoustic properties of such words in speech show objectively that "dog" consists of only one single acoustic segment. Thus, the child who does not hear three segments in "dog" is quite correct because they do not exist acoustically. The three segments of "dog" are only recognized by someone who has the abstract concept of phoneme and understands the phonemic system which is the foundation of alphabetic writing.

Another approach to the study of children's awareness of linguistic features is to test their responses to technical linguistic vocabulary in what DeStefano (1972) has called the "language instruction register"—"word," "sound," "letter," "number," "reading," "writing," and so on. Reid's study of Scottish five year olds found that these children exhibited considerable confusion about the meaning of such linguistic terms. Reid concluded that they had "a great poverty of linguistic equipment to deal with the new experiences" (p. 58) in reading instruction. Her findings were confirmed by Downing (1970) in a replication study conducted with a similar sample in England. This replication was extended to test children's interpretation of two technical terms that are used in reading instruction to refer to features of speech. Following appropriate pretraining in the experimental procedure, five year olds were asked to say whether each of a series of auditory stimuli was "a word." Five types of stimuli were used: a nonverbal sound, a meaningless vowel phoneme, a single word, a phrase, and a sentence. No child's category for "a word" coincided with the usual concept of a spoken word held by teachers. Some children made only random guesses; some excluded nonverbal sounds; and some thought that only the word, the phrase, and the sentence were each "a word." The experiment was repeated at three intervals during the first school year, but although responses were progressively in the direction of interpreting "a word" as a chunk of meaningful speech, no child achieved by the end of the year what teachers would usually consider to be the concept of "a spoken word." Similar results were found for the term "sound," although some pupils understood it as meaning a phoneme by the final testing session (Downing, 1971-1972).

Downing and Oliver (1973-1974) produced an improved instrument for testing children's understanding of the term "word." Subjects (Canadian) at the preschool, kindergarten, and grade-one levels were presented with eight classes of auditory stimuli to which they responded "yes" if they believed that each was a single word and "no" if they did not. All of the children, regardless of age, confused isolated phonemes and syllables with spoken words. Also, children up until the age of 6.5 years, in addition, tended to confuse nonverbal sounds, phrases, and sentences with words, but these confusions disappeared with older subjects. It was found also that children between the ages of 5.6 and 6.5 years tended to exclude long words from their conception of the spoken word, but children younger than 5.6 years and older than 6.5 years did not.

This last-mentioned finding from the experiment of Downing and Oliver was made possible because two of their categories of auditory stimuli were short words such as "cat" and long words such as "hippopotomus." This distinction between long and short words was introduced because of an earlier finding by Meltzer and Herse to be reviewed later when the possible cause of this curious age phenomenon will be discussed.

Papandropoulu and Sinclair (1974) asked young French-speaking children "to say a long word and a short word." The youngest (4.5-5.5 years) responded with the names for long and large objects or words for actions that take a long time as examples of "long words." Lundberg and Tornéus (1978) presented Swedish preschool children with word pairs—one long and one short. The words were spoken to each child who was then shown a card on which the two words were printed, but the experimenter did not reveal the relation between the printed word and the spoken word. The child was required to point to one of the words nominated by the experimenter. There was a clear tendency for these children to use a semantic strategy in selecting the printed word. For example, if the spoken word was the name of the larger object, then the child pointed to the longer of the two printed words.

Other researchers have focused their attention more on children's conceptions of features of written language rather than speech. Ehri's chapter in this book is an example. An early study in this area was the one by Meltzer and Herse (1969). They asked American kindergartners and first graders to cut off with scissors a word from a sentence printed on a card. Occasionally a word was cut off, but often it was two words and sometimes a part of a word. A variety of different testing procedures has demonstrated the same phenomena in children's understanding of the technical features of written language: Clay (1972) in New Zealand; Kingston, Weaver, and Figa (1972) in America; and Turnbull (1970) in Australia.

Meltzer and Herse found that grade-one children believed that a written word is comprised of only a limited number of letters. Downing and Oliver found a similar tendency to accept short spoken words as being "words" but not longer spoken words at about the same age level, though children younger and older than this group did not reject the longer words. Meltzer and Herse speculated that the tendency to reject longer words as being "words" may be caused by the predominance of short words in the basal readers used in first grade. The findings of Downing and Oliver suggest that this misconception transfers to children's beliefs about spoken words. Ferreiro and Teberosky also state that teaching methods in grade one can cause misconceptions about written language. Often their Argentinian subjects regressed in development of concepts of writing when they entered first grade.

How the development of these featural concepts and metalinguistic ability is related to achievement in learning how to read is still being debated.

Francis (1973), in her study of English primary school boys and girls, found that her highest correlation was between reading achievement and knowledge of a technical linguistic vocabulary, even with general vocabulary knowledge controlled. She concluded that factors independent of a general ability to deal with abstract concepts were involved in learning technical vocabulary, and that these were closely related to the reading process. She believes that, under current methods of instruction, children learn these technical concepts by groping their way through mostly unplanned experiences of hearing the technical jargon used by their teachers. Francis writes:

> It was as though the children had never thought to analyze speech, but in learning to read had been forced to recognize units and subdivisions. The use of words like *letter*, *word* and *sentence* in teaching was not so much a direct aid to instruction but a challenge to find their meaning (p. 22).

A very complex set of special technical terms and their underlying featural concepts is used by reading teachers. For instance, consider the following passage in which some of these linguistic concepts have been relabeled with nonsense words instead of the conventional school jargon:

> This is how you sove the zasp "bite". It is tebbed with the rellangs fly, ear, milk, wow. The last rellang is the holy wow. When you have a holy wow at the end of a zasp the ear says ear not ook like it does in the zasp "bit".

That this may not be such an uncommon experience for beginners in the cognitive phase of learning to read is shown when we translate the nonsense words into standard terminology:

This is how you write *the* word *"bite". It is* spelt *with the* letters bee, eye, tea, ee. *The last* letter *is the* silent ee. *When you have a* silent ee *at the end of a* word *the* eye *says* eye *not* i *like it does in the* word *"bit"*.

Egorov (1953) wrote that "The conceptual difficulties of this initial period of reading instruction are serious. Therefore, the teacher must take special care to avoid adding to the pupils' difficulties by introducing any unnecessary compli- -cations. For example, a common teaching error in the pre-primer period is flood- ing young beginners with too many new concepts, such as 'sentence', 'word', syllable', 'sound', 'letter', and so on." Egorov's pedagogical recommendations were based on extensive observations of children's behavior in first-grade class- rooms in the U.S.S.R.

It seems rather obvious that children's efforts to make preliminary fixes on the nature of the task of how to perform the skill of reading are not helped when instruction is overloaded with jargon that is not comprehended. Neverthe- less, once the jargon of the language instruction register is understood by the learner, it becomes very efficient equipment for thinking about the reading task.

The research by Ferreiro and Teberosky shows clearly that children begin their construction of the concept of written language long before they reach the age for beginning school. The same must be true of the metacognitive and meta- linguistic abilities involved in that construction process. In his chapter in this book, Mattingly proposes that this preschool metalinguistic ability is an impor- tant prerequisite for learning how to read. He states the following hypothesis:

> The child who is still actively acquiring language at the time he begins to read will be relatively mature phonologically, so that the orthogra- phy will correspond to a considerable extent with his morphophonemic representations. Having access to these representations, he will be linguistically aware, and the orthography will seem to him a plausible way of representing sentences. The analysis of a sentence on the basis of its lexical content will present no problems, since he has continued to use this analytic procedure in the course of learning new words. Moreover, he will, as a linguist, see that reading is a source of fresh data. . . . Thus the linguistic curiosity that has motivated his continuing language acquisition will motivate his learning to read as well.

Thus, preschool experiences with both speech and writing influence the child's early preliminary fixes on the task of learning the skill of reading. Interaction with adults and other children in games and discussions of everyday life experi- ences with speech and writing in the environment will influence the development of metacognitive and metalinguistic ability. Readiness for task awareness thus will show considerable individual differences. Mattingly mentions that "the course of language acquisition varies considerably. At one extreme is the indi- vidual who virtually abandons language acquisition as soon as he has developed the relatively modest body of analytical strategies needed to get by; at the other extreme, the 'word-child' who continues indefinitely to add to his grammatical knowledge." We have already quoted Mattingly's description of the happy state of the linguistically aware "word-child." In contrast, Mattingly writes "the child who is no longer actively acquiring language will surely find learning to read very difficult and unsatisfying. His morphophonemic representations will be less mature than they might be, so that the discrepancies between the orthography and the morphophonemic representations will be substantial. More seriously, these representations, being part of his grammatical knowledge, will have become less accessible to him; he will be lacking in linguistic awareness. As a result, the orthography will seem a mysterious and arbitrary way to represent sentences." Mattingly's chapter is of crucial theoretical significance to the area studied in this book.

In addition to such linguistic analyses, there is a need to study other variables in the culture and the family that may influence the child's conception of speech and writing. Two of the studies referred to earlier in this chapter in the section on concepts of the functional features of literacy also provide evidence for the influence of the child's environmental experiences on the development of featural concepts. Downing et al. (1975), in their comparison of Indian and non-Indian subjects in Canadian kindergarten classes, used another instrument produced by Evanechko et al. called the Technical Language of Literacy test, which investigated concepts of written language, such as "letter," "number," "sentence," and so on. Children had to circle the item nominated by the experimenter. The Indian kindergartners were found to be significantly more immature than the non-Indian subjects in their development of featural concepts. In the second study (Downing et al., 1977) of more than 700 kindergartners in a city in western Canada, this same test revealed a significant relationship between socioeconomic class and the level of understanding of featural concepts. The advantage held by the higher socioeconomic group persisted even after a year of kindergarten education. Furthermore, these results were quite highly correlated with more traditional tests of reading readiness, notably those of letter-name knowledge and auditory discrimination that are well known to be the best predictors of later attainments in reading. This correlation suggests that the test of featural concepts and the more traditional reading readiness tests probe the same underlying basic factor—the child's cognitive clarity or confusion about the featural concepts used in reading instruction.

Another study that showed a significant relationship between reading readiness and the development of featural concepts is that by Calfee, Lindamood, and Lindamood (1973). Children from grades ranging from kindergarten to grade twelve were tested to determine their ability to match phonetic segments by use of a sequence of colored blocks to represent auditory stimuli. They were tested also for reading and spelling achievements. The results showed that more than 50 percent of the total variance in reading ability could be predicted from the student's ability to perform this acoustic segmentation task.

Phoneme segmentation ability has been shown to be significantly related to reading achievement in several other studies. Liberman (1973) tested first-grade children's ability to tap out the number of phonemes in a word, and subsequently related their segmentation ability to scores on a word-recognition reading test administered in second grade. One half of the lowest third of the class in reading had previously failed the phoneme segmentation test, whereas none of the subjects in the top third of the class in reading had failed the segmentation test. Zifcak (1981) too reported a highly significant relationship between the reading success of first graders and their performance on Liberman's phoneme segmentation task. Helfgott (1976) and Treiman (1976) used other methods of testing young children's ability to segment phonemes and similarly found a significant relationship between phoneme segmentation ability and reading achievements.

The link between a wider range of metalinguistic abilities and reading achievement was demonstrated in an investigation by Evans, Taylor, and Blum (1979). They developed a battery of seven tests of various concepts of language features and administered them to 60 American first graders. Evans et al. analyzed their scores by stepwise multiple regression on standardized reading achievement test scores. They found that the children's performances on the featural concept tests were predictive of reading achievement.

In the previous section of this chapter we described how the LARR test was developed as an improvement on the earlier tests used in the two studies by Downing et al. (1975, 1977) on the influence of cultural and subcultural variables on children's development of task awareness. A similar improvement was made in the third test in the LARR battery—Technical Language of Literacy. This tests the child's knowledge of technical terms used in describing written or printed language such as "letter," "word," "sentence," and so on. For example, the child is asked to draw a circle around each part of a display that is an example of such stated technical terms. As was mentioned in the previous section of this chapter, the research on the reliability and validity of the LARR test also provides data for consideration in regard to task awareness theory. This work will be reported in the next section.

Task Awareness and Skill Integration

The richness of the research on the metalinguistic aspect is very satisfying but it must not divert us from the broader range of concepts involved in task awareness. Integration of a complex pattern of subskills is the key feature of skill performance, and the skill learner's preliminary fix is of this integrated whole. Therefore, it seems appropriate to probe task analysis along a broader front.

Earlier in two preceding sections of this chapter the three parts of the LARR test were described. It may be recalled that the three parts are: (1) Recognizing Literacy Behavior; (2) Understanding Literacy Functions: and (3) Technical Language of Literacy. As part of their program of research on the validity of the LARR test, Downing, Ayers, and Schaefer (1978) carried out a structured interview survey of children who were also administered the LARR test. The interview schedule comprised four parts:

I. Recognition of Acts of Reading and Writing

This part had 20 items. The first three required the child to show "the part that
people read" on a page from a children's story book, a food package, and a
newspaper. All three items contained pictures as well as print. In items 4-7, an
eraser, a pen, a pencil, and a ruler, respectively, were displayed and the child was
asked "Can someone write with that?" Items 8-14 consisted of photographs of
people either reading or not reading and the interviewer asked, "Is she (or he)
reading?" Finally, questions 15-20 inquired, "Is she (or he) writing?" about a set
of photographs of people either writing or not writing.

II. Concept of the Purposes of Reading and Writing

This part consisted of 20 photographs of people engaged in various activities that
either did or did not illustrate a purposeful use of reading or writing. For example, question 4 showed a woman combing her hair and the interviewer asked, "Is
this woman enjoying a story?" Question 5 asked, "Is this girl finding out what
music to listen to?" and the picture was of a girl reading the description on a
record container.

III. Concepts of Features of Printed Materials

There were 17 items in this part. The page from the story book was displayed
again. Questions 1, 2, and 3 required the child to show the interviewer "the
picture," "the printing," and "the writing," respectively. Questions 4-7 employed
a technique originally described by Clay. Three small white cards had to be
placed on the printed page so that only a stipulated feature was visible, for
example, "the top line," "just one word," "just one letter," and so on.

IV. Visual Perception

This part consisted of 20 drawings similar to (but not identical with) those
employed in what has been described as the most predictive subtest in The
Marianne Frostig Development Test of Visual Perception (Frostig, Lefever &
Whittlesey, 1964). However, the task was different from that required of the
child in the Frostig test. The latter asks the child to trace geometric shapes
that are present in each picture. In the adaptation used in this present investigation, the interviewer displayed the drawings one at a time and the child was
asked only to give a verbal "yes" or "no" response to questions such as "Is there
a circle in this picture?" This was scored as either correct or incorrect, but the
child's spontaneous additional comments were also noted.

At every stage in the interview, and whenever a new task was introduced,
pretraining was given to ensure that the child understood the type of response

Table 3-1. Subjects' scores on each part of the interview.

	Part I	Part II	Part III	Part IV
Number of questions	20.0	20.0	17.0	20.0
Mean number correct	17.8	15.8	8.5	18.4
Range of scores	6-20	4-20	1-17	10-20
Standard deviation	2.1	3.0	3.5	1.5

that was required. At the beginning of the interview, a "yes/no" game was played to prepare the child to give "yes" or "no" responses as were required for most of the questions. In Part III, the use of the three white cards was taught by demonstration, and then practice was given in isolating obvious items in the picture on the page of the storybook. In Part IV, the concepts of "circle" and "square" were demonstrated before the test items were displayed.

Further details of the interviewing method and the interviewers and their training can be found in the article by Downing, Ayers, and Schaefer (1978).

Schools were selected in a Canadian city by a method of stratified random sampling on the basis of socioeconomic class. Three hundred and ten children were interviewed—about one half of the population of kindergarten in the city's schools.

The results are summarized in Table 3-1. In comparing the scores on the four parts of the test, it should be noted that Part III had fewer items than the other parts. The mean number of items correct expressed as a percentage of the total possible score for each part was as follows: Part I, 89; Part II, 79; Part III, 50; and Part IV, 92.

The test of visual perception (Part IV) had the highest mean number of correct responses, 18.4, which is 92 percent of its 20 items. Furthermore, the standard deviation was very small. Thus, this test provides very little discrimination of individual differences in children. Most children easily perceived these graphic shapes. Spontaneous comments of the children also raised doubts as to whether the test is a perceptual one at all. It is more likely a very simple reasoning test. Nevertheless, it still provided a foil for the three other parts of the interview. Parts I, II, and III were predicted to be related to the LARR test, while Part IV was not.

The order of difficulty of the three other parts of the interview was I, II, and III. The mean number of correct responses for Part I was 17.8, which is 89 percent of the total score. The spread of scores (SD = 2.1) was wider than for the perceptual test of Part IV, but Parts I and IV are much closer in difficulty level than the other parts of the interview. Two items produced a much lower proportion of correct responses, 40 and 43 percent, respectively. These were the only two items showing a person drawing a picture. Thus, a large proportion of the sample confused "writing" with "drawing." The same confusion was found in an earlier open-ended interview study (Downing, 1971-1972) when some children drew pictures when asked to "write." The data from the present study indicate that about nine out of ten children aged five can recognize reading and writing behavior when they see it, but only about six out of 10 make a clear distinction between writing and drawing activities.

Part II of the interview, which investigated these kindergarten children's understanding of the purposes of reading and writing, proved more difficult. The mean number of correct responses was 15.8, which is 79 percent of the possible total score. The standard deviation (3.0) was also larger than that for Part I.

Part III, which examined the child's concepts of features of written language, was the most difficult section of the interview. The mean number of correct responses was 8.5, which is 50 percent of the possible total score. The spread of scores (SD = 3.5) also was the widest. The level of difficulty of the individual items was erratic but generally high. The first question, however, was readily answered by 96 percent of the sample. This required the subject to show the interviewer "the picture" on the page from a children's book. Among the comparatively better known concepts were: "printing" (74 percent), "letter" (69 percent on item 6; 75 percent on item 14), "bottom line" (70 percent), and "number" (71 percent). However, these proportions certainly do not encourage one to assume that beginners are always familiar with these concepts, and other technical concepts were much less well understood. For example, such commonly used teacher language as "sentence," "first word," and "first two words" were understood by only 15, 34, and 22 percent, respectively. These results give more detailed substance to the findings of the research on children's linguistic concepts reviewed earlier in this chapter. These children, like those in Reid's original study, had "a great poverty of linguistic equipment" to cope with the technical terminology of reading instruction. The large, carefully selected sample of kindergarten children in this investigation overcomes the corresponding weakness of previous research on this problem. The conclusion that young children's understanding of the functions and features of written language cannot be taken for granted seems inescapable.

Ayers and Downing (1982) have reported in a separate article the results of administering the LARR test to these same 310 children in kindergarten and then a year later in first grade, testing their reading achievement on the Cooperative Primary Reading test, form 12A (Educational Testing Service, 1967). Half of the classes were randomly assigned Form A and the other half of the classes form B of the LARR test. The Cooperative Primary Reading test was chosen because it is not limited by any ceiling effect through being too easy. It challenges the superior readers but, at the same time, progresses from recognition of simple words, through sentences, to interpretation of paragraphs. It could thus also be readily used for obtaining subscores for words, sentences, and paragraphs, as well as for the total score. In addition, a score was obtained for "inferences" which represented the highest level of comprehension, according to the test's authors.

Multiple regression analyses were conducted for each of the forms to determine the tests that made a significant contribution to the prediction of the various reading subscores and total reading score. Also, to determine if prediction varied across classrooms because of small numbers in the classes, Spearman rank-order correlations between the three tests in the LARR battery and two of the reading scores, word plus sentence score and total score, were calculated. Examination of the data by analysis of covariance was also conducted despite the fact that the children had not been assigned randomly to the classes. This was done in the hope that such analysis might provide some clues about the usefulness of LARR subtests for predicting under different classroom conditions.

The multiple regression analyses showed that, with one exception, the simple correlations of test 3 (Technical Language of Literacy) with the various reading subscores and total scores were as high as the multiple correlations. Consequently, these data are not included here.

In one case, test 1 (Recognizing Literacy Behavior) increased the correlation of test 3 alone from 0.48 to 0.51 in predicting the word subscore of the Cooperative Primary Reading test. The correlations of form 3A with sentence subscore, paragraph subscore, inference subscore, and total score were all approximately 0.60. For form B the correlations varied between 0.44 and 0.48. The differences in the sizes of the validity coefficients for the two forms is due in part to the lower reliability of the form B tests, as will be shown below in Table 3-4. Validity coefficients in the high forties are quite modest. (Clearly, form B needed the strengthening to be described in the next section of this chapter, which produced the improved reliability shown in Table 3-4.) However, those in the sixties for form A compare favorably with those usually obtained with other reading readiness tests, especially with a diverse cross section on socioeconomic status. (Compare, for example, Engin, 1974; Henderson & Long, 1970; Lessler, Schoeninger & Bridges, 1970; Randel, Fry & Ralls, 1977; Thackray, 1965, 1971).

The above correlations are based on the combined groups and may not reflect the prediction for single classroom groups which is what is usually involved in the practical situation. Table 3-2 shows the Spearman rank-order correlations between the three LARR tests and (1) the score for words and sentences and (2) the total score for each classroom group. It shows, with few exceptions, a consistency of prediction for both the simpler reading tasks involved in the word and sentence score and the combination of simple and difficult tasks in the total score. It also shows that test 3 is generally the best predictor, although, in two classes, test 1 is best and, in one class, test 2 (Understanding Literacy Functions) is best. But what is interesting is the size of the correlations for the classroom groups. Many are quite large, with several over 0.70. For test 3 there is only one group for which the prediction is not significant. For tests 1 and 2, however, less than half the groups have significant correlations.

The results of the analysis of covariance in which the Cooperative Primary Reading test scores at the end of grade one were adjusted for the LARR stanines obtained at the end of kindergarten are shown in Table 3-3. The data show that two classrooms were above average in reading skill in grade one (numbers 31 and 63), three below average (numbers 41, 73, and 64) and the remaining five were about average in the results achieved in reading as measured by the Cooperative Primary Reading test.

The above average, average, and below average groupings were used to classify the predictive validity coefficients in Table 3-2. When this was done, it was determined for both the above and below average groups that two of the LARR tests correlated significantly with reading in grade one, while for the average group, only test 3 had significant predictive validity. Although it is not possible to explain this finding, it does confirm that all three tests should probably be retained in order to ensure adequate predictive validity in a variety of situations.

The reliabilities and intercorrelations of the three LARR tests are reported in Table 3-4. The data indicate that the reliabilities of tests 1 and 3 are sufficiently high for individual prediction, and, while test 2 reliability is adequate for group prediction, it would need improving to be used for individual prediction.

It should be noted that the LARR test used in this first study was not as strong as it could have been. Subsequently, all of the items in both forms A and B of the three tests in the LARR battery were subjected to analysis, and the poor items were revised or replaced for a revised edition. In making this revision, the tests also were shortened to reduce testing time, while, at the same time, attempting to maintain adequate reliability. Table 3-5 shows the reliabilities and intercorrelations using the revised edition with a similar group of kindergarten children tested in the same city. Here the reliabilities for tests 2 and 3 proved adequate for individual prediction. Test 1 showed a need to be lengthened by about four items. Furthermore, the means and standard deviations show that form B has become about equal to form A in reliability. The intercorrelations, however, were much reduced from those in the original battery indicating that the tests were more homogeneous. This should result in improved prediction with more than one test contributing significantly to multiple R.

There is certainly sufficient information provided to indicate that test 3 is a useful predictor of reading achievement in grade one as measured by the part scores and total scores of the Cooperative Primary Reading test. Further studies should confirm this finding with similar reading achievement tests. The extent and conditions under which tests 1 and 2 would be useful in making decisions about children being taught to read has yet to be fully determined. There are indications, however, that with the lower intercorrelations in the revised battery and with the apparent usefulness of tests 1 and 2 in predicting with both high- and low-achievement groups, these tests should be retained in the battery. These encouraging results led the authors to produce the published edition of forms A and B of the LARR test (Downing, Ayers & Schaefer, 1983), which contains further technical improvements and better illustrations by a professional artist.

The validation work comparing the data from the structured interviews with those obtained from the LARR test has not yet been reported. However, the two sets of data are presented here in succession since they provide evidence on the importance of the several types of concepts involved in task awareness. The LARR test data indicate that metalinguistic ability in kindergarten, as measured

Table 3-2. Spearman Rank-Order Correlations of LARR Tests 1, 2, and 3 with Cooperative Primary Reading test (C.P.R) Part Score for Words and Sentences and with Total Score*

Teacher No.	N	C.P.R. Word and Sentence Score			C.P.R. Total Score			LARR Test Giving Significant Prediction
		LARR 1	LARR 2	LARR 3	LARR 1	LARR 2	LARR 3	
11	13	-0.07	0.28	0.38 (0.10)	-0.06	0.38 (0.10)	0.46 (0.06)	3
21	16	-0.08	0.07	0.55 (0.01)	-0.08	0.35 (0.10)	0.52 (0.02)	3
31	11	0.24	0.63 (0.02)	0.75 (0.004)	0.27	0.59 (0.03)	0.71 (0.01)	2 and 3
41	16	0.63 (0.004)	0.11	0.51 (0.02)	0.71 (0.001)	0.17	0.54 (0.02)	1 and 3
42	17	0.38 (0.07)	0.02	0.70 (0.001)	0.31	0.10	0.67 (0.002)	3
51	8	0.20	0.78 (0.01)	0.68 (0.03)	-0.04	0.53 (0.09)	0.41	2 and 3
52	8	0.41	0.23	-0.30	0.30	0.31	-0.41	
63	19	0.47 (0.02)	0.18	0.61 (0.003)	0.47 (0.02)	0.32 (0.09)	0.71 (0.001)	1 and 3
64	10	0.78 (0.004)	0.11	0.73 (0.008)	0.73 (0.008)	0.27	0.64 (0.02)	1 and 3
71	12	0.37	-0.19	0.51 (0.04)	0.35	-0.16	0.65 (0.01)	3
72	16	0.00	-0.03	0.65 (0.003)	0.12	0.16	0.73 (0.001)	3
73	10	0.28	-0.22	0.80 (0.003)	0.27	-0.26	0.79 (0.003)	3

*Levels of significance are shown in brackets.

Table 3-3. Ancova Cooperative Primary Reading Test Mean Scores by Teacher for Grade 1 Adjusted for Kindergarten LARR Test 3, and for Total of Tests 1, 2, and 3.

Teacher Number	Mean	COOPERATIVE PRIMARY READING TEST	
		Adjusted Mean* (LARR Test 3 only)	Adjusted Mean** (Total LARR Tests 1, 2, & 3)
11	36.38	35.93 (–0.45)	38.20 (+1.82)
21	34.69	35.53 (+0.84)	35.34 (+0.65)
31	31.64	34.84 (+3.20)	34.62 (+2.98)
41	25.31	19.58 (–5.73)	18.62 (–6.69)
42	32.59	32.29 (–0.30)	31.41 (–1.18)
63	36.74	43.49 (+6.75)	46.64 (+5.90)
64	26.20	24.25 (–1.95)	24.61 (–1.59)
71	35.92	35.40 (–0.52)	36.10 (+0.18)
72	36.56	36.90 (+0.34)	37.48 (+0.92)
73	26.20	20.80 (–5.40)	20.91 (–5.29)

*Main effect $F_{1,9}$ = 4.12; significance level = 0.0001; and multiple R = 0.72
**Main effect $F'_{1,9}$ = 3.33; significance level = 0.001; and multiple R = 0.63

by test 3, is most strongly associated with subsequent reading achievement in grade one. Other metacognitive abilities such as understanding the purposes of literacy in kindergarten are not so strongly and not so clearly associated with later achievements in reading. But the structured interview data show that metalinguistic tasks are much more difficult for kindergarten children than the other metacognitive tasks. Also, the metalinguistic tasks are more sensitive to individual differences. Testing at a younger age may find higher correlations between tests 1 and 2 and later reading achievement.

Table 3-4. Intercorrelations, KR 20 Reliabilities, and Summary of Scores for LARR Tests 1, 2, and 3 End of Kindergarten, May-June, 1977.

Test and Form	Test and Form					
	1A	2A	3A	1B	2B	3B
1A	(0.90)					
2A	0.62	(0.84)				
3A	0.66	0.62	(0.93)			
1B				(0.84)		
2B				0.52	(0.77)	
3B				0.63	0.37	(0.90)
No. of Items	28	24	41	28	24	41
$\overline{X}$	21.3	11.7	20.0	20.2	12.1	19.0
s	5.9	5.1	8.8	4.9	4.4	7.6
N	154	151	146	140	137	141

Table 3-5. Intercorrelations, KR 20 Reliabilities, and Summary of Scores for LARR Tests 1, 2, and 3 End of Kindergarten, May-June, 1978.

Test and Form	Test and Form					
	1A	2A	3A	1B	2B	3B
1A	(0.76)					
2A	0.34	(0.95)				
3A	0.45	0.23	(0.87)			
1B				(0.82)		
2B				0.22	(0.84)	
3B				0.45	0.23	(0.91)
No. of Items	22	20	29	22	20	29
$\overline{X}$	18.4	13.6	19.1	15.5	11.6	18.3
Range	3-22	3-20	4-28	7-22	3-19	3-29
s	3.2	6.1	5.7	3.9	4.4	6.4

Conclusion

In this chapter, the evidence that reading belongs to the class of behavior that psychologists call "skill" has been reviewed. It has been proposed that, since reading is a member of the skill category, we can apply psychological research findings on skill learning in general to learning how to read in particular. One such finding is that the initial step in learning a skill or a subskill is to make a "preliminary fix" about what one has to do to make a good performance. This is "task awareness," also referred to as the "cognitive phase" in skill development.

Task awareness involves a variety of metacognitive abilities, such as metalinguistic knowledge and understanding the purposes of literacy. Our review of the research evidence to date shows that there is an important relationship between metacognitive awareness of the reading task and success or failure in learning to read. The precise causal nature of this relationship is not yet determined. It remains controversial because so much of the evidence is from correlational-type research. But many investigators would accept that there is a connection. The problem of controversy is whether the metacognitive development is a cause of success in learning to read or vice versa.

For example, Gambrell and Heathington (1981) in a recent interview study assessed adult good and poor readers' metacognitive awareness of task and strategy variables in reading. They reported that, in contrast to the good readers, the adult disabled readers had a limited understanding of reading as a cognitive process. But was their inadequate task awareness the cause of their poor reading or did their lack of experience of reading cause their vagueness about what one does in performing the skill of reading?

The present author takes the view that task awareness and reading achievement are interactive, but that task awareness is, nevertheless, a cause of success in learning how to read. Francis has stated this position well. She writes that children learn technical linguistic terms "in the course of learning to read" (Francis, 1975, p. 152), but, when she spells out in greater detail what happens in the "course of learning to read," both the interaction between task awareness and progress in reading and the causal effect of task awareness on reading achievement seem to be indicated:

> For many children, the ability to analyze and compare seen words seems to precede that with spoken words . . . , but this may simply reflect the easier experience of comparing items presented concurrently and continuously (words on paper) rather than those presented sequentially and intermittently (heard words). What is made conscious, however, is not a knowledge of structuring such as is captured in linguists' theories of syntax and phonology, but, simply the knowledge that both written and spoken language do have recognizeable regularities, and can usefully be taken apart and put together again the better to convey and grasp intended meanings. This is a further understanding of literate behavior. The underlying coding remains obscure . . . but the outcome of unconscious information processing in relation to text is the development of some knowledge of system in spelling, even with such a complex orthography as that of English. This knowledge guides performance in both reading and writing, not in the sense that it guarantees accuracy, but in the sense that it helps the child to make appropriate decisions and provides the framework for the growth of greater precision in performance (p. 149).

Ehri (1979) expressed this interaction hypothesis in words which accord well with this present author's conclusion:

> Although alternative causal relationships between lexical awareness and learning to read may be distinguishable logically, they may not be all that separable and mutually exclusive in reality. Rather lexical awareness may *interact* with the reading acquisition process, existing as both a consequence of what has occurred and as a cause facilitating further progress. For example, the beginning reader may learn first the printed forms of sounds he recognizes as real words. In this case, lexical awareness helps him learn to read. Once known, these familiar printed landmarks may, in turn, aid him in recognizing the syntactic-semantic functions of unfamiliar printed words so that he can mark these as separate words in his lexicon. In this case, decoding written language enhances lexical awareness. If this picture of the process is more accurate, then there exists truth in both positions. Rather than battling over which comes first, it may be more fruitful to adopt an interactive view and to investigate how a child applies his knowledge of spoken words to the task of reading printed language, and how enhanced familiarity with written words changes his knowledge of speech, enabling him to accommodate better to print (p. 84).

This present author would generally extend Ehri's interaction hypothesis from lexical awareness to task awareness. This seems to fit well with the psychological evidence that we have reviewed on the course of development in all skills.

CHAPTER 4

Students' Perceptions of Reading: Insights From Research and Pedagogical Implications

Jerry L. Johns

Overview

After providing an historical orientation to the origins of research in linguistic awareness, this chapter reviews studies relating to students' perceptions of reading and provides a number of practical suggestions for helping beginning and intermediate-grade readers develop a perception of reading in which meaning is the central focus. Research and instructional activities that have implications for 〝 disabled readers are also presented.

When asked how he viewed reading, a second-grade student said that reading was "stand up, sit down." When the youngster was asked to explain what he meant, he said that the teacher had him stand up to read; he would continue to read until he made a mistake, then the teacher would tell him to sit down. Hence, he perceived reading as a stand-up, sit-down process.

Educators, psychologists, and learning theorists generally accept the view that children's thought processes differ from those of adults. Piaget's (1959) theory of the development of children's thinking makes it clear that children progress through definite stages in their thought processes. The student's view of reading mentioned above is an example of a perception of reading that greatly differs from reading perceptions that are held by adults. Students in various stages of learning to read do their best to make sense of their instruction; however, they may develop perceptions about reading that impede their progress or make the task of reading a somewhat mysterious activity.

Beginnings of Research in Linguistic Awareness

Consider, for example, the work of Vernon (1957), who conducted an extensive review of research on reading disability. This review led her to conclude that "the fundamental and basic characteristic of reading disability appears to be cognitive confusion" (p. 71). The confusion seems to lie in why certain printed letters should correspond to certain sounds. In short, the disabled reader seems to be confused about elements that compose the reading process, and he or she tends to remain in this state of confusion.

Based on the work of Vernon and others (Edwards, 1958; Vygotsky, 1934), a new area for research began to evolve. Goodacre (1971) identified the exploration of children's perceptions of reading as an area for thought-provoking work. Her observation was an accurate prediction of things to come.

Strands of Linguistic-Awareness Research

Since 1971, a large number of research studies related to linguistic awareness have been conducted throughout the world. These studies originated from a variety of scholarly fields, theoretical orientations, or just simple curiosity about how students approach the task of reading. A casual review of the literature reveals that studies have been conducted in areas such as "concepts about print," "word consciousness," "linguistic awareness," "phonemic segmentation," "units of print," "metalanguage," "cognitive clarity," "metalinguistic ability," "print awareness," "knowledge of word structure," "syntactic awareness," "word boundaries," and "schemata for reading." Despite the variance in the terminology used in these studies, it is possible to identify the major strands of research within the general field of inquiry called linguistic awareness. These strands, while not necessarily discrete, provide an overview of the major topics explored in the area of linguistic awareness.

Eight strands of research have been identified by Hutson (1979). Her organizational scheme provides one way to approach, in a systematic manner, the large number of studies on linguistic awareness that have been conducted since the late 1950s. A slightly modified version of her topics follows:

(1) the nature and functions of reading
(2) the conventions by which we represent language in print
(3) definitions of "word," "sound," and "sentence," and differentiation of one term from another
(4) understanding of the relationship between words and sentences
(5) the relationships between words and sounds
(6) the development of metalinguistic judgment and the ability to hold language as an object of thought and to perform certain logical operations upon it
(7) the ability to manipulate units of speech intentionally
(8) the development of expectations about the literary forms one will encounter in oral and written language.

The first strand of research may well be the most global of the eight areas since it deals with general notions or perceptions about reading rather than with the more specific aspects or technical terms used in the teaching of reading. This area also provided the stimulus for much of the subsequent research that has centered on linguistic awareness.

The general plan for the remainder of this chapter, then, is to systematically review children's perceptions of reading, and to use the insights from this strand of research to offer pedagogical implications.

Research on Students' Perceptions of Reading

The major thrust of this section is to summarize the literature on students' perceptions of reading and to suggest how these perceptions may influence reading performance.

Early Explorations with Disabled Readers

One of the first inquiries into students' perceptions of reading was reported by Edwards (1958). He sought to find out students' perceptions in their approaches to reading by asking them what they had thought "good" reading was when they first started school. The sample was composed of 66 disabled readers from grades two to four who ranged from normal to superior in intelligence. Using interviews with groups of four (or fewer) students, Edwards asked the students to remember what their teachers and parents meant when they described the students' reading as "good." Although no statistical tests were used in this loosely controlled study, a significantly large number of students responded with the perception that "good" reading was a matter of speed and fluency. Glass (1968) found a similar emphasis on speed of reading in his interviews with poor readers from fourth grade through college. Edwards suggested that this view of reading may promote speed at the expense of reading for meaning.

Because students in the Edwards study were disabled readers, he noted that these students might have been more productive if the correct perception of the reading process had been taught directly and thoroughly. Suggestions for teaching the purpose of reading were given in a later article (Edwards, 1962). He concluded that beginning readers who are delayed in understanding that reading is for meaning could develop ineffective reading habits; moreover, these habits could possibly produce retardation in reading.

Johns (1970) echoed these thoughts when he found that 10 of 12 severely disabled readers said "I don't know" to the question, "What is reading?" He concluded that one of the contributing factors to students' reading problems may be their failure to understand what is involved in the reading process. Although the investigations by Edwards and Johns had the usual limitations of such informal interview techniques, they both suggested that students' perceptions of reading were different from those of adults; moreover, that this difference might influence students' reading achievement.

Studies with Preschoolers and First-Grade Students

In 1963, Denny and Weintraub began an investigation in five first-grade classrooms in three school systems to explore students' perceptions of reading. The

111 first-grade students came from widely divergent socioeconomic backgrounds: rural, middle class, and lower class (Denny & Weintraub, 1966). Each student was interviewed individually and asked three questions about reading. Responses were taped, analyzed, and classified into logical categories. The two investigators had a 90 percent level of agreement in the assignment of responses to the categories. An independent judge achieved an 82 percent level of agreement with the investigators' classifications.

In summarizing the responses to the three questions ("Do you want to learn how to read?", "Why?", and "What must you do to learn how to read in first grade?"), Denny and Weintraub (1966) noted that:

> A fourth of all these entering first-graders could express no logical, meaningful purpose for learning to read and a third of the children had no idea how it was to be accomplished. The need for helping pupils see a reason for learning to read and for gaining some insight into how it is going to be accomplished becomes apparent. Most research on learning supports the proposition that it helps the child to learn if he knows the reason for a learning situation and sees a purpose in a task. Inasmuch as reading is not nonsense learning, but a complex mental process, it may be important to identify it as such and to help beginners establish purposes for wanting to learn to read (p. 447).

In another report, Weintraub and Denny (1965) presented an analysis of 108 first-graders' responses to the question, "What is reading?" The analysis revealed that 27 percent of the responses were of the vague or "I don't know" variety; 33 percent of the responses were object related, such as "Reading is when you read a book"; and 20 percent of the responses were of a cognitive nature or described reading as a cognitive act that "helps you to learn things." The remaining 20 percent of the responses were distributed almost evenly across three categories: (1) value terms—"I think reading is a good thing to do"; (2) mechanical descriptions—"It's words and you sound them out if you don't know them"; and (3) expectations—"It's something you have to learn how to do."

Within the limitations of the study, which included a relatively small sample size and many different socioeconomic levels, Denny and Weintraub noted that students came to school with greatly disparate perceptions of the reading process. In addition, 27 percent of the students failed to verbalize an intelligible perception of the reading process. Based on their findings, Denny and Weintraub stressed the need to develop well-planned and carefully executed experiences to help students build an understanding of reading. Dictated stories, poems, and actual experiences were included among the activities they suggested to help students build an accurate perception of the reading process.

A number of other studies have explored students' perceptions of reading. Perhaps one of the most frequently cited studies is the one conducted by Reid (1966). Her study involved students five years of age from a classroom in an Edinburgh, Scotland school. The study had several purposes, one of which was to explore the students' perceptions of reading. The 12 students, seven boys and five girls, were randomly selected. The students were interviewed individually three times during their first year in school. Although there was a core of standardized questions, the order of the questions was varied as each individual interview progressed, and additional probing was done when necessary.

In discussing the findings of the interviews with regard to students' perceptions of reading, Reid noted that, although most of the students were aware that they could not read, they had very few precise notions about reading. They were not even clear whether one read the pictures or the other marks on the page. In short, there was a general vagueness about the nature of reading. Like previous researchers, Reid speculated that a conscious and careful effort to develop an awareness of what reading is might make a difference in students' progress in reading.

A replication and expansion of the Reid investigation was conducted by Downing (1970). He selected 13 students, six boys and seven girls, from a school in Hertfordshire, England. In addition to the interview technique that Reid used, Downing used an experimental method and certain "concrete" stimuli (for example, pictures of someone reading and toy buses with different route numbers). The three research methods produced findings that complemented those found by Reid. Specifically, Downing concluded that students had difficulty in understanding the purpose of reading and had only a vague perception of how people read. He also raised the issue of whether or not teachers were making unwarranted assumptions about students' perceptions of the nature of reading.

In addition to the findings of Reid and Downing, a study was conducted with preschoolers in the United States by Mason (1967). The 178 children, aged three to five, represented a racially, socioeconomically, and intellectually stratified sample of children from Clayton County, Georgia. Four questions were posed to each child during individual interviews, the first of which was "Do you like to read?"

By the time approximately half of the children had been interviewed, the investigators realized that most of the children had said they could read—a startling statement since the children were preschoolers. In addition to the four questions, the interviewers asked the remaining 80 children, "Can you do it [read] all by yourself?" Over 90 percent of these children gave affirmative responses. Unlike the students in the Reid and Downing studies, preschoolers in the Mason study believed that they could already read. This finding may be due, in part, to the younger age of the children involved in the Mason study and the different questions asked.

Mason's implication that "one of the first steps in learning to read is learning that one doesn't already know how" (p. 132) appears to be a modest one, especially in light of the major research findings about students' perceptions of reading. There seems to be considerable evidence to suggest that students need to understand what reading is.

Studies with Elementary Students

Tovey (1976) obtained perceptions of the reading process from 30 students, grades one through six, in a small, midwestern, industrial community. Five students from each grade were selected by their teachers. Individual interviews that also included reading activities were used in the assessment process. Tovey found that 43 percent of the responses to the question, "What do you think you do when you read?" expressed the idea that reading is looking at and pronouncing words. Only 28 percent of the responses indicated that reading had something to do with meaning, while 29 percent of the responses were vague or irrelevant (e.g., spelling, breathing). In his interpretation of these and other findings, Tovey argued that students tended to perceive reading as a word-calling process in which unknown words should be sounded out. He proposed that this view needs to be replaced by one in which students think of reading as reconstructing meaning from print.

In one of the largest studies undertaken to explore students' perceptions of reading (Johns & Ellis, 1976), 1,655 students from grades one through eight were selected from several public elementary and middle schools located near a large, midwestern, industrial area of the United States. The sample was assumed to represent the generally expected ranges of intelligence and reading achievement. An informal assessment of students' backgrounds revealed socioeconomic status ranging from upper-middle class to lower class. Most of the 826 boys and 829 girls were Caucasian.

An individual interview format was used to gather students' responses to three questions:

(1) What is reading?
(2) What do you do when you read?
(3) If someone didn't know how to read, what would you tell him/her that he/she would need to learn?

The responses to the questions were recorded on audio tape and later transcribed and classified into five categories:

Category one: no response, vague, circular, irrelevant, or "I don't know."
Category two: responses that described the classroom procedures involved in reading or the educational value of reading.
Category three: responses that characterized reading as decoding or involving word-recognition procedures.
Category four: responses that defined reading as understanding.
Category five: responses that referred to both decoding and understanding.

In summarizing the responses to the first question, "What is reading?" Johns and Ellis noted that 69 percent of the students gave essentially meaningless responses (categories one and two). Only 15 percent of the students gave responses that associated comprehension or understanding with reading; furthermore, over two-thirds of these responses were from students in grades seven and eight.

Results from the second question, "What do you do when you read?" indicated that 57 percent of the students' responses were classified as meaningless. Only 20 percent of the students indicated that they sought meaning when reading; moreover, nearly two-thirds of these responses were from students in grades six, seven, and eight.

Replies to the third question, "If someone didn't know how to read, what would you tell him/her that he/she would need to learn?" provided the following results: 36 percent of the responses were meaningless; 8 percent of the responses referred to comprehension or understanding; and 56 percent of the responses emphasized word recognition or decoding. Over one half of the students seemed to know that pronouncing words was a part of reading, but they overemphasized this aspect at the expense of comprehension—the heart of reading.

Johns and Ellis concluded that, while older students had a somewhat more accurate understanding of the reading process than did younger students, the vast majority of students (over 80 percent) did not include meaning in their responses. Most of the students' meaningful responses described reading as a decoding process or as an activity involving the use of a textbook and occurring in the classroom or school environment.

Although the Johns and Ellis study is a large one, it has several limitations. First, it is quite possible that some students had perceptions of reading and were unable to verbalize them. The three predetermined questions used by Johns and Ellis may have given the students an inappropriate stimulus for expressing their perceptions of reading. In addition, it is possible that the questions may have been too abstract, especially for the younger students. Reid (1966) notes, however, that the absence of a term in a student's response (such as meaning) can indicate that "it is not one of the terms he habitually thinks with" (p. 58).

A second limitation is due to a possible warm-up effect for the three questions. The number of vague, irrelevant, or "I don't know" responses (category one) dropped from 33 percent for question one, to 22 percent for question two, to 14 percent for question three. It is possible, therefore, that students' perceptions of reading were somewhat distorted or underestimated.

To obtain a more accurate idea of students' perceptions about reading, the data for elementary students (grades one through six) from the Johns and Ellis study were reanalyzed. One provision of the reanalysis was to exclude all students whose responses were placed in category one, that is, responses of the vague, irrelevant, or "I don't know" variety. A second provision was to collapse and reorder the remaining categories. The sample for the reanalysis was then composed of 1,122 elementary students in grades one through six.

In this reanalysis, responses to the first question, "What is reading?" centered on classroom procedures and/or the educational value of reading. Representative responses included "Reading books out loud," "Switch to a different class," "Do worksheets," "Read in books," and "You read to your teacher and to the kids in your class." Category two contained responses that focused on word recognition, such as "Sounding out words," "Learning your vowel sounds," and "Words, sounds, and letters." The third category included responses that focused on meaning or understanding. A typical response included "Recognizing words and understanding what the author is trying to tell." Figure 4-1 graphically displays the results from the question, "What is reading?"

Responses for the total group were distributed as follows: 64 percent in category one; 25 percent in category two; and 11 percent in category three. Trends from grade one to grade six were noted in categories two and three. With few exceptions, word-recognition responses decreased through the grades while meaning responses increased. The vast majority of responses at all grade levels, nevertheless, was in category one.

The same three categories of results for question two, "What do you do when you read?" are shown in Figure 4-2. A trend similar to that found in the responses to question one was found for question two: 52 percent of the responses were in category one; 30 percent in category two; and 18 percent in category three. Through the six grades there was a gradual decrease in the number of word-recognition responses and a gradual increase in responses that identified meaning as part of reading. This increase was most noticeable in the fifth and sixth grades.

Figure 4-3 shows quite a dramatic shift in students' responses to the third question, "If someone didn't know how to read, what would you tell him/her that he/she would need to learn?" The majority (64 percent) of the responses dealt with word recognition or decoding while only 5 percent of the responses concerned meaning or comprehension. Almost one-third (31 percent) of the responses were concerned with classroom procedures. Over half of the first-grade students mentioned classroom procedures in their responses to the third ques-

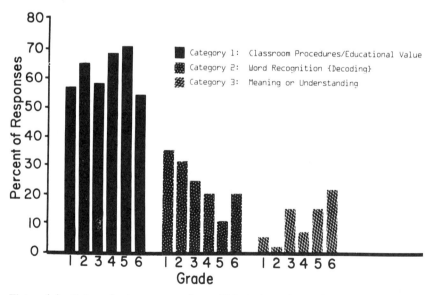

Figure 4.1. Categorized responses from 683 students in grades one through six to the question, "What is reading?"

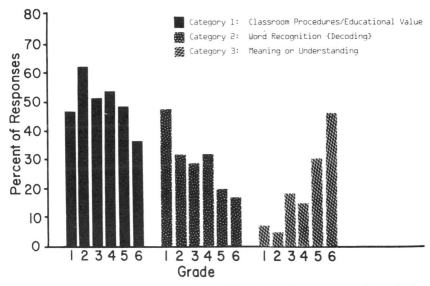

Figure 4.2. Categorized responses from 855 students in grades one through six to the question, "What do you do when you read?"

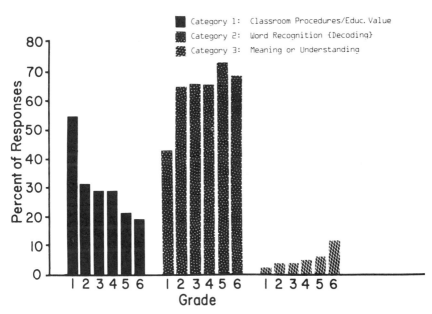

Figure 4.3. Categorized responses from 951 students in grades one through six to the question, "If someone didn't know how to read, what would you tell him/her that he/she would need to learn?"

Table 4.1. Comparison of Original and Reanalyzed Data in Percents to Three
Questions About Reading for Students in Grades One Through Six

| | Category One | | Category Two | | Category Three | |
| Question | Classroom Procedures or Educational Value | | Word Recognition (Decoding) | | Meaning or Understanding | |
	Original	Reanalysis	Original	Reanalysis	Original	Reanalysis
What is reading?	36	64	16	25	15	11
What do you do when you read?	35	52	22	30	21	18
If someone did not know how to read, what would you tell him/her that he/she would need to learn?	22	31	56	64	8	5

tion. These first-grade students often mentioned taking turns reading or asking
the teacher for help.

The reanalysis of the Johns and Ellis data revealed some differences in the
distribution of responses when compared with the original data. Table 4-1 contains a summary of the findings from both studies to permit comparisons. In the
reanalysis, the largest shift in responses took place within category one (classroom procedures). When students verbalized their ideas about reading, they very
seldom gave responses that could be classified as reconstructing meaning from
print. A very rough estimate would be that less than 15 percent of the students
expressed a perception of reading in which the focus was on comprehension.
When students described what they do when they read, about 25 percent of
them named word-recognition strategies such as sounding out words, saying
words, and the like. It is worth noting that when students explained what someone would need to learn in order to read, the emphasis on learning words and
sounding techniques was mentioned by over 55 percent of the students.

Generally, over 30 percent of the students mentioned classroom procedures
when sharing their perceptions of reading. Responses such as "Switch to a different class," "Do worksheets and workbooks," and "Follow every line with your
marker" were not uncommon.

Synthesis of Research Findings

As a group, the studies cited in this review focused on students' perceptions of
reading. Interviews were used to obtain this information. Some of the interviews
were highly structured while others were loosely structured. Students who par-

ticipated in the studies ranged from preschoolers to students in the eighth grade. They lived in different countries and in various areas of the United States. Taken as a group, these studies have certain limitations in common. The studies relied, for the most part, on interview questions that were usually evaluated and interpreted by only one professional; moreover, some of the questions may not have adequately assessed what the students knew about reading. Since many of the students were of preschool age or in the early years of schooling, it is possible that they knew more about reading than they could communicate orally. Given these limitations and the problems that are often inherent in using the interview technique to explore a process as complex as reading, only the most cautious of conclusions should be drawn. The following conclusions appear to be warranted by a cautious interpretation of the data.

When asked to answer questions about reading, preschoolers and first graders appeared to be confused about what the process entailed. Students above the fourth grade expressed a somewhat better understanding of reading; however, most students did not know (or were unable to verbalize) that an effort to reconstruct meaning from printed materials is essential in reading. Students seemed to acquire numerous perceptions relating to reading as they progressed through school; however, even though most of these perceptions were merely the tangential aspects of the reading process, they tended to remain the most influential or dominant factors when students shared their perceptions about reading.

Impact of Students' Perceptions of Reading on Their Reading Ability

If one accepts the idea that students differ in their perceptions of reading, it is worth investigating whether this factor influences their reading abilities. The available research evidence does not answer this question with any degree of certainty. Correlational studies that show relationships do not necessarily demonstrate cause and effect. The available research, nevertheless, is promising.

Although Muskopf (1962) did not find a significant correlation between first-grade children's perceptions of reading and their reading achievement, Edwards (1961) found a very slight correlation with fifth-grade students. In addition, both researchers cautioned that the tests used to measure students' perceptions of reading were of questionable validity.

Johns (1974) conducted a study with 103 fourth- and fifth-grade students to determine if there were differences in the perceptions of reading held by "good" and "poor" readers. He classified as "good" readers the 36 students who were reading at least a year above grade placement according to the comprehension section of a standardized reading survey test. The 29 students classified as "poor" readers were reading at least one year below grade level on the same standardized test. Perceptions of reading were measured by students' responses to the question "What is reading?" and classified as meaningful or nonmeaningful. Although Johns found that the "good" readers differed significantly in their perceptions of reading from the "poor" readers (χ^2 = 6.04, $p < .05$), he mentioned several factors that may have affected the results (i.e., the method used to classify readers and their responses). In addition, he noted that less than half of the "good" readers gave responses that were judged to be meaningful.

As part of a larger study, Schneckner (1976) tested 30 first-grade and 30 third-grade students with a standardized reading survey test. He asked them five questions to assess their perceptions of reading. Schneckner reported that there was a significant positive correlation between reading achievement and perceptions of reading for first-grade students. There was no relationship between these same two variables for third-grade students.

Canney and Winograd (1979) contended that there appeared to be a relationship between students' comprehension performance and their perception that reading entails some meaning-focused activities. They cited their study and referred to the work of others (Cohen, 1974-1975; Golinkoff & Rosinski, 1976; Weber, 1970a).

From the large amount of research reviewed, it is appropriate to conclude that many students in the beginning stages of reading instruction are unable to express a meaningful perception of reading. A tempting implication is to say that this lack of linguistic awareness is likely to cause difficulty in reading. While there is some evidence of a relationship between students' reading ability and students' perceptions about reading, the evidence does not produce a convincing case to establish a cause-and-effect relationship. It is possible that the acquisition of a good perception of reading is a consequence of the individual student's reading achievement. Perhaps Ehri (1979) is correct when she hypothesizes that linguistic awareness may interact with the reading-acquisition process so that the awareness exists as both a consequence of what has occurred and as a cause of further progress in reading.

Pedagogical Implications

The review of research presented in this chapter should be approached cautiously; nevertheless, teachers may wish to ensure that their students have repeated exposure to a wide variety of oral and written language activities. Such activities may help students to develop a good perception of reading quite naturally. Weaver and Shonkoff (1979) believe that a curriculum rich in language activities seems to foster the development of linguistic awareness. While this may be true, it is also clear that many students, even after several years of instruction that supposedly contained at least some of these rich language activities, developed a perception of reading in which decoding played the dominant role. Meaning was largely ignored. It may be worthwhile, therefore, to cite some of the specific language activities that are likely to foster linguistic awareness in such areas as: knowing what reading is; knowing what makes up a letter, word, sentence, or story; and knowing the conventions of print (e.g., left to right). Although it is not known whether children's perceptions about reading are likely to foster progress in reading, Berman (1963), Bruner (1973), and Burton (1956) have asserted the necessity of defining and understanding a task before attempting to achieve that task.

Implications for the Initial Stages of Reading Instruction

Preparation for reading instruction must give primary attention to the notion that reading, writing, talking, and listening should be treated as a unity (Department of Education and Science, 1975). The process of instruction in most classrooms, however, often fragments the language arts. Attention also should be given to creating a smooth transition between the period of learning to talk and the period of learning to read. These two notions suggest many other important ideas for reading instruction, such as showing that talk can be written down, that print is meaningful, and that communication between an author and a reader should take place during reading.

Using the Language-Experience Approach

Perhaps the most natural and meaningful way to maintain a unity among the language arts in the teaching of reading is through the use of the language-experience approach. This approach, used by teachers throughout the world (Au, 1979; Pienaar, 1977; Wilson, 1979), places emphasis on those activities that allow students to increase facility with oral language so that reading becomes an extension of speaking and listening. Reading, in the language-experience approach, is built around the student's experiences. The central focus of the curriculum is the student. It is not necessary to "fit" the student into a workbook or program that is ill suited to his/her needs. Interest is ensured because the student is translating his/her ideas into print. Understanding is guaranteed because familiar experiences and concepts are represented in print.

Although there are numerous sources for teachers who desire to implement the language-experience approach (Allen, 1976; Lee & Allen, 1963; Stauffer, 1980), an overview may be in order.

(1) The teacher provides a stimulus or uses a recent experience that can be discussed with the student. This can be done in a group or individually.

(2) The student dictates a story to the teacher. The story should be in response to the stimulus or experience. It can be of any length and should be recorded exactly as the student dictates it, paragraphing and punctuating it appropriately.

(3) When finished, the teacher reads the story back to the student asking him/her to indicate whether there is something he/she wants changed.

(4) After the teacher rereads the story, the student thinks of a title. In addition, the date of the dictation is recorded.

(5) The story is given back to the student. The student reads the story with the teacher's help. Then he/she reads the story alone, underlines all the known words, and illustrates the story.

(6) The next day the student reads his/her story again. The words that he/she knows are underlined. If he/she knows the words two days in a row, they will be underlined twice.

(7) The student then reads the story aloud to the teacher. The teacher notes whether the student actually knows the underlined words. Those words the student knows both days are put on small cards with the date of the story on the back. These words are then put in a "word bank" of known words. The word bank can be a cigar box, an oatmeal box, an envelope, etc.

(8) The words in the word bank should be used for work with word-analysis skills. The student can do many things with his/her words and should do something each day (Garton, Schoenfelder & Skriba, 1979).

(9) Any word the student forgets should be removed from his/her word bank. The student can find the word in his/her story because the date is on the back of the card. Thus, unknown words may become known again with relatively little effort.

(10) The same procedure should be followed with each story. The stories may be kept in a notebook.

An example of how the language-experience approach enabled a student to express the relationship between speaking and writing came from a kindergarten classroom. As the teacher typed the young girl's dictation, the girl expressed this thought, "Oh, you're typing what I'm saying, aren't you?" Recognizing that talk can be written down to be read helps to establish the connection between speech and print. This kindergartener has apparently made that connection. Agnew (1982) noted that the language-experience approach is well suited for helping students "understand both the function and the technical features of reading" (p. 451).

Activities to Stimulate Speaking and Listening

In addition to using the language-experience approach, there are numerous other activities that teachers can use to foster the student's reading ability.

Help students learn poems and nursery rhymes. Through such activities students learn the grammar of language and the rhythm of sentences. They will become familiar with sentence patterns that are used in reading written language. As students become familiar with a rhyme or poem, leave out a word and have them say it. This will promote the notion that language is predictable and may help build good feelings about language. Familiar poems and nursery rhymes may be printed on poster boards so they can be read by students.

Share books and tell stories. The vast number of outstanding books for students gives teachers an excellent opportunity to read and discuss books that can promote many possible benefits in the reading program. Oral reading by the teacher can (1) demonstrate that print is meaningful; (2) serve as a model for reading; (3) show that reading is pleasurable; (4) build an interest in reading; (5) stimulate language development; (6) help students move naturally into reading; (7) create favorable attitudes toward reading; (8) build rapport; (9) encourage students to listen actively; and (10) stimulate students to react to what is read.

Books without words, such as *A boy, a dog, and a frog* (Mayer, 1967), provide an opportunity for students to use their language to interpret a story. The student's "story" can be recorded or written down. Sharing these stories with a small group can help promote language and reading skills.

Develop class stories. A pattern like the following may be used:

Today is *(Monday)*.
The date is *(May 25)*.
The weather is *(sunny)*.
There are *(14)* boys and *(12)* girls.
(Susan's) birthday is today.

Through such stories students can be helped to see that the printed text tells a story or gives information. The visual symbols convey a language message. Although students may not be able to initially identify specific letters or words, they will perceive that the various shapes of letters and words act as cues to the various sounds that are made when reading. As the student's linguistic awareness grows, he/she will see that there is a correspondence between the order of the sounds spoken and the left-to-right sequence of words as printed. Framing specific words while the story is being read may help students develop a concept of a word. There are numerous sources (Baird, 1965; Lamb, 1971; Menagh, 1967; Rasmussen, 1962) to help teachers foster students' creative expression and language learning. Some of these activities should help students realize that reading is the process of reconstructing meaning from the printed page. In addition, many of the activities can easily be adapted for use by parents of preschool and primary-grade children.

Helping Middle-Grade Students Improve Their Perceptions of Reading

To help middle-grade students improve their perceptions of reading, the following activities are suggested. The activities center on several statements about reading. After completing the activities, students should have a better understanding of reading. They can demonstrate this understanding by repeating these statements about reading in their own words and by integrating them in reading tasks. When students understand these statements, they could try to help a younger child who does not have a good perception of reading to develop one.

Statement One

Reading occurs when meaning is reconstructed from written symbols. Suggested activities include the following:

(1) Present students with the following: *Hun sog voy ado sinco.* Ask them if they can read it and, if they think they can, ask them what it means. Emphasize that in order for reading to take place there must be meaning. Even if they pronounce the words, this is not reading because these words do not make any sense.

(2) Show students these words: *home everyone went and party the over o'clock was ten at.* Ask students if they have read this. If they say "yes," go back to the definition above and ask them what meaning they got from "reading" those words. Discuss why even though these are familiar English words there wasn't any meaning and point out that this was not reading. Words need to be put into a meaningful sequence to make sense. Encourage the students to arrange the words into a meaningful sentence that can be read.

(3) Present students with the following: *Yo voy al cine.* Ask them if they can read the sentence. Tell them that the language is Spanish and tell them how to pronounce the words. Ask them if they can read the sentence now. By now they should begin to realize that they are not reading unless they know what the words mean in English. After they know how to interpret the Spanish words, they will be able to read the sentence because in their minds they will understand the meaning. (It means "I am going to the movie.") If some students can read Spanish, use the French equivalent: *Je vais au cinema.*

(4) Discuss how reading is related to speaking. When we speak, our purpose is to say something that will have meaning for a listener. If our listener does not get any meaning from our speech, it would not make much sense to keep on talking to him/her. Therefore, it does not make much sense to keep looking at a group of words if they have no meaning. You either have to find out what they mean or quit trying to read them. Have students pretend they all have lost their voices and have to communicate with each other for a while by writing down what they want to say and having their would-be listeners read what they have written. Discuss how books they read are similar to their notes: authors of books have something meaningful to say that they want students to know. Through these activities students should realize that reading is reconstructing meaning from print; reading is not just pronouncing or saying the words.

Statement Two

Reading is a way of satisfying our needs; it has a purpose. Suggested activities include the following:

(1) Ask students what reading they had done recently outside of school. Ask them if they had a purpose for that reading. If they cannot think of any

real purpose that they had, try to show them that there was one. Perhaps it was simply to satisfy curiosity or for entertainment. Explain that there are many purposes for reading besides reading in school.

(2) Have students make a list of all the kinds of reading materials they can think of and the purpose for reading each one.

Examples: comic books for entertainment

labels on cans to know what you're buying

encyclopedias to find information

(3) Have students list specific information that can be acquired through reading. Have others try to think of reading material that supplies information and where that material can be found.

Examples: What is a yegg? Look in a dictionary.

What is the largest state in the union? Look in a geography book, an almanac, or an encylopedia.

(4) Have students think of different things they sometimes write for others to read, the purpose they had for writing them, and the purpose the reader had for reading them.

Examples: an invitation to a party

a thank-you note

These activities should lead students to the realization that reading is a way for people to satisfy their needs. There are many different purposes for reading: to learn how to do something, to learn how to find something, for entertainment, etc. In the above activities, the teacher should supplement students' contributions with additional purposes.

Statement Three

Reading is when you use your experiences and knowledge to gain meaning from written symbols for some purpose. Because not everyone has had the same experiences and because people have various purposes for reading, different people can get different meaning from the same written symbols. Suggested activities include the following:

(1) Have students list things they like to read that another person their age would probably not be interested in. List things they like to read that some-one several years younger would not want to read.

(2) List things students read that their parents would not be interested in and vice versa. With each of the above lists, discuss why people would have dif-ferent feelings about the same reading materials and what these feelings might be.

Statement Four

Reading is reacting. Suggested activities include the following:

(1) List some different types of reading material and have students state their

reactions. Discuss the different kinds of reactions people often have from reading and suggest ones that students do not mention (e.g., sympathy for a character, dislike for a character, suspense, excitement, fear, boredom, wishing things had happened differently, or confusion).

(2) Read the whole class the same story and have them write down their reactions to it. Have them compare their reactions to see how they are similar and different. Suggest things that the students might not think of, such as, "What would you have done if you were the main character? How would you have felt?"

As a culminating activity have students keep a log of their reading activities for a week. Have them record the different kinds of reading material that they read and some specific examples of each different type. They should list the purpose they had for reading the item, what need was satisfied by reading it, and their reaction to the material read.

Through these activities students should see that reading is a meaningful process in which the reader does much more than just say the words. The reader must actively participate in the reading process to reconstruct meaning and to satisfy his/her needs.

Implications for Older Students and Disabled Readers

Reading is a highly complex process that involves a dynamic interplay of many factors. A number of authorities in the field of reading (Harris & Sipay, 1980; Malmquist, 1971) believe that several interrelated factors or combinations of factors cause, or contribute to, most reading disabilities. Thus, tackling the problem of reading disability is not an easy task.

One factor involved in reading—in fact, in any type of language activity—is a knowledge of syntax, which enables the reader to use syntactical rules to go from the surface structure (visual forms) to the deep structure (meaning). Some research seems to support the notion of a relationship between a student's level of language development and his/her reading ability, although it must be remembered that correlation does not necessarily indicate a cause-and-effect relationship. Other factors may influence both variables.

Research on Language Development and Reading

Chomsky (1972), in her investigation of developmental linguistic stages during the acquisition of syntax, studied the reading backgrounds of children aged five to ten years. She found positive relationships between reading exposure and linguistic stages: children at the top linguistic stage were characterized by having read (or listened to) a larger number of books, having taken out more books on regular visits to the public library, and having read more complex materials. Several studies based on Chomsky's work have provided further evidence that good readers have a stronger knowledge of syntax than do poor readers. Van Metre (1974), using some of Chomsky's measures of linguistic maturity, compared similar groups of bilingual and monolingual third-grade students. The high groups of both bilinguals and monolinguals, who scored in the top 25 percent on a standardized reading test, scored significantly better in linguistic maturity than the low bilinguals and low monolinguals, who had scored in the bottom 25 percent. Goldman (1976) used one of Chomsky's constructions to compare the listening and reading comprehension of 96 third-, fourth-, and fifth-grade students. Since the low-skill group performed significantly below the high-skill group on the listening task, Goldman concluded that low-skill readers, in addition to having problems arising from the graphic input, also have less well-developed skill in syntactic analysis.

A third study (Mavrogenes, 1977, 1978) based on Chomsky's work involved 20 ninth- and tenth-grade students who achieved a mean grade equivalent of 5.0 in vocabulary and 4.9 in comprehension on a standardized reading test. The students were tested on their ability to understand Chomsky's five exceptional structures, which represented the five developmental stages in the acquisition of syntax. In a comparison of Chomsky's group (age 5-10) and this group of disabled readers (age 14-15), the difference between mean scores was not significant. In addition, 22 competent readers (age 14-15) were also exposed to Chomsky's structures. On the same standardized reading test, these students scored a mean grade equivalent of 11.5 in vocabulary and 12.6+ in comprehension. Using Chomsky's structures, the difference between the mean scores of the disabled readers and those of the competent readers was significant. One conclusion is that the high-school disabled readers have a low level of language development.

In his 12-year longitudinal study, Loban (1976) showed that decreased language ability accompanies low reading ability at all levels. In both written and oral language, low readers were low in average number of words per communication unit, syntactical elaboration of subject and predicate, number of grammatical transformations, proportion of mazes to total speech, height and range of vocabulary, use of connectors, number of dependent clauses, and use of adjectival clauses.

Additional studies have provided evidence of the relationship between language development and reading (Takahashi, 1975; Kuntz, 1975). Vellutino (1977), in his overview of research on why some students cannot read, argued in favor of a "verbal-deficit" hypothesis for the most convincing explanation of reading disability. He cited many research reports that suggested disabled readers were not as proficient as other readers in their "knowledge of words, syntactic facility, and verbal fluency" (p. 345). Therefore, to help disabled readers improve their overall language and reading abilities, the following activities are suggested.

Expose Students to Good Literature

Use literature that has not been adapted to a lower level. Exposure to rich language is based on the nature of language acquisition that occurs naturally as students mature and develop in an environment where they are adequately exposed to language and where they can use such inputs in their own ways. Hoskisson and Krohm (1974) recommend "assisted reading" using a tape recorder, a listening post, and reading couples. This kind of reading gives students the full context of written language while they learn to read, just as their environment should have provided them with the full context of spoken language while they were learning to speak.

Listening while reading has its roots in the basic assumption of the psycholinguistic-information-processing theory of reading: "Reading programs must enhance the strategies natural to the reading process.... They need practice at the strategies natural to the skill of reading" (Cooper & Petrosky, 1976, p. 200). Such exposure would provide an antidote to "controlled" reading and reading "made easy" as students' "structural understanding of their reading material could be enhanced by greater exposure to the more complex syntax of informed adult speech" (Stotsky, 1975, p. 43).

Several research studies support the benefits of having students follow along as good literature is read to them. Cohen (1968) reported on 580 second graders from seven New York City schools with large minority populations. The experimental group was read an interesting story each day of the school year and participated in many kinds of follow-up activities. The goals were to strengthen the vocabulary of socially disadvantaged children, to stimulate their desire to read, and to increase their reading achievement. The experimental group made significantly higher gains than the control group in vocabulary, word knowledge, and reading comprehension.

Schneeberg (1977) described a similar situation at the fourth-grade level in an inner-city Philadelphia school. In this study, the experimental group read and listened to over 70 books during two school years, for two to four hours a week, and also participated in follow-up activities. These students gained an average of 2.5 years on the reading portion of a standardized achievement test. Normally, educationally disadvantaged students progress 0.65 to 0.70 grade-equivalent units per year (Tempo Center for Advanced Studies, 1971).

Provide Instruction in Sentence Combining

Combs (1977) reported on two groups of seventh graders, one of which received sentence-combining instruction. Although the two groups were similar on a pretest in reading rate and comprehension, the experimental group, after sentence-combining practice, experienced significant gains in reading comprehension.

There are several ways to incorporate sentence combining into the instructional program. One way is to use commercial sentence-combining programs. O'Hare (1975) has a complete workbook called *Sentencecraft*. If a somewhat easier program is desired, Strong's (1973) *Sentence combining: A composing book* might be appropriate. In addition, sentence-combining exercises may be adapted from classroom books and materials.

Besides sentence combining, there are other ways of providing practice for the development of syntactic understanding. Sentences can be lengthened in the manner suggested by Christensen (1967).

Another possibility uses exercises that give practice in structures such as expanded verb forms, relative clauses, participial phrases, causal clauses, question forms, negative sentences, passives, connectives, and indirect discourse (McCarr, 1973). Sources for these games and related activities can be found in Bailey (1975), Hurwitz and Goddard (1969), and Shipley (1972).

Conclusion

In 1968, Clymer noted that a clear perception of reading is not just an academic concern. Students' perceptions of reading, based on the research reviewed in this chapter, are often vague, meaningless, or concerned with only fragments of the reading process. Meaning, the ultimate goal of reading instruction, is frequently missing from students' perceptions. It is replaced by notions about doing workbook pages, saying words, and paying attention to teachers.

Perhaps one of the most unresolved issues is whether efforts should be made to teach students a meaningful perception of reading. Ample research has been conducted in various countries throughout the world to suggest that students' perceptions about reading are not well developed, at least as those perceptions have been measured and investigated. Vygotsky (1934) noted that the direct teaching of concepts is impossible and fruitless; however, since reading is regarded as a language-based process, teachers may want to provide both general and specific activities that have the potential to help students unlock the mystery of reading.

CHAPTER 5

Children's Thinking about Language and Their Acquisition of Literacy

Anne Sinclair and Ioanna Berthoud-Papandropoulou

Introduction

Some of the many currents of research and theory in the field of literacy development are concerned with questions of metalinguistic thinking. By their very nature, studies concerning the perceptual (or motor) processes involved in the activities of reading and writing do not address the question of underlying (explicit or implicit) linguistic knowledge. Similarly, research dealing with educational methods and practice does not often address this aspect either, as emphasis is put on comparisons between different methods, and on the evaluation of the pupil's performance. Other studies on literacy concern the role of socio-affective factors such as motivation and deal mainly with the content of the material proposed for beginning readers. Researchers who investigate the development of the child's knowledge of the written system (as a system in itself without particular links to other systems) tend to point to the differences between text and spoken language and leave aside questions of knowledge of a general nature about the object of thought, "language." In contrast, researchers who view reading and writing as activities that are based on a certain type of transposition of spoken language naturally consider both production and comprehension capacities, as well as metalinguistic knowledge. This brief sketch of the different trends in studies of literacy development serves to clarify the perspective chosen for this chapter. We will describe and discuss research results that belong to the last two trends described, thus leaving aside perceptual processes, motivation, and instruction methods, and concentrating on conceptual components.

In this chapter we will attempt to examine various reflections or types of knowledge of a fundamental nature about natural language, either as a very general object of thought, or more specifically about its spoken or written form, or concerning the written system itself, as a particular system with its own properties and characteristics. Our aim will be to point to certain parallels or similarities in the reflections children make in these different domains, and to show how reflections in one domain may be transposed or applied to another. Examining parallels is a classical method for establishing links between different types of behavior, and we hope that this approach may allow us to conclude by discussing mutual interdependencies between "thinking about language" and the development of literacy. This approach may also serve to elucidate general questions concerning knowledge about natural language, independently of its particular form or use.

In a deliberate attempt to focus on the links between thinking about language and development of literacy, we will limit our discussion in various ways. At present, the study of metalinguistic thinking or awareness includes a large collection of heterogenous phenomena; many different activities and many types of explicit or implicit knowledge are classified within it (self-corrections, awareness of syntax, semantics, social rules, comprehension, and explanation of ambiguity, jokes, riddles, etc.; see Clark [1978] for a review).

First of all, we will consider only reflections or knowledge of a general nature that children apply to both oral and written examples of language, and reflections that seem to establish links between the two systems (spoken and written).

Second, as reading procedures, literacy development, and perhaps also certain aspects of metalinguistic thinking depend (at least in part) on the type of language, the type of orthography, and the type of script, we will refer to results from studies we are most familiar with, concerning children who speak French and Spanish, languages that are alphabetically written. We will not consider syllabaries or logographic writing systems.

Third, we will take into account results of research performed with children at the ages of three to six or seven, that is, when they generally cannot yet read and write in an adult sense or are just learning to do so. No matter how we view it, literacy develops slowly in the sense that performance in reading and writing generally improves with age. In our chapter, we use the terms "be able to read" or "be able to write" as implying an understanding of the alphabetic principle, as well as the capacity to apply this understanding in reading and writing behaviors (and thus having at least some knowledge of our conventional letter shapes and the sounds they represent). When we say a child can read, we thus mean that he is capable of reading aloud and understanding a short—possibly very short—written display, composed of words that are part of the child's spoken vocabulary, but have not been encountered in their written form previously. The child must be able to use his knowledge of phoneme-grapheme correspondences and carry out phonological recoding in a manner that permits access to lexical meaning. Extracting meaning from longer texts is a different matter in which many other factors intervene: the child's reading proficiency, his knowledge of language and the world, the nature of the text, and, last but not least, the child's capacity to use the information or show that he has understood it. To write, the child must be able to write a few words (not learned by rote, such as his name or a well-known brand name) using the alphabetic principle (though he may make many orthographic errors).

Even from this particular perspective, and taking into account the scope of our discussion, one general problem should be mentioned. It is difficult to distinguish (both theoretically and operationally) one of the following notions from another: principles underlying behavior, implicit knowledge, awareness, etc. The degree of the subject's consciousness and the role these components play in behavior are even more difficult to define, particularly when one is dealing with young children who often cannot express what they think and know. In this chapter, we have taken the widest possible definition of "thinking about language" or "thinking about the written system," and our discussion will include explicit verbal comments as well as principles underlying responses to certain questions as they are deduced by the researcher. For children's ideas about orally presented language, we will limit our discussion to explicit comments and judgments, and will not take into account production and comprehension capacities. (The links between metalinguistic knowledge and comprehension and production raise highly complex problems we cannot go into here.) For reflections on the written system, on the other hand, we will consider both judgments (and comments) as well as interpretations of the child's productions (writing attempts) and reading procedures.

Relationships Between Signifier and Signified

Piaget (1923) and Vygotsky (1934) have pointed out that, in the young child's mind, words and the things they can refer to entertain particular relationships which are not the same as those conceptualized by adults. Berthoud-Papandropoulou (1978, 1980) has investigated this problem in more detail. One hundred and sixty French-speaking children, aged four to twelve, were individually asked several questions, all dealing with the linguistic unit "word." Different studies were carried out to explore various aspects of metalinguistic conceptualization. One study's aim was to obtain children's definition of the term "word." In order to probe the children's ideas, the experimenter started a conversation that led to the question: "Do you know what a word is?" In another study, the experimenter uttered seven or eight simple, frequently used words one by one (some content words, an article, a conjunction, a pronoun, etc.) and asked the children whether each was a word or not, and why. In a third study, children were asked to give examples of words with certain properties: a *long* word, a *short* word, and a *difficult* word, and were encouraged to justify their answers. They were also asked to invent a word and to produce an utterance containing it. A last task was a word-counting task: the experimenter uttered a sentence and asked the child to count the words. The child was then asked: "What were those (n) words?" Several sentence types were presented. The experiments were conducted with a supple subject-oriented interview method. The results show a slow development of the concept of word towards a capacity of conceiving the word both as a meaningful constituent element of larger units and as a unit which is itself built up of smaller elements.

Young children (four, five, and six years old) do not differentiate, when asked to think about what words are, between words and the things they represent. In a sense, for children at these ages, a particular word and the object referred to form an inseparable unit. From this point of view, "strawberry" is a word "because it grows in the garden" and "train" is a long word "because there are lots of carriages." Similarly, when asked to say a short word, children at these ages mention small objects or actions that are rapidly carried out: for example, "primrose," "the ball jumps." These responses all show that for young children words share at least some of the characteristics of their referents.

This general characteristic of the child's view of language is also evident in early writing behavior. Ferreiro and Teberosky (1979) carried out a study of literacy development in which children between the ages of four and six were asked to carry out several different tasks of production and interpretation of written material. They tested the following groups of Spanish-speaking children in Buenos Aires, Argentina: a group of 30 five-year-old children from a shanty town (lower class), whom they tested several times during their first school year (longitudinal study); and groups of four-, five-, and six-year-old children from both the middle class and the lower class, 75 children in all. They used Piagetian theoretical constructs and Piagetian interview methods. Their results show that children go through several conceptual stages before grasping the alphabetic principle, and that there are differences between the shanty-town children and the middle-class group. (We will return to these social class differences later.)

According to Ferreiro and Teberosky (1979), before children know how to write (at the ages of four and five), they will often think that some of the properties of the referent determine some of the properties of the written symbolization. For example, they will propose to write "elephant" with more letters than "butterfly," because "elephants are bigger." Similarly, for a picture with three dogs in it, they propose to write three letters, one for each dog. These children are representing some quantitative characteristic of the referent (size, quantity, and number) rather than the sounds of the corresponding words. Thus, according to Ferreiro and Teberosky (1979), the first link children make between letters and meaning is in the domain of measurement and number and not that of sound.

The children's particular conception of words is also evident in other results. If they are asked to give examples of words, they most often give proper names or nouns that refer to people ("a man," "a little girl," "granny," etc.) and natural entities (animals, fruit, and flowers) (Berthoud-Papandropoulou, 1980). As these words designate natural referents, and sometimes even unique ones, we may suppose that they are chosen because they illustrate the closest possible link between word and referent. Once more, evidence of this viewpoint appears if we examine early reading and writing behavior. When the child begins to identify and name isolated letters (which does not imply linking them to sounds), he will often associate one letter with his own name or with the name of a member of his family. For example (Goodman, 1980, p. 10), "When Mark was as young as four, he looked at the word TIMEX, pointed to the M, and said: 'that's my word'."

Or: a three year old holds up a plastic M and says "that's Mommy," and then picks up a P and says "that's Papa" (personal observation). Subsequently, the child's own name seems to be the first stable written form that he produces. Besides profound reasons having to do with the construction of the system, this is partly due to the behavior of family members and institutions who will often mark the child's place or his possessions with his name.

These few examples may suffice to show that the child's view of the relationship between signifier and signified, where the two terms are neither coordinated nor differentiated, manifests itself in similar ways in different kinds of language activity.

Linguistic Units

The child's conception of words is also evident if we ask him to segment a spoken utterance into units. His particular conception of the relationship between signifier and signified leads him to consider that only certain words are indeed "words," while others are not. Berthoud-Papandropoulou (1980) and others (such as Karpova [1966]) have shown that segmenting utterances into words is a difficult task for young children. Ferreiro (1978, 1979) asked prereaders to identify the different parts, separated by blanks (i.e., words) of a written sentence after having read the sentence aloud for them, in order to find out what meaning they attributed to what for adults are different words. Even though these studies were performed in different countries and in different languages (in French, Russian, and Spanish), there are striking similarities in the results. We will present some of them, without going into the details of the results of the experiments. When asked to count the words in a sentence, at the lowest levels (four and five years of age), the children center their attention on the reality referred to and not on the verbal form of the sentence. For example: "How many words in: *six enfants jouent?*" (six children are playing): "Six . . . Maria, Jimmy" (and the child gives six proper names). Similarly, when confronted with the written sentence PAPA PATEA LA PELOTA—daddy kicks the ball (subjects are told what is written on the display; capital letters are used for written displays throughout) at least some young children will point to PAPA as saying "daddy kicks the ball," to PATEA as saying "daddy is sick" (for example), and for LA "daddy writes," PELOTA finally meaning "daddy goes to bed" (Ferreiro, 1979).

Some other comments made by young children in the word-counting task also show that they are centering their attention on the reality referred to. For example, "How many words in *le cochon a mangé beaucoup* (the pig ate a lot)?": "19 words, because he eats a lot." Or: "How many words in *la rose est belle* (the rose is beautiful)?": "Seven, because it has a lot of leaves." Similarly, some of Ferreiro's (1978, 1979) subjects, once they have attributed "daddy kicks the ball" to the first written fragment (PAPA), will say that the other written fragments are "mamma," "arco" (goal), and "judadores" (players), for example. Ferreiro concludes: "If written text is thought to be a representation of the objects or persons (or of their names) referred to in the utterance and not a representation of the enunciation, there is nothing strange in adding details which simply complement the meaning and do not alter it" (Ferreiro, 1978, p. 34).

At a later stage, when children begin to turn their attention to the message (utterance or written sentence) itself, rather than to the reality referred to, both Ferreiro's (1978, 1979) and Berthoud-Papandropoulou's (1980) experiments show that utterances and written sentences are felt to have two main parts, subject and predicate. Consequently, young children do not isolate the verb. For example: "How many words in *le garçon lave le camion* (the boy washes the truck)?": "Two words, the boy, and that he washes the truck."

Similarly, in a written sentence, young children often indicate two parts (words) as the places where the subject and the predicate are written, or possibly, as Ferreiro and Teberosky (1979) remark, a noun and a verb with its object, since the sentences presented do not allow a distinction between the two interpretations. For example: LA NENA COME UN CARAMELO (written display: the girl eats a candy). "Where is *the girl* written?": the child shows LA NENA. "And this (the experimenter shows COME), what does this say?": "Eats a candy." "And does it say *a* somewhere?": "I don't know." "And how does it go altogether?" "The girl (child shows CARAMELO) eats a candy (child shows all the rest, from right to left)" (Ferreiro & Teberosky, 1979, p. 166).

Formal Properties of Verbal Signifiers

Words, although in the child's mind they may be closely linked to referents, are nevertheless composed of sounds and are thus audible, pronounceable, and so on. Even three year olds are globally aware of this characteristic of speech. For example, three year olds report only one cause for incomprehension of a verbal message: messages that are inaudible. Thus, they report only one strategy for dealing with incomprehension, i.e., to speak louder or to ask the other person to speak louder (Sinclair, 1980). Five year olds can give relatively good descriptions of how they actually produce certain phonemes and/or syllables (movements of the body, mouth, tongue and lips—see Zei [1979]). It has also been shown that three year olds can discriminate admissible but nonexisting sound sequences from inadmissible sequences in English (Messer, 1967). Such intuitions or capacities do not imply phoneme analysis, but do show that even very young children have some awareness of language as composed of certain types of sounds. Similarly, three year olds know that one cannot read a blank page, or that one cannot read with one's eyes shut, and at least some of them do not accept a series of invented letter shapes as "good for reading," but do systematically accept as legible written displays composed of our conventional letter shapes (observations from ongoing research). Of course, one need not know anything about the function or meaning of letters to realize that reading is carried out from visual input, or to be able to distinguish letters from "nonletters."

Thus, very young children already seem to possess knowledge of some formal characteristics of both written text and spoken language. However, to learn more about what they think characteristic sequences of sounds or shapes should have to qualify as "possible language material," it is necessary to avoid situations where they can attribute a meaning to the material. We have already pointed out that children under seven years of age, when confronted with linguistic material for judgment or discussion, concentrate most often on the meaning of the material (the referent, or the utterance's truth, plausibility, appropriateness, comprehensibility, etc.) rather than on the form of the utterance. Apparently, thinking about meaning overrides thinking about formal characteristics.

For written language, it is easy to induce children to concentrate on form; all one has to do is present written material to subjects who cannot yet read and not read the displays to them. For spoken language, the problem is more difficult. We finally chose the following approach. We told the child that a puppet who speaks Italian was going to try and say, in Italian, "le chien a mangé le bifteck" (the dog ate the steak). The child was to tell us if the doll had expressed this event ("told the same story," "said the same thing," and "said it right"), and, if so, how did he know; if not, why not, etc. We then presented various utterances for judgment, the first of these being "cane cane cane cane," the second "cane bistecca," and the third "mangiato bistecca," followed by several other possibilities. These questions are comprehensible to children as young as four years of age, provided that they have some conception of the existence of other languages besides French, which is always the case in Geneva kindergartens. We tested 20 children aged four and five in a kindergarten, 20 children aged six and 20 children aged seven in a primary school, as well as older subjects, all monolingual French speakers. Only three of our subjects (four and five year olds) did not answer our questions as we intended, claiming that the Italian doll did not say the same thing, because "she speaks a different language," thus refusing to consider the possibility that the same message may be expressed in two different languages. All the other four-, five- and six-year-old subjects (37/40) immediately rejected the first utterance as telling the same story. The reasons for their rejection, at these ages, have primarily to do with the repetitive quality of the utterance. For example: "No, it can't be, she always says the same thing"; "She repeats the same thing over and over, no"; "She says *cane cane cane* - no." Some assign a meaning to "cane," guessing correctly that it means "dog" (or guessing incorrectly that it means something else), but this does not, at these ages, provoke a change in their responses. "She only says *dog dog dog*, that's not the same as the French doll." It is only after the age of six that children will also discuss what is missing in this utterance, what it actually might mean, how it might be interpreted, what parts should be added or taken away, etc. (For example: "*Cane* is O.K., the rest not. The beefsteak is missing: he ate the beefsteak."). Young children all stress the repetitiveness and thus the lack of variation. Many of them, therefore, go on to accept the second utterance ("cane bistecca") as telling the same story, either because "it is not always the same thing," or because they identify one or both nominal elements ("cane"-"bistec-ca") as corresponding to the French words "chien" and "bifteck." On the other hand, all these children immediately accepted an Italian utterance composed of different elements and of comparable length as expressing the same meaning, without understanding that meaning. All these judgments are made immediately and with conviction, despite the subjects' lack of knowledge of Italian (hesitations, conflicts, and alternative responses appear at later ages).

As we have said, there is no problem in getting children to think about the formal properties of written text whose meaning they do not know. One simply presents, as several researchers have done (Ferreiro & Teberosky, 1979; Lavine, 1977), cards with different graphic displays (geometrical drawings, letters in different combinations and dispositions, numbers, etc.) and asks the children whether what is on the cards "can be read," "is good for reading," etc. These authors have shown that children who cannot yet read consider only some displays as legible, mentioning formal characteristics which show similarities to the comments they made about our incomprehensible spoken material. For example, confronted with AAAA, children aged five and six will say "You can't read that, they're all the same." (However, they will accept displays like AERTZ, IUOP, etc.) This argument is the same as that used by children of the same ages to reject our "cane cane cane cane" example. The authors mentioned above have also shown that a minimal quantity of letters must be present for the child to accept the display as "readable": displays with one or two letters cannot be read. In our research on the judgment of incomprehensible spoken material, we did not uncover a similar criterion of quantity, most likely because we did not present utterances of less than five syllables in length. Four, five, and six year olds did, however, use the criterion of length in our task when asked to judge if a short (but varied) utterance in Italian could express the same meaning as a long one in French. (Those who accept the utterance use criteria we will not go into here; those who refuse it do so for reasons having to do with length.)

It thus seems to be the case that, concerning both written and spoken language, children abstract some formal properties (and not others) of the material they hear and see, and consider these properties to be essential for the sequences of sounds or shapes to be elements of a language system. The abstracted properties have to do with quantity and variety. Clearly, such ideas are not taught, but are, at least in part, the result of hypotheses constructed by the children themselves. Moreover, some of the children's notions cannot be attributed to the influence of what they observe in their environment. For example, the large majority of subjects (45 out of 56 children) tested by Ferreiro (1979) accept a written sentence with no blanks between the words as "readable," and often they declare this form to be "better" than the sentence with blanks. Surely blanks between words are more readily perceptible than letter variation or mean word length. Interestingly, some of the properties the children abstract are profound ones: one can read and understand text without blanks, but a series of identical letters or similar sounds simply cannot have meaning at all.

The origin of these criteria, applied to both written displays and oral examples, remains mysterious to us. We must note, however, that these criteria are never explicitly formulated as general rules, and that they become evident only when the child is confronted with constructed examples that he rejects. The rejections are motivated and explained, the underlying principle itself is not.

Links Between Sounds and Letters

What do children think about sounds and letters and correspondences between the two before they are capable of reading and writing in an alphabetic way? As we have pointed out in the previous section, where we discussed the quantity and variety of elements present, even very young children (aged three) have some intuition and knowledge concerning the substance of spoken and written language. However, these intuitions are but a first step towards establishing stable phoneme-grapheme correspondences.

Once the child had made the hypothesis that there is a link between letters and sounds (and not letters and the properties of the referent, or letter and letter name), Ferreiro's and Teberosky's Spanish-speaking children went through a stage where they believed this link to be syllabic, i.e., they thought that one letter corresponds to one spoken syllable. Ferreiro and Teberosky noted an important difference between the two groups of children they tested (middle-class children in kindergarten and children from a shanty town). At the age of four, the differences between the groups are not large: at the age of five, however, larger differences appear and no child from the shanty-town group goes beyond the "syllabic hypothesis," whereas many of the middle-class children do. At the age of six, the differences are even larger, with many of the shanty-town children still persisting in syllabic writing behaviors. Ferreiro and Teberosky's data thus seem to show that the syllabic hypothesis is more difficult to overcome for the shanty-town children.

The idea that one letter stands for one syllable is, of course, most evident in spontaneous (or requested) writing behavior, for example, four letters or letter-like shapes for "mariposa," two for "papa," etc. When attempting to read, the hypothesis is contradicted by information from the environment. If the child knows how the written display should be read and attempts to read in this way, large amounts of letters will be "left over" or "unread." Other knowledge the child possesses may also contradict the syllabic hypothesis: he may, for example, have learned to write his own name correctly, as well as a few other words. Conflicts inevitably arise and, according to Ferreiro and Teberosky, it is these conflicts that induce children to revise their hypotheses and subsequently gradually come to reinvent and understand the alphabetic principle.

It has long been known that children find it easier to segment phonetic strings into syllables than into phonemic segments (Bruce, 1964; Rosner, 1974; Liberman et al., 1974; Rosner & Simon, 1971; Gleitman & Rozin, 1973, etc.). Segmenting capacities do not appear until the age of five or six (depending on the studies): it is not surprising that analysis of phonetic strings, at least in this form, does not begin earlier. If children consider "train" to be a long word and "strawberry" to be a word "because it grows in the garden," they are not yet considering words as being composed of some specific substance. Segmenting into words is a task of a different nature, as morphosyntactic criteria intervene. Results depend largely on task instructions and the researcher's requirements. For example, Holden and MacGinitie (1972) conclude that five and six year olds find it easier to segment into words than into syllables, but in Berthoud-Papandropoulou's (1980) counting task, articles are not included systematically as separate words until the age of eleven.

It has been surmised that syllabic segmentation is easier than phonemic segmentation (see Liberman et al., 1977) because acoustically no genuine segments exist. We impose our apprehension on signals which are in reality continuous. The syllable peak, however, has a correlation in physical amplitude. These facts are reflected in other psycholinguistic results. Studies of speech perception with adult subjects have shown that phoneme detection is at least highly dependent on syllable identification (Segui, Frauenfelder & Mehler, 1981) if not preceded by it (Savin & Bever, 1970). Illiterate adults reportedly fail to carry out phonemic segmentation, and children who are "failing" readers perform well when taught logographic writing (Rozin, Poritsky & Sotsky, 1971) and word-syllabic reading (Gleitman & Rozin, 1973). Moreover, Gelb (1952, p. 203) points out that "Almost all the writing introduced among primitive societies in modern times stopped at the syllabic stage." It has also been shown that the history of writing systems (see Gelb, 1952, or Cohen, 1958) goes from a direct representation of meaning (which bypasses sound entirely) to representation of syllables, culminating in the construction of alphabetic systems. There is thus plenty of evidence to support the fact that children's segmenting capacities are in part dependent on the nature of the sound stream itself and, in part, dependent on general properties of human cognition.

Ferreiro's and Teberosky's (1979) subjects thus transposed their analysis of the sound stream directly to the written system. Clearly, believing that one letter stands for one syllable is an idea constructed by the child: the alphabetic system itself has no specific properties that suggest this type of decomposition, and we may be fairly certain that older children or adults never make comments that would induce the child to believe this. It is, however, quite possible that once children have made a link between sound and letter, the discrete nature of text (words separated by blanks and letters most often separate entities) suggests some kind of segmentation of the oral stream, and that they simply carry out the only segmentation of which they are capable at that point. It is surely not a coincidence that segmentation capacities and beginning writing seem to develop concurrently (see also Sinclair, 1982).

Here, we are thus dealing with specific analysis of some of the properties of language (the sound system and how it may be segmented) and use of that analysis for other activities (reading and writing), and not with general ideas or under-lying principles that manifest themselves in different tasks or areas.

Discussion

The data we have discussed show that the capacity of grasping the alphabetic principle and of applying it uniformly is preceded by a slow development of various conceptual substages, even though it is as yet impossible to organize these substages into a coherent developmental sequence. Within such a framework, phonological analysis, establishing stable phoneme-grapheme correspondences, and carrying out phonological recoding appear to be characteristics of covert knowledge in a broad sense, rather than a particular skill acquired through association and practice.

The various parallels we have described show that certain ideas or concepts manifest themselves in both areas we have considered, namely, "thinking about language" and the development of literacy. The comparisons we have carried out are certainly not exhaustive; moreover, the experiments described were carried out in different countries, by different researchers. On the one hand, this makes it difficult to carry out comparisons, but on the other hand, it renders the parallels themselves more striking.

Despite the absence of a clearly established developmental sequence, the various observations reported suggest the following tentative interpretations. A gradual dissociation of the different observable forms of language (utterances, written words, or phrases) from their meaning underlies the progress we have described. From the adult and theoretical point of view, this dissociation also appears necessary in order to grasp the alphabetic principle (language must be apprehended as composed of sounds, independently of meaning). Moreover, the meaning of a written text is of a fundamentally different nature from that of spoken language (which is accompanied by immediate gestural, pragmatic, situational cues, etc.; see Olson, 1977). To understand this difference, it may be necessary to link words and meaning in a conceptually new way, not quite the same as that used in speaking and comprehending.

The first parallel discussed (relationships between signifier and signified) points to an initial lack of dissociation: for very young children, words and their referents have certain properties in common. Both the children's responses to metalinguistic questions (concerning long and short words, for example) and some of their early writing attempts (more letters for "elephant" than "butter-fly," for example) can be attributed to this confusion.

The initial lack of dissociation between word and referent also explains, in part, the behaviors we have discussed in our second parallel (linguistic units). The children's particular view of word-referent relationships leads them to establish classifications of "words" versus "nonwords" of a particular type: when counting words in utterances, as well as when attributing meaning to written fragments, they consider only some elements to be "words."

Our third parallel (formal properties of verbal signifiers) shows that young children nevertheless formulate at least some ideas about the properties of shapes and sounds and their combinations in longer sequences. (We have called these "formal properties.") Variety and quantity appear as properties considered necessary for an example (visually or orally presented) to be "language." These ideas, however, only appear when the experimental situation does not allow the child to consider meaning, which shows that he need not, to answer our questions, operate a dissociation between verbal samples and their meaning. Certainly, ideas about such properties participate in some way in the gradual change of perspective children undergo concerning the question of meaning: at present, however, it is unclear how.

Our fourth point (links between sound and letter) shows that once children have made the hypothesis of a link between sound and letter, independently of meaning, their analysis of spoken language becomes directly useful and pertinent for reading and writing activities.

What, finally, can we say about the difficult question of mutual interdependencies between thinking about language and the development of literacy? As we have taken the widest possible definitions of "thinking about language" and of "literacy development," our discussion will perforce remain of a general nature.

It is possible to speculate that it is precisely for the acquisition of such activities as reading and writing that ideas about language (or metalinguistic knowledge) will become necessary. Mattingly (1979a) points out that speaker-hearers are not conscious of (or have no access to) performance mechanisms, but that they do have access to grammatical knowledge. The fact that languages such as English transcribe morphophonemic knowledge is directly exploited in reading (which requires access to morphophonemic representations) but is not exploited in listening. He concludes: "... it is my contention that written language, far more than speech, places a direct demand on the individual's acquired knowledge of language, what has been called grammatical knowledge; and that such knowledge ... must be accessible to the reader, as it presumably is to the language learner." (See also Chapter 2 of this book.)

Conversely, we could hypothesize that reflections or ideas about the written system—its material permanence and its discrete properties rendering it a more suitable object for reflection than spoken language—stimulate and guide children's metalinguistic thinking. Children abstract some of the properties of written text: that it is composed of a stock of letters that appear in different combinations, that it can be read aloud, that it has meaning, etc. Perhaps the children gradually come to transpose these abstractions to language in general, including spoken utterances.

Our writing system represents some aspects of sounds and bypasses meaning entirely. To use alphabetic reading and writing techniques, new conceptions, where words and their meaning are coordinated and dissociable, must be constructed. Exactly how these new conceptions are elaborated is at present unclear, but, certainly, ideas about language as well as reflections on the written system play a part and interact. Further research, particularly with subjects exposed to different writing systems and with subjects from an illiterate group, will help elucidate this complex problem.

Acknowledgment

The research on which this paper is based was carried out with the help of the Fonds National Suisse de la Recherche Scientifique, Grant No. 1-956-0. 79.

CHAPTER 6

Cognitive Development and Units of Print in Early Reading

Alan J. Watson

Introduction

With the excitement of one who has made a major discovery, he pointed to his name, *Peter*, printed on the corner of his playgroup painting. Running his finger rather haphazardly across the letters, he pronounced, "Pe-ter Gor-don Mey-ers." Peter, nearly four, had made significant progress towards an awareness of the functions and purposes of written language. He realized that print provides clues for reading and that components of his printed name in some way represent speech sounds. However, the nature of the units of which print is composed and the one-to-one correspondence of their visual and auditory elements were yet to be grasped. Understanding the units of print seems to be an aspect of the young child's language awareness which is critical in learning to read. This chapter will review theory and research on attaining concepts of the units of print and will present evidence which tests a cognitive developmental view of this task.

The units of print are a particularly difficult constellation of concepts for the child to grasp. Even after a year's reading instruction, school beginners did not reliably distinguish a word, a letter, and a sentence (Reid, 1966; Downing, 1970). Part of the difficulty of the task seems to be that the unit of reading, chamelion-like, takes so many different forms in different situations. Sometimes the unit might be a letter or a group of letters, while, at other times, it might be a word or a phrase or a sentence. Furthermore, the phonemic elements of speech are particularly difficult to identify. Although spoken words have acoustic peaks which help the child form syllables (Peter did this quite well), there are no acoustic criteria for distinguishing phonemes (Liberman, Shankweiler, Fischer & Carter, 1974) and, even when the phonemic elements of speech have been identified, there is the problem of ambiguity in their representation. There is no simple correspondence of phoneme and grapheme. In addition, the alphabetic system requires the child to manipulate meaningless units at a time when the child's focus in regard to language is upon its communication purpose.

Thus, the elusive nature of the unit itself, the difficulty of identifying its phonemic elements, the ambiguity inherent in its representation, and its meaninglessness for the communication-oriented child suggests that to appreciate the units of print is a demanding cognitive task which calls for mental flexibility of a high order. It is the contention of this study that it calls for the most advanced level of conceptual reasoning the child can produce, and that the basis of the failure of many children to understand the units of print is their, as yet, inadequate reasoning capacity.

Since manipulating and making sense of print is likely to require an appreciation of its units, it is not surprising that a wide range of language-awareness tasks have exhibited a considerable relationship with reading progress in a number of different studies (Francis, 1973; Ehri 1975, 1976; Liberman, 1973; Liberman & Shankweiler, 1979; Fox & Routh, 1980; Snowling, 1980). Although there is little controversy over the *existence* of an important relationship between reading and language awareness, there is considerable controversy about the *explanation* of that relationship. Three major hypotheses merit close examination. One hypothesis, developed from linguistic theory (Mattingly, 1972, 1979a), proposes that the same underlying language acquisition processes are used in learning to speak and to read but that reading calls for an awareness of language developed initially, at least, by enriched experience with oral language. A second hypothesis is an information-processing view (Ehri, 1979) which suggests that the relationship is the direct result of specific experience of print, e.g., learning to read produces language insights. A third hypothesis, a cognitive developmental view favored by the author, proposes that the child's developing reasoning capacities support both language awareness and learning to read in a way not required as the child learned to speak.

Language Acquisition

According to Mattingly (1972), speech is a primary linguistic activity while reading is a secondary linguistic activity dependent on the learner's awareness of the primary activity. The aspect of that primary linguistic activity which is critical in beginning reading is awareness of or "having access to" the appropriate units of one's morphophonemic representation (Mattingly, 1979a, p. 14). He argues that, although the same biologically based language-acquisition processes are used to learn both speaking and listening as well as reading and writing, the need for this access accounts for the greater difficulty involved in reading and writing. Linguistic awareness is acquired as an extension of the early grammatical development which supports speaking and listening. There have been very few studies of preschool children's language awareness, and none seem to have traced its relationship to reading progress, but a large range of both correlational and training studies, reviewed by Golinkoff (1978) and Rozin and Gleitman (1977), attest to the importance of the relationship of language awareness (especially phonemic awareness) and reading among school-age children. In some more recent studies it has been found that severely retarded readers, in contrast to normal readers, could not perform phonemic analysis (Fox & Routh, 1980), could not use the surface code (rhyme) to assist memory (Byrne & Shea, 1979), could not do an auditory oddity task or provide a matched rhyming word (Bradley & Bryant, 1978), and could not match appropriate graphemes and phonemes (Snowling, 1980).

Despite this very general suggestive support, the language acquisition hypothesis has several limitations. First, the account of how linguistic awareness is acquired as a nonfunctional precursor to reading is unconvincing. Since morphophonemic knowledge is not needed in speaking and listening, it is unlikely to be acquired before reading is begun because language is not learned independently of its functions (Halliday, 1973). As MacGinitie (1979) remarked of this aspect of Mattingly's view, "the human propensity to acquire grammar beyond the point of functional value has been rescued from the extinctive fate of other sabretoothed characteristics by the invention of writing. It seems unlikely" (p. 4).

Second, the cultural influence on learning to read is underrated by this account. Mattingly has convincingly argued for the existence of important differences in the nature of primary and secondary language activity. A critical cause of these differences is the effect of the cultural invention, orthography. This seems to imply significant differences in the learning conditions of the primary and secondary language activities. The discussion of these conditions, however, which proposes the same learning processes for both does not do justice to the differences established. For example, it is maintained that learning to read will be motivated by the same linguistic curiosity that motivated oral-language acquisition. No account is taken of social effects which seem likely to influence a cultural skill, reading, rather differently and much more than they did a more biologically based skill, listening (Waterhouse, 1980). The explanations of the nature of reading and of the way it is learned do not sit together very comfortably and the effect of social influences on learning to read are largely overlooked.

Third, the account is very difficult to subject to empirical testing. Although there is evidence of a very general kind which seems to support some of its main contentions, their formulation as yet lacks specific testable hypotheses. Is the existence of oral-language-learning mechanisms a necessary but not a sufficient condition for the acquisition of linguistic awareness? How do the language-learning mechanisms relate to other processes, e.g., perception and memory, which have long been considered important to reading?

Despite these limitations, the language-acquisition hypothesis has served an important seminal role in current reading theory. The delineation of primary and secondary linguistic activity, and the concept of linguistic awareness as an explanation of the difference between the two and of difficulties in learning to read, has drawn together much recent knowledge in a fairly coherent and plausible way. The theory, however, as Mattingly acknowledges, is largely conjecture and needs to be developed considerably and tested empirically before it can be used as a firm basis for classroom practice.

Information Processing

While the language-acquisition hypothesis gives little place to social influence in learning to read, an information-processing view gives a much greater place to social effects. Language insights are gained as a direct result of experience with print, usually in learning to read (Ehri, 1979). It is a knowledge-based rather than a mechanism-based explanation. The orientation of this perspective is towards the detailed interpretation of empirical data rather than towards broad theoretical proposals. The general concept, linguistic awareness, is avoided and, instead, content specific insights, e.g., about words or about phonemes, are studied. They are viewed as bearing no necessary relationship to one another.

A range of evidence indicates that learning to read is commonly a critical variable in the development of word and phoneme awareness. Interview studies with school beginners have suggested that print awareness typically develops as a result of reading instruction (Reid, 1966; Downing, 1971-1972; Francis, 1973). Ehri (1976) found that, of two age-matched kindergarten groups, readers had a considerable advantage over prereaders on a task requiring word awareness and were not significantly different from older grade-one readers. This is interpreted to mean that reading ability, rather than age and its correlates, is the critical factor in accounting for word consciousness. Additional evidence for this view is provided by the finding (Ehri & Wilce, 1980b) that the way children spelled words influenced the way they were segmented into phonemes, and it is concluded that word learning has a causal influence on phonemic awareness. In summing up the print experience view, Ehri (1979) states that, as the results with beginning readers are correlational and since there are some children with a well-developed word consciousness before learning to read, experience with print is "not a necessary condition for acquiring lexical awareness but only a sufficient condition" (p. 84).

There is, however, a difficulty with this conclusion. Whether children of the same age and grade are readers or nonreaders may well be influenced by their underlying general cognitive or language development. No control was maintained for these factors which are indeed correlates of age but which act quite independently of it within a narrow chronological range. To assert that something is a "sufficient" condition implies that it is, in itself, enough to produce the result. No other conditions are required if that condition exists. If, however, a certain level of cognitive development is necessary in addition to the child's experience of print to produce an understanding of its units, then experience of the written word cannot be a sufficient cause of print awareness. This is an example of the problem of causality in developmental data raised by Wohlwill (1973). If there are "normal developmental processes" which have a momentum of their own, independent of specifiable external agents or conditions, it is only possible to isolate necessary, not sufficient, causes of developmental or developmentally related variables. Whether there are important general developmental influences on language awareness is the subject of this study. The possibility that they exert a significant influence has not been excluded by Ehri's data or by the other studies surveyed.

One study which sought to test the print experience and cognitive development hypotheses of phonemic awareness compared literate and illiterate adults from a poor agricultural area of Portugal who were similar in childhood experiences and environment (Morais, Cary, Alegria & Bertelson, 1979). While the illiterate subjects were unable to delete or add a phone (a single sound) at the beginning of non-words and fared little better adding a phone to words, the literate subjects had little trouble with either task. Ability to manipulate beginning sounds of words is not attained spontaneously as a result of general cognitive development and any incidental experience of print they may have had (this is not documented). The experience of print in learning to read seems to have produced that capacity. This conclusion, however, does not rule out cognitive development as an important influence, since specific instruction may have no effect before a certain level of cognitive development is attained. The question of whether reading experience is a sufficient cause of language awareness requires a comparison of groups which are equated for reading achievement, not age. The numerous studies which have found that backward readers have low language awareness by contrast with normal readers of the same age and intellectual level (Rozin & Gleitman, 1977; Golinkoff, 1978) confound language awareness and reading experience. Poor readers' lower language awareness may be due to their limited experience with reading by comparison with that of normal readers. Do children who are at a similar level of reading possess a similar degree of print awareness?

Snowling (1980) tested this hypothesis by asking normal and backward readers, matched on reading and I.Q. (rather than on chronological age), to say whether visually presented nonsense words, e.g., *dorn-dron* and *sint-snit,* were the same or different. Retarded readers were significantly poorer than normal readers, indicating that, despite having had about as much experience of reading, they were less able to use phonemic segmentation. This is evidence against the view that reading experience is a sufficient cause of phonemic awareness. There seem to be other factors of great importance for its development. The data suggest that the retarded readers perform in a qualitatively different way to the normal readers. Normal readers show a distinct developmental trend in visual-auditory matching, but retarded readers show no significant changes with reading age. Retarded readers reach the same reading level as normal readers with far less capacity for phoneme segmentation. Finding it difficult to decode unfamiliar words, it seems that they depend more on other methods, e.g., sight vocabulary. Snowling interprets this in terms of Vellutino's verbal-deficit hypothesis (Vellutino, 1977), but it could well imply a cognitive processing deficit. Retarded readers are able to operate at the "associative" level (White, 1965, 1970) to learn whole words as part of sight vocabulary but have great difficulty operating at the "decision" or "reasoning" level required to master phonemic awareness tasks which are mentally much more demanding. A similar mental organizing deficiency has been shown in a phonemic differentiation task by retarded readers who were also matched with normals for reading level and I.Q. but not age (Bradley & Bryant, 1978). The normal readers were far superior in discerning phonemic similarities and differences in groups of words, e.g., "bed, red, fed, had," and in producing a rhyming word. As in the Snowling (1980) study, there was a clear developmental trend evident for the normal readers, but no such trend was shown by the backward group.

Both studies are subject to the limitation of being based upon cross-sectional data, but the suggestion is clear. Normal readers develop a phonemic awareness which poor readers, who are at the same reading level, do not develop. These abilities do not appear to be the result of experience with written language since the groups were matched for reading level. Indeed, in regard to the quantity of time spent with print, the older backward readers have probably had much more. A plausible explanation is that there are different processing capacities brought to the learning situation which have a profound effect on phonemic and word-awareness learning. This is the contention of the cognitive developmental view.

Cognitive Development

This explanation claims that an underlying cognitive developmental capacity is called upon by both language awareness and by reading that learning to speak and listen did not require. It does not deny the involvement of the child's lan-guage-acquisition processes in language awareness and in reading. Nor does it imply that print experience and instruction are of little importance for language awareness. To the contrary, they are seen as critical for reading progress. Instruc-tion, however, exerts an effect in the degree to which its structure, timing, and content match the child's developmental status. In themselves, oral-language mechanisms and print experience are inadequate explanations of the relation-ship between language awareness and learning to read. This view was suggested by Williams (1979) in a response to the language acquisition theory of Mattingly.

> It seems to me more reasonable to assume that every "normal" human being has enough of whatever it takes to learn to speak and listen, and that there is some other capacity, not equally distributed among all people, that (when environmental conditions permit) leads to facile reading acquisition and reading performance (a cognitive, non-linguistic ability, perhaps?) (p. 4).

The extent of individual differences across groups of children who have had the same amount of schooling (Calfee, Lindamood & Lindamood, 1973) implies the importance of factors other than instruction and print experience. The difference between the language awareness of good and poor readers matched for reading and I.Q. (Snowling, 1980; Bradley & Bryant, 1978) indicates that reading competence does not explain these individual differences in language awareness. The nature of these factors is suggested by a study which analyzed qualitative differences in error patterns of kindergarten and grade-one responses to a word-awareness test (Holden, 1977). The subjects had to detect which word had been added to the second of two orally presented sentences when the word added made a homophone change meaning, e.g., "John leaves after dinner," "John rakes leaves after dinner." They showed a developmental trend from larger unit responses (sentences or incorrect sentences) in kindergarten to smaller unit responses (the homophonous word or a phrase containing it) in grade one. This move from global to more discrete error units is not readily explained by memory improvement nor by specific experience or instruction since no one instructs response errors. A developmental pattern in analysis of utterance from global to discrete with a later emergence of the ability to differentiate words from the semantic matrix seems to exist. The data, however, suffer from being cross sectional, and the interval of one year does not seem well chosen when the greatest change over that period is in the number of correct responses (43 to 85 percent) which can tell nothing about the qualitative changes being studied. A finer-grained examination of the ability starting during kindergarten and followed for a year or so with measures four or six months apart would provide a more convincing demonstration of this developmental change. Although Ehri (1979) interprets the increase in correct responses to show the effect of print experience, it is not possible to isolate the effects of instruction and cognitive development in this improvement. Probably both are operating. However, the qualitative change in error patterns suggests a developmental component in the child's emerging lexical awareness.

Corroboration of this finding is provided by the two studies of phonemic awareness (Bradley & Bryant, 1978; Snowling, 1980) in which print experience is controlled by matching normal and backward readers on reading achievement. Both found that normal readers underwent a developmental change towards use of phonemic awareness while backward readers did not. These developmental patterns, taken together with Holden's (1977) evidence for a developmental component in lexical awareness, suggest that language awareness is a capacity influenced, not only by instruction, but also by underlying developmental change. This is fully consonant with a cognitive developmental viewpoint, but whether the development can be explained as an expression of the simple-to-complex organization of the task or also as an expression of the child's underlying mental structures requires further evidence. Some means of identifying possible underlying reasoning capacities is needed to give a basis for this investigation. The most comprehensive current developmental theory of cognition, that of Jean Piaget, provides a starting point.

Piaget's (1967) explanation of the emergence of a unit of number in terms of operative thought may provide an important clue to understanding the child's conception of units of print.

A whole number is in effect a collection of equal units, a class whose subclasses are rendered equivalent by the suppression of their qualities. At the same time, it is an ordered series, a seriation of the relations of order. Its dual cardinal and ordinal nature thus results from a fusion of the logical systems of nesting and seriation, which explains why true number concepts appear at the same time as the qualitative operations (p. 53).

To gain the concept of a unit of print the child seems to need to do several things simultaneously. He or she must classify, for example, all letters as having something in common, despite their evident differences. He or she must also recognize that these letters participate in some complex form of serial relationship to one another, e.g., *T* before *o* before *m* in the boy's name. As well as doing these with both the spoken and written forms of the letters, the child must also match these two elements in a one-to-one correspondence. Although attaining the concept of units of print is more complex than gaining the unit of number (Elkind, 1974), it seems to possess basic similarities and does appear at about the same time as operational thought.

Studies of Piagetian operations and reading have shown low to moderate correlations of conservation and classification with reading but a stronger relationship of seriation with reading (Waller, 1977). Another aspect of operativity, perceptual regulation, has also displayed a significant and consistent relationship with reading (Elkind, 1976). Perceptual regulation, which refers to the ability to see beyond the more salient aspect of a complex visual field, is also claimed by Elkind to underlie linguistic awareness. The findings in regard to the relationship of seriation and perceptual regulation to reading suggest that cognitive development is an important underlying capability for learning to read and for language awareness.

Two studies support this suggestion, the first with school-age subjects and the second with preschool children. Holden and MacGinitie (1973) examined the relationship of word awareness to a battery of measures including operativity (single seriation), I.Q., and readiness skills in 50 kindergarten and 50 first-grade subjects. The strongest predictor of word awareness for the whole sample was age ($r = 0.57$), which also shared heavy common factor loadings with word awareness while no other variables loaded heavily on more than one factor. This relationship (reflected in the marked superiority of grade one over kindergarten pupils) and a rapid increase in word awareness at about age six is clearly not to be attributed to age per se (Ehri, 1976). Yet its interpretation is difficult. Although it is likely that experience of print through reading instruction begun in grade one is a major influence, as Ehri (1979) claims, the substantial relationship between seriation and word awareness suggests that cognitive development may also be important. Kindergarten seriation correlated 0.48 with word awareness and the combined kindergarten and grade-one correlation of the two was 0.46, despite a relatively low reliability of 0.54. When corrected for reliability, the combined scores correlation was 0.68. Comparison with word awareness in its correlation with I.Q., 0.54, and in its correlation with Gates-MacGinitie Readiness Skills test, 0.60, both well-established predictors of school tasks whose reliabilities can be expected to be greater than that of seriation, indicates that operativity cannot be dismissed lightly as an influence in the development of word awareness. Its basis in a wide-ranging developmental theory in contrast to the pragmatic basis of I.Q. and readiness gives it a potential explanatory power which those measures lack. Cognitive development as well as reading instruction may be the basis of the marked improvement in word awareness detected at about six years of age.

However, the data present a difficulty for this interpretation. The cross tabulation of word awareness and seriation, in which half of those low in seriation are high in word awareness, suggests that operational competence is not necessary for word awareness. This effect, however, is almost entirely confined to grade-one pupils where 68 percent are assessed as low in single seriation, a task which has been found to be mastered by the majority of children in Western society by age 7 (Inhelder & Piaget, 1964; Elkind, 1964; Lunzer, 1970; Nixon, 1971). Since the average age of Holden's and MacGinitie's subjects is 7 years 2 months, the high-low cutoff point of 27 (maximum 30) seems too stringent. There must, therefore, be some reserve about the conclusion, drawn from this cross tabulation, that language awareness precedes seriation operativity. The kindergarten cross tabulation provides a somewhat more dependable guide to the relationship. Hence, despite the rather high cutoff point for seriation, 78 percent of those low in seriation are also low in word awareness. In this comparison, few children achieve a high level of awareness without high seriation.

In sum, the relationship between word awareness and seriation found by Holden and MacGinitie suggests that cognitive development may provide an underlying capacity which enables the child to benefit from print experience during reading instruction. Individual differences in word awareness after a year or two of reading instruction may be caused partly by differences in conceptual reasoning ability. However, the evidence is as yet inconclusive and needs corroboration and extension. No control has been exercised for the influence of oral language and the direction of the relationship needs to be clarified. Above all, the use of cross-sectional data is a severe limitation in the study of developmental phenomena, since change is not directly studied but can only be inferred.

Hiebert (1980, 1981), who studied preschoolers' print awareness and its relationship to operativity (including seriation), oral language, and home experience, found that operativity made a unique contribution to print awareness. The best predictor model for the measures of print awareness was a combination of operativity, home teaching, and vocabulary which accounted for 56 percent of the variance. Although it is claimed that linguistic and cognitive variables seem "inextricably interwoven" (Hiebert, 1980, p. 321) in the development of print awareness, this is not a conclusion based on the data presented. Hiebert's regression procedure shows that the major influence on print awareness comes from operativity and that each of the three variables contributes independently of the other two. Within the limits which regression procedures impose upon the interpretation, cognitive development must be seen as an explanatory variable of considerable consequence for awareness of print.

Finally, it is necessary to summarize this cognitive developmental view of language awareness to provide a basis from which hypotheses can be developed for empirical investigation. Although experience of print (especially reading instruction) is of great importance for the development of linguistic insight, wide individual differences in the language awareness of those who have had similar print experience indicate that other factors are also highly important in its development. Print experience does not seem to be a sufficient condition for the development of language awareness. Striking differences exist between normal and backward readers matched for reading level (e.g., Snowling, 1980). Qualitative differences in the reading of these two groups and developmental progress in awareness and reading by the normal but not the backward readers suggests that a developmental effect important to language awareness is at work. Furthermore, changes in word awareness, resulting in errors not readily explained as a result of improved memory or of instructional experience (Holden, 1977), also suggests that a developmental factor is operating. Studies of Piagetian variables which have shown a moderate to strong correlation of operativity (especially seriation) and reading indicate that these conceptual reasoning capacities may provide a basis for the study of the influence of cognitive development on language awareness.

Thus, the developmental change evident in the qualitative analysis of pupil responses and the relationship of Piagetian variables to linguistic insight provides the basis of a substantial case that cognitive development is an important underlying factor in the attainment of several aspects of language awareness. Findings in regard to developmental variables based on cross-sectional data, however, are subject to a severe limitation. Static observations, insensitive to the dynamics of change, are likely to confound growth with individual differences and to assume that interindividual differences are stable over time. Longitudinal research, the "lifeblood of developmental psychology" (McCall, 1977), is vitally important for an understanding of the relationship between the development of the cultural skill, reading, and the biologically based capacities of thought and language. This is especially so since the more biologically based skills are not unitary phenomena but complex jigsaws of varied components and control mechanisms, and the cultural skills may be so flexibly achieved as to have different components and arrangements in different individuals (Waterhouse, 1980).

An Empirical Investigation

A study was undertaken to overcome the limitations of cross-sectional data and to extend present findings by a longitudinal examination of the relationship of conceptual reasoning (operativity), oral language, print awareness, and reading during the first three school years (kindergarten, grade one, and grade two). The hypothesis to be investigated was concerned with the extent of the effect of operativity on awareness of units of print. Its influence was compared with the effects of competing explanatory variables, oral language, and reading experience.

The hypothesis states that the effects of operativity on print awareness are partly distinct from the effects of oral language and reading ability and that operativity exerts a causal influence on print awareness.

To test this hypothesis, it is first necessary to examine the relationships within and between the several variable groupings. The evidence for multiple mental abilities which develop in different ways with greater and lesser degrees of continuity and consistency (Bayley, 1970) suggests that there may be varying patterns shown within the abilities under study. These varying developmental patterns may have important implications for attaining an awareness of units of print and for learning to read. Language capacities have been found to be more developmentally consistent than reasoning processes which could be expected to undergo important developmental change in the age period of the study, five to seven years. The different correlations of seriation and word awareness in kindergarten (0.48) and in grade one (0.26) found by Holden and MacGinitie (1973) suggest that there may be changing patterns of relationship from year to year between predictor and dependent variables. However, the lack of previous research makes it impossible to formulate specific predictions of how these interrelationships may change with time. This preliminary exploration will prepare for an examination of the overall pattern of effects as expressed in the hypothesis.

Method

Sample

The subjects were selected by chance procedures from kindergarten entrants to three adjacent public schools in a homogeneous middle-class southern area of the city of Sydney, Australia. The average age was 5 years 10 months. In the first year the sample consisted of 148 subjects (69 boys and 79 girls) with a mean age of 5 years 10 months and a standard deviation of 3 months. In the second and third years, pupils who moved were followed to their new schools as far as possible, and the sample size was kept at 142 and 137, respectively.

Materials

The various measuring instruments used were as follows. Reading was assessed with a group-comprehension measure, the Paragraph Understanding test (Hall & Pacey, 1978), and with an individual test, the St. Lucia Graded Word Reading test (Andrews, 1969), both devised for use in Australian schools. The Units of Print test, a group measure devised for this study, sought to assess the child's appreciation of the units of which print is composed. It presented a series of examples of letters, numbers, words, or sentences with distractors and asked the child to nominate, for example, which one was a letter. In this way, the child's knowledge of the correspondence between the written and the spoken forms of the items was tested. One problem, which became apparent in pilot testing of an early version of the measure, was the possibility of confounding reading ability with print awareness. To avoid this difficulty, words and sentences were made as simple as possible and the test was given after at least a full year of reading instruction. Two tasks, adapted from Piaget's work, were used as indices of operativity: seriation, and perceptual regulation. Seriation was measured in a two-part test assessing single seriation (Lunzer, Dolan & Wilkinson, 1976) and multiple seriation (Watson, 1979). Perceptual regulation was measured with the Perceptual Ambiguity test developed by Elkind and Scott (1962) which tests whether the child can look beyond the most salient aspect of an ambiguous line drawing. Two aspects of oral language, vocabulary and grammatic prediction, were tested. The Grammatic Closure test from the Illinois Test of Psycholinguistic Abilities (Kirk, McCarthy & Kirk, 1968) was used to measure syntactic prediction, and the Peabody Picture Vocabulary test (Dunn, 1959) was used to assess lexical knowledge.

Procedures

The operativity and oral-language tasks were administered at the end of kindergarten and grade one by a team of undergraduate research assistants trained until a criterion level of competence was attained. Each measure was administered by a different tester or group of testers. The language awareness and reading measures were given at the end of grades one and two by the researcher and an assistant to groups of some 25 subjects seated in rows in a classroom and separated by screens. Precautions were taken to avoid contamination of results by ensuring that no tester administered individual criterion and predictor measures to the same child.

Data Analysis

Internal consistency, as assessed by Cronbach's coefficient alpha, was used as an index of reliability for all measures and was calculated on the basis of the responses of the full sample (Goldstein, 1979).

The computations were made with the commercially prepared SPSS programs (Nie et al., 1975; Hull & Nie, 1981). The correlation table was arranged partly in the form of a MTMM (multitrait-multimethod) matrix (Campbell & Fiske, 1959) with time difference replacing method difference. The diagonals underlined (Table 6-1) represent stability values for the predictor and dependent variables. Comparison of these values and those with different traits assessed at different times (in the square blocks surrounding the diagonals) and with different traits assessed at the same time (values in triangles immediately above or to the right of the square blocks) gives an indication of divergent validity in a way similar to that suggested by Campbell and Fiske (1959).

Path analysis, a method for representing and attaching quantitative estimates to a theoretically structured model, was used to assess the pattern of causal

Table 6-1. Correlation Matrix of All Variables

	Kindergarten				Grade 1							Grade 2		
	PR	Ser	GrPr	Voc	PR	Ser	GrPr	Voc	UOP	RC	WR	UOP	RC	WR
Kindergarten														
Perceptual Regulation	(0.64)													
Seriation	0.10	(0.80)												
Grammatic Prediction	0.20	0.13	(0.97)											
Vocabulary	0.23	0.04	0.43	(0.90)										
Grade 1														
Perceptual Regulation	0.29	0.27	0.23	0.07	(0.79)									
Seriation	0.27	0.46	0.12	0.10	0.14	(0.79)								
Grammatic Prediction	0.28	0.34	0.69	0.41	0.31	0.33	(0.99)							
Vocabulary	0.35	0.14	0.47	0.60	0.07	0.12	0.54	(0.99)						
Units of Print	0.13	0.48	0.15	0.12	0.30	0.37	0.39	0.19	(0.99)					
Reading Comprehension	0.19	0.28	0.24	0.20	0.26	0.35	0.44	0.25	0.45	(0.96)				
Word Recognition	0.21	0.31	0.28	0.07	0.18	0.28	0.46	0.07	0.39	0.50	(0.95)			
Grade 2														
Units of Print	0.28	0.37	0.09	0.10	0.23	0.20	0.31	0.07	0.47	0.25	0.34	(0.98)		
Reading Comprehension	0.19	0.28	0.31	0.09	0.30	0.35	0.45	0.18	0.46	0.57	0.57	0.46	(0.95)	
Word Recognition	0.24	0.33	0.26	0.10	0.23	0.22	0.47	0.14	0.37	0.52	0.85	0.34	0.62	(0.95)

Reliability coefficients (Cronbach's alpha) are shown in parentheses.
Stability coefficients are underlined.

effects within the data (Kerlinger & Pedhazur, 1973; Wolfle, 1980). Recursive equation models (with no feedback loops) were drawn up, and standardized regression coefficients were used to estimate the values of the path relationships. Associations shown in the path analysis were decomposed using Duncan's fundamental theorem (Wolfle, 1980) to allow an assessment of the extent of direct, indirect, and spurious effects. Although there are limits to causal attribution on the basis of path analysis (Rogosa, 1979), the difficulty of applying experimental methods to the study of developmental variables (Wohlwill, 1973) suggests that the technique is a highly suitable one for this study.

Results

The issue of central concern, as discussed earlier, was the relationship of operativity, oral language, and reading to awareness of print. The predictive efficiency of the variables was assessed in several stages. First, the relationships within variable groupings were gauged by inspection of a simple correlation matrix. Second, the prediction of dependent variables was evaluated from the correlation matrix and from cross tabulation of some promising relationships. Third, the pattern of causal effects was examined using path analysis.

Relationships Within Variable Groupings

The Pearson correlations for predictor and dependent variables are presented in Table 6-1. Examination of the language variables, grammatic prediction and vocabulary, reveals a consistently strong reliability and a high stability (0.69 and 0.60).

The application of the divergent validity criteria of Campbell and Fiske (1959) that require lower correlations between unrelated constructs indicates that both possess sound discriminant validity. Furthermore, although distinct from each other, the two are also quite closely and consistently related, correlating with each other (0.43, 0.41, 0.47 and 0.54) more strongly than they do with any other variables.

The operativity variables, perceptual regulation and seriation, are not as reliable or as stable as the language measures. The reliability of perceptual regulation moves from rather low (0.64) to moderate (0.79), thus limiting its predictive potential. The low stability coefficient (0.29) is likely to be, in part, a reflection of this but may also indicate a surge of developmental change in this period and considerable individual differences in the development of this capacity (McCall, 1977). Discriminant validity is fair, with the stability coefficient exceeding 10 out of the 12 values in the comparison of its correlations with other constructs. Seriation is a more reliable (0.80 and 0.79) and more stable (0.46) measure. The stability again seems, in part, to reflect individual differences of development shown by considerable rank-order change over this time period. Seriation also shows sound construct validity with its stability value exceeding all 12 relevant comparison values. The relationship of the two operativity measures in these data gives little basis for their combination, but evidence shows that only a slight relationship between operative tasks can be expected before these capacities have settled more firmly into the response repertoire (Flavell, 1977; Tomlinson-Keasey et al., 1979; Arlin, 1981).

Units of print shows high reliability (0.98 and 0.99), moderate stability (0.47), and sound discriminant validity. The two reading measures, which show strong reliability and stability (especially word recognition) and a satisfactory discriminant validity, are closely related and their combination as a composite reading measure seems well justified.

Prediction of Dependent Variables

The kindergarten language variables, although highly reliable, reveal little relationship to units of print, either at grade one (0.15 and 0.12) or at grade two (0.09 and 0.10), but grade-one language predicts units of print more strongly (Table 6-1). By contrast, seriation, despite its lower reliability, is the strongest kindergarten predictor of units of print with a correlation of 0.48 at grade one and 0.37 at grade two. The predictive strength of seriation at grade one is lower but still significant (0.37 and 0.20). Perceptual regulation displays a mainly significant but rather low relationship with units of print. To some degree this is influenced by its low reliability, but it contributes much less than seriation to the prediction of print awareness and reading.

To provide a further estimate of the importance of seriation for units of print, the relationship was examined more closely by cross tabulation of these variables. Seriation, which showed a tendency to cluster at three points on the kindergarten distribution, was broken into three levels, and units of print was broken into two, as shown in Table 6-2.

Table 6-2. Cross Tabulation of Units of Print and Seriation

		Units of Print Grade 1					Units of Print Grade 1		
		Low	High	Total			Low	High	Total
	Low	22	7	29		Low	4	0	4
		(76)	(24)	(20)			(100)	(0)	(3)
Seriation	Mid	32	50	82	Seriation	Mid	26	24	50
Kindergarten		(39)	(61)	(58)	Grade 1		(52)	(48)	(35)
	High	8	23	31		High	32	55	87
		(26)	(74)	(22)			(37)	(63)	(62)
	Total	62	80	142		Total	62	79	141
		(44)	(56)	(100)			(44)	(56)	(100)

Percentage values are shown in parentheses: Within the matrix are the row percentages; beside the matrix are Seriation subgroup percentages; below are Units of Print subgroup percentages.

The cross tabulation of kindergarten seriation with grade-one units of print shows that high seriation has a strong tendency to go with high print awareness a year later (74 percent of those high in seriation are high in units of print) and that low seriation is strongly associated with low print awareness a year later (76 percent of those low in seriation are low in units of print). Of those high in units of print, 91 percent were mid or high in seriation, while, of those low in units of print, 87 percent were low or mid in seriation. Although the support is not complete since there is no empty cell, this suggests that attainment of at least a moderate level of seriation may be necessary for gaining a high level of print awareness. Yet, since a large proportion of those at a mid level in seriation are still low in units or print, a moderate level of seriation is not a sufficient condition for attaining high print awareness. Other factors apparently also have an important effect.

These contentions are supported by the cross tabulation of grade-one seriation with grade-one units of print. Of the 29 children low in seriation at the end of kindergarten, only four remain low at the end of grade one, all of whom are also low in units of print. (An examination of the result sheets showed that all four were part of the original 22 in the "low-low" cell). However, it may be observed that high seriation does not necessarily produce high units of print, since, of those low in units of print, 52 percent are high in seriation. A moderate level of seriation seems to be a necessary but not a sufficient condition for a high level of units of print.

Pattern of Effects

The path analysis has two analytic goals, each concerned with linking the data to reveal the pattern of its longitudinal relationship: first, to assess the relative effects of operativity and oral language on print awareness, and second, to evaluate the effect of reading achievement on print awareness in relation to operativity and oral language. This will allow longitudinal testing of the hypothesis that operational thinking is causally implicated in the development of an awareness of the units of print. The first analytic goal is addressed by the path diagrams shown in Figure 6-1.

Units of print at the end of grade one is considered to be dependent on operational thought and oral language in kindergarten and grade one directly, and for the kindergarten measures, indirectly through the grade-one measures (Figure 6-1A). Although preschool children have been shown to possess some degree of print awareness (Hiebert, 1981), this has not included the ability to distinguish reliably between the various units of print, an insight which children typically develop in their first year of formal reading instruction (Reid, 1966; Downing, 1971-1972; Francis, 1973). Thus, it is assumed that the more fundamental cognitive abilities, thinking and language, possess temporal and causal precedence over the more culturally specific capacity, awareness of units of print at grade one. Another assumption of the model is that the causal flow between Opl and Ll moves from Opl to Ll, since there is evidence that new language capacities

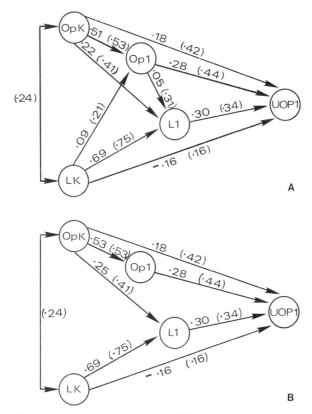

Figure 6-1. Path analysis showing effects of kindergarten and grade one operativity and of kindergarten and grade one oral language on grade one units of print. Single headed arrows show the direction of effect. Strength of effect is shown by the path coefficients on each arrow. Simple correlations (total associations) are shown in parentheses. Double-headed arrow joins two exogenous variables. A. Full model; B. reduced model.

developed at 6 or 7 years of age are under the influence of logical operations (Sinclair de Zwart, 1969; Beilin, 1975; Tenezakis, 1975). The greater strength of the path OpK to Ll than the path LK to Opl seems to support this. However, several path coefficients, including Opl to Ll, do not attain significance and may thus be subject to deletion in model trimming.

The criterion for determining if paths may be deleted is whether, using the path coefficients of the new model, the original correlation matrix can be reproduced with few discrepancies and none exceeding 0.05 (Kerlinger & Pedhazur, 1973). All paths susceptible to deletion (on the basis of their failing to meet the 0.05 level of significance) were tested in this way and it was found that the paths Opl to Ll and LK to Opl could be deleted with few discrepancies from the original correlations and none greater than 0.02. As a precaution, the assumption of the opposite causal direction for the two variables, Ll and Opl, was made and tested. Again, the path could be deleted from the model without substantial alteration to the correlation matrix. The omission of path LK to Opl in the more parsimoneous model (Figure 6-1B) is in accord with the hypothesis that opera-tivity exerts an influence distinct, in part, from that of oral language.

The decomposition of the associations of the predictor variables with grade one units of print (Table 6-3) reveals the strength of the total effect of operativity

Table 6-3. Decomposition of Associations of Kindergarten and Grade-One Operativity and of Kindergarten and Grade-One Oral Language with Grade-One Units of Print

Variables	Type of Effect	Decomposition			
Kindergarten	Total Association	0.42			
Operativity and	Total Effect		0.41		
Grade-One Units	Direct Effect			0.18	
of Print	Indirect Effect			0.23	
	Through OP Gr 1				0.15
	Through L Gr 1				0.08
	Joint Association		0.01		
Grade-One Opera-	Total Association	0.44			
tivity and Grade-	Total Effect		0.28		
One Units of	Direct Effect			0.28	
Print	Indirect Effect			...	
	Spurious Effects		0.14		
	By Op K			0.10	
	By Op K through L Gr 1			0.04	
	Joint Association		0.01		
	Lost in model trimming		0.01		
Kindergarten Oral	Total Association	0.16			
Language and	Total Effect		0.05		
Grade-One Units	Direct Effect			−0.16	
of Print	Indirect Effect			0.21	
	Through L Gr 1				0.21
	Joint Association		0.10		
	Lost in model trimming		0.01		
Grade-One Oral	Total Association	0.34			
Language and	Total Effect		0.30		
Grade-One Units	Direct Effect			0.30	
of Print	Indirect Effect			...	
	Spurious Effects		−0.02		
	By Op K			0.05	
	By Op K through Op Gr 1			0.04	
	By LK			−0.11	
	Joint Association		0.04		
	Lost in model trimming		0.02		

(0.41 and 0.28) by comparison with that of oral language (0.05 and 0.30). This is especially marked in kindergarten, where it is interesting to note that the indirect effects of operativity (through grade-one operativity and language) are greater than its direct effects. If kindergarten operativity were left out, undue importance would be attached to the effects of grade-one operativity and oral language; the total association of grade-one operativity with units of print is reduced by 32 percent (0.14) and of grade-one language with units of print by 27 percent (0.09), due to the spurious influence of kindergarten operativity. Kindergarten language seems to be acting as a suppressor variable in its considerable indirect influence on units of print through grade-one language. This is, in part, offset by the positive indirect effect of kindergarten operativity on units of print through grade-one language. The moderate zero-order relationship of grade-one language and units of print is influenced by counterbalancing spurious effects which would not be noticed without the present decomposition. Again, the importance of kindergarten operativity for units of print is made clear. Its longitudinal direct effects seem to operate independently of the effects of oral language, and its longitudinal indirect effects seem to operate in part through oral language. The direct effects of oral language, although weak in kindergarten, are much stronger and about the same as those of operativity in grade one. Thus, both operativity and oral language, in accord with the hypothesis, exert independent effects upon units of print. The model, however, takes no account of reading ability which is developing rapidly at this time and is likely to be an important explanatory variable. Perhaps, as Ehri (1979) claims, it is even an alternate hypothesis to explain the development of language awareness, i.e., experience of print, as shown by reading ability, is a sufficient condition for acquiring lexical insight. To provide a test of the impact of reading ability, a reading achievement measure is added in the next model (Figure 6-2). For this model, units of print was measured one year later in grade two and grade-one reading results were used in an attempt to overcome the problem of feedback loops in a recursive equation model (Wolfle, 1980). The assumptions of these additions to the earlier model are that reading has an effect on later awareness of print and that grade-one reading follows kindergarten and grade-one operativity and language.

This is justified on the grounds that the earlier established abilities are more likely to influence the newly developing capacity, reading, rather than the other way around, at least in its beginning. Grade-one units of print is likely to be exerting an effect but is omitted to avoid feedback loops, as any interaction with grade-one reading cannot be analyzed in this model. Any effect it might have, however, would be likely to reduce the present model's estimation of the effect of grade-one reading. If the analysis is consistent with the cognitive developmental hypothesis without grade-one units of print, its inclusion would strengthen that support.

The model, which has been trimmed in a similar way to the previous one, shows that kindergarten operativity has a direct effect on grade-one units of print over and above the effect of reading. In addition, the decomposition of the relationship of grade-one reading and grade-two units of print (Table 6-4) reveals that 41 percent of that association is spuriously caused by operativity and lan-

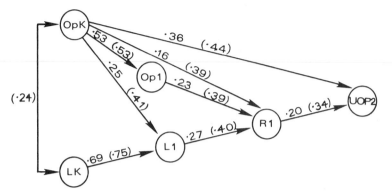

Figure 6-2. Path analysis showing effects of kindergarten and grade one operativity, kindergarten and grade one oral language and grade one reading on grade
two units of print. Single-headed arrows show the direction of effect. Strength
of effect is shown by the path coefficients on each arrow. Simple correlations
(total associations) are shown in parentheses. Double-headed arrow joins two
exogenous variables.

guage. In this spurious effect, the total effect of operativity (OpK + Opl) is 0.12
(35 percent) of the variance of grade one reading and grade two units of print. If
operativity were omitted, the effect of reading on print awareness would have
been markedly larger and an undue importance attached to a spurious component.

    In sum, the path analysis gives support for the interpretation that both opera-
tivity and oral language have a causal effect on units of print and that each is
partly distinct from the other in this effect. Operativity at kindergarten plays a
dominant causal role both one year and two years later in its relationship with
print awareness. It also exerts a causal influence on grade-one operativity and language in their relationship to units of print. The addition of reading achievement
to the model shows that reading has an additional causal influence on later print
awareness but the effect of operativity on print awareness is felt above the influence of reading while oral language influences print awareness through reading.

Discussion

The principal finding of this study is that the conceptual reasoning (operativity)
of the five- to seven-year-old beginning reader seems to have direct and indirect
causal effects on the child's understanding of the units of print. These effects
operate over and above the influence of oral language and the influence of reading ability. Both oral language and reading seem to make important contributions towards the emergence of language insights but no explanation of print
awareness seems adequate which overlooks the impact of the child's underlying
cognitive development.

Table 6-4. Decomposition of Association for Reading Grade One and Units of Print Grade Two

Variables	Type of Effect	Decomposition		
Reading Grade One	Total Association	0.34		
and Units of Print	Total Effect		0.20	
Grade Two	Direct Effect			0.20
	Indirect Effect			...
	Spurious Effects		0.12	
	By OpK			0.06
	By OpK through Op1			0.04
	By OpK through L1			0.02
	Joint Association		0.02	

Establishing a Causal Interpretation

The validity of the causal interpretation of the data requires careful consideration especially since causality has commonly been seen as attributable only on the basis of experimental results. There is, however, considerable doubt about the suitability of experimental methods to establish causality in developmental phenomena (Baltes & Nesselroade, 1979; Wohlwill, 1973). Granted that there are "normal developmental processes" which have a momentum of their own involved in important variables of this study, simple cause-and-effect analysis limited to the activity of an externally imposed condition (i.e., an experimental manipulation) is inadequate and causal modeling procedures are called for.

There are three conditions for establishing causality from path analysis: first, two phenomena must show reliable covariation; second, one must be demonstrated to have temporal precedence; and third, the possibility that both variables are jointly dependent on some other determinant(s) must be excluded. These conditions, in regard to the present data, will be discussed in turn.

The simple correlation of kindergarten seriation and grade-one units of print, 0.48, in this study is remarkably similar to the relationship found by Holden and MacGinitie (1973) between kindergarten word awareness and single seriation, 0.48, and that found by Hiebert (1980) between preschoolers' logical reasoning (including seriation) and five different indices of print awareness in which the correlations ranged from 0.47 to 0.60. With three somewhat different measures of operativity and several different indices of language awareness administered independently by different researchers in America and Australia to preschool and beginning school pupils, there is a very close similarity of results. Further verification of this is desirable but the relationship seems to be a consistent one and cannot be ignored in the explanation of language awareness.

The temporal precedence of kindergarten operativity over grade-one units of print depends less on the earlier assessment of operativity than on the finding from three studies that, before formal reading instruction, children were unable to distinguish reliably between print units such as a word, a letter, or a number (Reid, 1966; Downing, 1971-1972; Francis, 1973). The more fundamental cognitive processes are assumed to have precedence over the more culturally specific capacity, awareness of print.

The exclusion of other possible determinants, underlying both operativity and oral language, is critical for the causal interpretation of this path analysis. The control of oral-language mechanisms is provided by the design of the study. Although the measures are not complete or absolute indices of oral language (or of operativity), there are sound theoretical and empirical grounds for their choice, and the data indicate that they possess good reliability and construct validity. The results show that operativity influences print awareness quite independently of oral language at kindergarten and thus language seems unlikely to provide a sufficient causal explanation.

Another proposed alternative explanation, print experience (Ehri, 1979), is controlled in two ways: first, by comparing subjects who have had very similar teaching exposure and, second, by monitoring reading ability. Although all subjects have had the same number of years of schooling, there is a wide range (from 4 to 23 out of 24) of scores in grade-one units of print. A further check of this was made with a subgroup, a class of 31 pupils, selected because it had been taught by the same method and with similar materials. From it, any pupil with less than 85 percent attendance was excluded. The range for grade-one units of print was 7 to 23. This, in full accord with the findings of wide individual differences in other studies (e.g., Calfee, Lindamood & Lindamood, 1973), suggests that factors other than print experience are important and that print experience is not likely to be a common underlying determinant of operativity and print awareness.

Reading achievement is sometimes considered, with good reason, to be a better guide to print experience than teaching exposure. The path analysis, however, shows that operativity has an effect on print awareness distinct from that of reading achievement, and it is concluded that experience of print in terms of reading progress, as measured in this study, is excluded as a general underlying determiner.

Although a case can thus be made for the exclusion of oral language mechanisms and print experience as causal effects underlying operativity and awareness of the units of print, no such exclusion process can ever be complete. More powerful measures of these variables could be found to change the results or other underlying causes could be discovered. However, within the context of the measures employed, the middle class sample and the limits of present theoretical understandings, the data support the conclusion that operativity has a causal effect on the child's knowledge of the units of print.

A Cognitive Developmental Theory

The results are consistent with a cognitive developmental explanation of the units of print. The five and six year old's emerging operational capacity, unformed when he or she learned to speak, is of great importance for gaining print awareness and learning to read. This does not mean, however, that understanding the units of print is gained spontaneously in the course of cognitive development or that development and print experience are competing explanations of linguistic insight. Although Piaget's concrete operations may well be attained as a result of the need to survive (Elkind, 1975), knowledge of the units of print does not appear to emerge in this way, since illiterate adults, who might reasonably be thought to have achieved concrete operations, could not segment speech, while literate adults, matched in background and experience, could do so (Morais et al., 1979). Specific experience of print, typically in reading instruction, is also highly important (Francis, 1973; Ehri, 1976). Since print is a comparatively recent and rather complex cultural invention (Rozin & Gleitman, 1977), it is not surprising that the learner usually requires at least some deliberate social initiation.

These findings are, therefore, in accord with an interactive theory of learning to read. How well a child learns will depend largely on the success of the match between the teaching—its content, structure, and strategies—and the developmental status of the child—his or her cognitive, language, and motivational state. There are important practical implications flowing from this on determining conceptual level and providing training where there are weaknesses. These will be considered in a later section.

Explaining Some Current Issues

The cognitive developmental view offers insight into three puzzling issues in current research and theory: the low relationship between general vocabulary and print vocabulary, the ease of listening and the difficulty of reading (Mattingly, 1972), and the bimodal distribution of lexical awareness scores.

In regard to the *first* issue, Francis (1973) found that the relationship between reading and vocabulary of units of print (0.41) continued to be considerable (0.34), even when general vocabulary skill was partialled out. She believes this indicates that factors independent of general conceptual ability are involved in learning the print units vocabulary and that these are closely related to reading. The present results afford a similar demonstration. When general vocabulary (Peabody Picture Vocabulary Test) was partialled out of the correlation of grade-one units of print and grade-one reading, the value changed very little (from 0.49 to 0.47). Although the vocabulary of print units seems to be a subset of general vocabulary ability, there is only a low correlation between the two and they are quite independent of one another in their relationship to beginning reading. The present results provide the suggestion that the difference lies in the much higher level of operational thinking which the vocabulary of print units requires. It seems that, in the preparation of the Peabody Picture Vocabulary Test, developmentally unstable influences on its items, such as would be exerted by operativity, have been largely omitted. Linguistic awareness, in the form of understanding units of print, seems to be another aspect of language development which, like relational terms (more and less), coordinated structures (tall and thin) (Sinclair de Zwart, 1969; Tenezakis, 1975), and the appreciation of riddles

(Shultz & Horibe, 1974), is especially dependent on the cognitive changes of middle childhood.

A *second* issue for which the cognitive developmental view offers an explanation is the connundrum raised by Mattingly (1972) of why "listening is easy, reading is hard" (p. 135). The many differences in the learning tasks and the linguistic awareness which Mattingly has argued is needed in reading seem to require a higher level of cognitive functioning. The data provide support for the suggestion of Williams (1979) that, while every "normal" human being has enough of the capacity required by speaking and listening, there is a nonlinguistic cognitive capacity, not equally distributed among the population, needed in learning to read. That capacity, as indicated by the operativity measures of the present study, seems to include aspects of conceptual reasoning or executive functioning. In White's (1965) terms, it is the "decision zone" or the "cognitive layer" of response which emerges as an important new dimension of ability during middle childhood and which the study has shown exerts a decisive causal effect on print awareness. This explanation avoids the considerable difficulties of Mattingly's suggestion that the language-learning mechanisms are prepared for reading by the accumulation of nonfunctional grammatic knowledge during early childhood.

These data also offer a feasible explanation for the *third* issue, the bimodal distribution of scores shown by different measures of word awareness among kindergarten and first-grade subjects in several studies (Holden & MacGinitie, 1973; Calfee, Chapman & Venezky, 1972; Ehri, 1979). Ehri notes that a single skill, either possessed or not possessed by the subjects, seems to underlie these tests but looks to further investigation for an explanation. Measures of Piaget's operations have typically shown the "on-off" bimodal distribution similar to these. The distribution of kindergarten seriation scores in this study reveals three modal points with 20 percent of subjects (29) clustered around one of these modes at the lower end of the scale. Of these 29 subjects, 76 percent are low in units of print (Table 6-2) and, of the four still low in seriation 12 months later, all are low in units of print. Although the units of print test does not show a bimodal distribution, due, it seems, to the difficulty range and the number of its component parts, the cross tabulation indicates a strong relationship between a failure to achieve seriation and low scores on units of print. Although there may be other factors at work, seriation seems to be an important ingredient in the explanation of the bimodal distribution of word-awareness scores.

Implications for Teaching

A central implication of the interactive theory proposed here is that good teaching makes a productive match of the child's developmental status, including the level of her or his cognitive processing, with the instructional content and strategy. Like the experimenter in the instructional approach to developmental cognitive research (Belmont & Butterfield, 1977), the teacher too has one critical task—to ascertain the conceptual level of the child and provide instruction in those terms. Sometimes teachers, through shrewd observation of children and a thorough knowledge of the subject, have structured the situation to produce good results. At other times, they have created unintended mismatches and poor learning or even failed outright to promote learning performance. All reading teachers would be helped by a more systematic knowledge of the cognitive processes called on in print awareness and reading. This research gives some insight into these processes and could provide the basis for the adaptation of Hunt's idea of a learning style conceptual level (Hunt, 1975) in terms of reading-related cognitive components. Hunt found that pupils with low conceptual levels (dependent on external standards and incapable of generating their own concepts) learned better with high-structure teaching while those with high conceptual levels (independent in their capacity to generate concepts) did well with either high- or low-structured teaching. It seems that the low-conceptual-level pupil lacks a degree of executive control which the teaching approach can make up in a similar way that researchers can serve as the executive for individuals in their experiments (Belmont & Butterfield, 1977). Although many pupils will learn the units of print incidentally, others, low in conceptual functioning, need deliberate teacher explanation of terms and daily instruction, e.g., through game or song, based on the various print units. The success of programs, such as those of Elkonin (1973a) and Williams (1980), will depend on the effectiveness of the match between the greater structure they offer and the learning needs of the children.

If it is possible to ascertain the conceptual level of the pupil and match instruction to it, perhaps it is also possible to raise a low level of conceptual ability by training and so strengthen print awareness. Does a cognitive developmental-based capacity lend itself to this manipulation? There is evidence in the work of Gelman (1969) and Siegler and Liebert (1972) for the contention that "rigid organizational limits on cognition can no longer stand" (Belmont & Butterfield, 1977), but this does not mean that limits do not exist, even though they may be more elastic than originally thought. Specific intervention to stimulate operational development may improve the language awareness and reading of those who lag behind their peers. However, it is likely that a high specificity of effect would be found, and therefore the training of seriation tasks directly involved in print awareness and reading is desirable. For this purpose, the recent development of an alphabetical seriation test (Weed & Ryan, 1982) is of interest. More structured teaching of the units of print using concrete materials devised by Elkonin (1973a) and adapted for use in English (Ollila, Johnson & Downing, 1974) may provide help especially valuable for those with low conceptual levels.

Conclusion

The evidence presented provides support for the view that understanding units of print, as well as drawing on existing language mechanisms, seems to call on new mechanisms not required while learning to speak and listen. These new mechanisms involve cognitive capacities which are part of the higher mental processes whose development gains major new impetus in the five- to eight-year period. Their causal effects, which are very difficult to uncover with cross-sectional or experimental data, are revealed by this longitudinal study to influence print awareness over and above the effects of oral language and reading achievement. The heavy demands which awareness of print makes on conceptual development seem to be an important reason why learning to comprehend print is harder than learning to comprehend speech.

The cognitive developmental perspective consistent with the findings offers valuable guidance for the subtle matching process which is at the heart of good instruction. In doing so, it calls for a more detailed study of the changing mosaic of interaction between the developing characteristics of children and the instructional patterns of teachers as children gain an understanding of the units of print and learn to read.

Acknowledgments

This study was supported in part by a grant from the Education Research and Development Committee of the Federal Government of Australia. The close cooperation of the New South Wales Department of Education, the Principals and staff of Gymea Bay, Gymea North, and Laguna Street Primary Schools, and a dedicated team of research assistants in carrying out this project has been greatly appreciated. Grateful thanks is also due to Dr. Herbert Marsh and Professor Donald Spearritt of Sydney University who gave valuable advice and help in analysis of the data and to Dr. Kenneth Sinclair, of Sydney University, who provided assistance and encouragement throughout the project and who made helpful comments on earlier drafts of the manuscript.

CHAPTER 7

How Orthography Alters Spoken Language Competencies in Children Learning to Read and Spell

Linnea C. Ehri

Introduction

Pictures that allow the mind to behold invisible aspects of reality may be worth much more than the 1,000 words proclaimed in the adage, particularly to children in the midst of constructing cognitive schema to make sense of their experiences, and particularly when the picture entails a system of symbols organizing an entire dimension of experience. For example, when children learn to read printed language, they become able to visualize what they are saying and hearing. When children learn to read clocks and calendars, they acquire a visual means of representing the passage of time. When children learn to read music, they become able to visualize what is sung or played on an instrument. In each case, a visual-spatial representational system is acquired by the mind for perceiving and thinking about experiences which cannot be seen and which have temporal duration rather than physical extent as a basic property. Acquisition of a spatial model offers several potential advantages. It enables the possessor to hold onto and keep track of phenomena which themselves leave no trace or have no permanence. It imposes organization upon the phenomena by specifying units, subunits, and interrelationships which might otherwise be difficult to detect or discriminate. However, some degree of distortion or inaccuracy may also result because properties of space may not be completely isomorphic with properties of the nonspatial modality, and also because the spatial system, being a cultural invention, carries no guarantee that it is perfectly conceived.

The purpose of this chapter is to consider the acquisition of one type of picture, printed language, and to develop and present evidence for two ideas: (1) that a visual representational system for speech is acquired when children learn to read and spell, and (2) that because print gets established in memory by being built onto the learner's knowledge of spoken language, acquisition may work various changes in children's competencies with speech.

First, it is important to review what aspects of speech are made visible in print. Written English includes two basic types of spatial symbols, one nested within the other. Horizontal sequences of letters separated by empty spaces symbolize words. Letters within words symbolize phonetic segments which are blended together in pronunciations. In the present chapter, spellings and letters will be referred to as symbols, while the aspects of speech being represented will be termed their referents. In the case of spellings, the referent is a composite which includes the word's pronunciation, its syntactic function, and meaning. In the case of letters (graphemes), the referent is the phonetic or phonemic subunit. Sentences are another aspect of speech represented in print as sequences of words initiated by a capital letter and ending in a period. Less will be said about sentences since words and letters are regarded as the primary units of the visual representational system to be established in memory.

Print maps speech systematically at both the lexical and the phonetic levels. Lexical symbols are in one sense more reliable than phonetic symbols since specific word spellings represent the same referents consistently whereas letters singly or in combination may symbolize more than one phonetic referent or they may lack any referent in speech. However, it is important to note that although there is variability in grapheme-phoneme relations across different word spellings, there is little variability within specific word spellings since letters are fixed by convention in English. To illustrate, although the symbol for /f/ may be F or PH in different words, it rarely varies in spellings of the same word. In this respect, phonetic symbolization and lexical symbolization are equally reliable.

The possibility that printed language aids in the perception and analysis of speech is suggested by several facts about spoken language. At the lexical level, speakers experience most words in the context of other words, and their attention is centered upon meanings, not upon linguistic structure. Moreover, there are no auditory signals segmenting fluent speech into word units. Hence, words as components of speech are neither salient nor clearly marked. Also, many words such as auxiliaries, past tense verbs, prepositions, and conjunctions depend for their meaning upon the presence of other words. If heard as isolated sounds without contexts, these words may not be recognized because they evoke no independent meaning. The pronunciations of some words may even change when they are lifted out of context and pronounced separately (e.g., "of" versus "gimme a piece *a* candy") and this may impair recognition. The fact that linguists cannot agree on how to define words as units in speech testifies that there is substantial ambiguity to lexical segmentation.

At the phonological level, the problem of segmenting pronunciations into phonetic units is even more formidable since sounds are folded into each other and in some instances co-articulated. Also, many sounds, especially those in consonant clusters, are very brief in duration, making detection difficult. Furthermore, the phonetic properties of sounds may be altered by adjacent sounds, making it harder to recognize recurrences of the same sound in various contexts.

Gelb (1952) claims that alphabetic orthography has been invented only once. This suggests that conceptualization of such a system is far from straightforward and thus has much to gain from a visual representational system.

Although printed language may provide the clearest view of speech, there may be other ways to gain awareness as well. At the lexical level, words can be substituted for each other, combined, and recombined in different orders across sentences. Speakers may be able to gain some awareness of words as separate units by manipulating and monitoring these features of spoken sentences. At the phonetic level, the speaker may gain some awareness of sound units by monitoring articulatory and acoustic cues associated with mouth movements as speech is produced slowly or carefully. This points out that structures in speech may not be inaccessible but may merely be harder to penetrate without print. However, in the absence of print, there may be little reason to undertake such an analysis.

Among linguists, there appears to be consensus that printed language has little formative influence on speech. Most linguists are not much interested in writing. In fact, many do not consider it a topic of study in linguistics since it is not language but only a reflection. Bloomfield (1933) describes print as "merely a way of recording language by means of visible marks." Further on, he says "In order to study writing, we must know something about language, but the reverse is not true" (p. 21). There are some exceptions among linguists, however. Gleason (1961) portrays the relationship between speech and writing as intimate and possibly interdependent. Although he pays more attention to the dependence of written language structures on spoken language, he acknowledges that the influence may run in the opposite direction, specifically in the form of spelling pronunciations and a literary form of speech. The idea of spelling pronunciations emerging from a knowledge of print is extended by Kerek (1976) who proposes that word spellings provide an alternative, psychologically compelling model for sound that competes with and may replace spoken pronunciations when the two differ. He proposes the iconic principle of "one graphic form—one phonetic form" (p. 326). According to this principle, when orthography is discrepant with speech, there is pressure to change pronunciations so as to maximize the iconic relationship. Kerek cites several examples of historical shifts in pronunciation which conform to orthographic patterns. The most extreme view is proposed by the maverick linguist Householder (1971) who argues for the primacy of writing over speech. Among other things, he points out that historically there are many more instances where spellings have changed the pronunciations of words than where pronunciations have altered spellings. Also, the law is on the side of spellings. It cares little if a person decides to pronounce his name differently, but court action is required to make an altered spelling legitimate.

In contrast to most linguists, psychologists have paid much attention to the relationship between print and speech, particularly those interested in how children learn to read. Their interest has centered not on spelling pronunciations or literary forms of speech, but rather on children's developing knowledge of structural units in speech. Opinion differs about whether print shapes or merely reflects what develops in speech. This is evident in alternative explanations for the large positive correlations between reading acquisition and children's ability to segment sentences into word units and words into phonetic units. One interpretation is that structural knowledge of speech is a prerequisite or readiness skill which matures between the ages of five and seven, making a child ready to learn to read (Mattingly, 1972, 1979a; Chapter 2 in this volume). Other interpretations are that linguistic segmentation skill is a consequence of learning to read or that the two interact during the acquisition process (Ehri, 1979). Print would be assigned a passive role in the former case, an active formative role in the latter case. Still another interpretation is that segmentation ability and reading skill are correlates, both emerging as a consequence of cognitive maturational factors such as the development of logical operational thinking skill (see Watson, Chapter 6 in this volume).

The strict prerequisite (passive) view and also the cognitive maturational view appear to be incorrectly based on evidence presented by Morais, Cary, Alegria, and Bertelson (1979). They administered a phonetic analysis task to Portuguese literate and illiterate adults who were thought to be similar in other respects (origin, childhood history, and employment). Whereas illiterates were unable to add or delete phones at the beginnings of nonwords, literates performed the tasks easily. That these differences were detected in adults indicates that awareness of phonetic units in speech does not develop spontaneously or as a result of cognitive maturation during childhood but rather requires special experiences such as learning to read an alphabetic system.

There is also disagreement among psychologists about the role of print as a representational system in memory. This is apparent in alternative theories about how children learn to read. Also these theories differ in whether and how print is regarded as influencing speech during reading acquisition. One view is that readers learn associations between visual forms of words and their semantic referents. This is termed a whole or sight-word strategy where words are stored in memory as visual Gestalts (Smith, 1972). The representational system in this theory is strictly lexical. Although reading experiences might be expected to improve learners' awareness of words as units of speech, they would be expected to have little impact on learners' awareness of phonetic units or their pronunciations of words since letters are not thought to be analyzed for their associations with sound. Any phonetic benefits of printed language might be attributed not to reading experiences but to spelling experience which is regarded as quite a different process.

Another view is that children learn to read by converting letters into sounds, combining the sounds, and then recognizing the word from its pronunciation (Liberman et al., 1977). This is the recoding (or decoding) view of word recognition thought to be fostered by phonics instruction. The visual representational system suggested by this view is a limited one in which letter-sound relations are stored in memory as part of a general system for recoding print into speech. Printed words remain outside the mind on the page always to be converted to a pronunciation each time they are seen. Proponents of this view might be expected to regard experiences with printed language as making learners highly sophisticated about the phonetic structure of words. Printed language might even be seen as changing the way learners say words when a discrepancy is detected between spellings and pronunciations. However, in actuality, these researchers have not drawn such conclusions, perhaps because they have declared speech primary and print as "parasitic on speech" (Liberman, Liberman, Mattingly & Shankweiler, 1980).

In sum, advocates of the sight-word and phonetic recoding views of reading acquisition have paid little attention to the possibility that printed language might become established in memory as a visual representational system for speech. Furthermore, they have attributed little, if any, formative powers to print. Assumptions about the primacy of speech have precluded recognition of the possible impact of print. Also, commitment to one level of analysis in print, either lexical or phonetic, and exclusion of the other has limited the scope of these views. Another limiting factor may be their exclusive concern with reading and not spelling skill.

Theory of Printed Language Acquisition and Evidence

An alternative view, one which has been guiding our research on how children learn to read and spell (Ehri, 1978, 1980a, 1980b), is that the full representational system offered by printed language is acquired and stored in memory during acquisition. Lexical representations in the form of alphabetic images are stored in memory. Also, letters are learned as symbols for sounds, both within specific word spellings and as general rules. As will be evident, this view carries several implications regarding the consequences of literacy for linguistic and cognitive processing since print itself is viewed as a tool of memory.

According to the theory, beginners' knowledge of language includes a lexicon of words acquired from their experience with speech. The lexicon is comprised of word units having several identities. Each word has a phonological identity, that is, how the word sounds and is articulated. Each word has a syntactic identity, how the word functions in sentences. Each has a semantic identity, what the word might mean in various contexts. In the course of learning to read, another identity is added to the lexicon: an alphabetic image of the word. This image is integrated with the other identities to form a unit in lexical memory. LaBerge and Samuels (1974) refer to this process as unitization. Alphabetic and phonological identities are unitized when letters are processed as symbols for sounds. Alphabetic, syntactic, and semantic identities are unitized when printed words are read and given meaningful interpretations in sentence contexts. As a result, alphabetic representations come to function as symbols for meanings as well as sounds.

Spellings are thought to enter memory not as unanalyzed visual figures but as sequences of letters bearing systematic relationships to acoustic and/or articulatory segments detected in the word's pronunciation. The first few times a printed word is seen, its component letters either singly or in combination are processed as symbols for component phonetic segments, the letter sequence enters memory, and it becomes a visual symbol for the sound structure of the word. The process by which letters enter memory as symbols for sounds is called phonetic symbolization. To store images, readers must be able to analyze words into the relevant phonetic segments suggested by letters they see in spellings. Likewise, they must be familiar enough with the sound-symbolizing function of the letters to recognize how each matches up with a sound in the pronunciation. To the extent that letters are grounded in sound, clear alphabetic representations are formed which can be used for reading printed words accurately and rapidly and also for producing approximately correct spellings.

Various studies have been interpreted as providing evidence that word spellings are stored in memory and that letters enter memory by being processed as symbols for sounds in pronunciations. The purpose of the next section is to review this evidence. (Fuller descriptions of this research can be found elsewhere: Ehri, 1980a, 1980b; Ehri and Wilce, 1979, 1980b, 1982.)

Spellings to Improve Memory for Sounds

One series of experiments was conducted to show that orthography can function as a mnemonic device, that letters can be used to preserve sounds in memory when they provide an adequate printed *symbol* for those sounds (Ehri & Wilce, 1979). It was clear from a previous study that young children have difficulty remembering nonsense syllables. A paired associate sound-learning task was designed to see whether memory for nonsense syllables might be improved when children were shown spellings of the sounds during study periods. Beginning readers (first and second graders) were given several trials to learn four oral, consonant-vowel-consonant (CVC) nonsense syllables such as MAV, REL, KIP, and GUZ. Recall of these responses was prompted by a variety of stimulus cues in four experiments. During study periods children saw either correct spellings of the sounds or misspellings, or they rehearsed the sounds, or they listened to oral spellings. We found that children learned the sounds fastest when they viewed correct spellings during study periods. Looking at misspellings made it especially hard to learn the sounds. These results are interpreted to indicate that spellings improve memory for sounds because they are retained as visual symbols helping to preserve the sounds in memory.

Individual differences were observed among beginning readers in their ability to make use of letters. Only some of the first graders benefited from spellings in the sound learning task (i.e., their learning was faster with than without spellings). These subjects were found to be superior to unsuccessful learners on several other measures of reading subskills: phonemic segmentation, decoding, and spelling. This indicates that spellings are effective as mnemonic devices primarily among children who possess the phonetic symbolization skill to make the letters functional.

Evidence was collected indicating that this letter symbolization process for sounds may underlie the storage of printed words in memory. Very high correlations, as high as $r = 0.75$, were obtained between children's scores on the spelling-aided sound learning task and their scores on a printed-word identification task.

Spellings to Shape Conceptualization of Sounds

Another set of studies was undertaken to show that the storage of spellings in memory affects the way learners conceptualize the sound structure of words (Ehri & Wilce, 1980b). We reasoned that if, when printed words are stored, letters are interpreted as symbols for sounds, then we ought to see some *variations* in learners' conceptualization of the phonemic structure of words depending upon how the words are spelled and which sounds are represented by letters. For example, the spelling of "pitch" may cause readers to think it has four phonemes whereas the spelling of "rich" may indicate only three phonemes. In both words, a phonetic element corresponding to the extra letter T is present in articulation. Whether or not readers think of /t/ as a separate phoneme may depend upon the presence of a letter symbol in the spelling.

The first experiment was designed to test the relationship between phonemic segmentation and spelling knowledge. Seven pairs of words were selected so that each pair shared the same target phoneme, yet the spelling of one member symbolized an extra phonetic element adjacent to the phoneme while the spelling of the other member symbolized only the phoneme. These word pairs, their shared phonemes, and the extra phonetic elements are listed in Table 7-1. A phonemic segmentation task was used to reveal how fourth graders conceptualized the phonological structure of these words. Children pronounced each sound separately and positioned blank counters in a row to represent each sound. A spelling task was employed to determine whether children knew all the letters in the words.

Based upon our theory, it was expected that, among children who knew the spellings, extra segments would be detected in words whose orthographic identities included letters symbolizing those segments but would be omitted in words lacking extra letter symbols. This is what we found. As evidenced in Table 7-1, sounds were discovered frequently in words whose spellings included a letter for that sound but were almost never detected in words whose pronunciations were parallel but whose spellings lacked the letter.

Children were able to spell most of the target words. However, there were some misspellings. To determine whether subjects were less likely to detect an extra segment if they did not know the word's spelling, the number of these cases was counted. There were 21 misspellings in which the extra letter was omitted. In 90 percent of these, the extra sound was also not detected in the segmentation task. This suggests that it is when children acquire orthographic symbols that they become aware of additional phonemes in the pronunciations of words.

It was evident from subjects' comments that spellings were influencing their

Table 7-1. List of Extra-Letter and Control Word Pairs, Phonetic Description, Frequency that Extra Phonemes Were Detected (Phonemic Segmentation Task) and Frequency of Correct Word Spellings (Maximum = 24 Subjects per Word)

Word Pairs		Sound Structure		Phoneme Detection		Spellings	
Extra Letter	Control	Shared Phoneme	Extra Phonetic Element	Extra Letter	Control	Extra Letter	Control
catch	much	/č/	t alveolar tap	15	1	19	24
pitch	rich	/č/	t alveolar tap	13	3	18	24
badge	page	/ǰ/	d alveolar tap	13	0	17	24
can you	menu	/u/	y glide	19	0	24	20
new	do	/u/	w glide	18	0	24	24
own	old	/o/	w glide	12	0	19	24
comb	home	/m/	b bilabial stop	6	0	20	24
			Means	13.7	0.6	20.1	23.4

segmentations. Some children remarked about their uncertainty whether you could really hear the B in "comb" or the T in "pitch." This was in spite of the fact that the experimenter never mentioned spellings, and subjects' attention was focused upon sounds in words, not letters. However, spellings were not the sole basis for segmentations. Only a couple of subjects allocated chips for silent E and for C and H separately. Most children ignored truly silent letters and they created only one sound in segmenting words spelled with consonant digraphs such as CH. This suggests that spellings interact with pronunciations to determine how the sound structure of words is conceptualized. Letters do not dictate sound structure. Neither are they ignored.

A second experiment was conducted to investigate whether phoneme conceptualization can be shaped by experiences with the printed forms of words, specifically whether, for identically pronounced words, the way readers conceptualize their sound structure depends upon which sounds they see symbolized in the words' spellings. Fourth graders practiced reading five pseudowords assigned as names of animals. Half of the subjects read spellings with extra letters, half saw control spellings. The pairs of words (extra letters underlined) were: *banyu-banu*; *drowl-drol*; *simpty-simty*; *tadge-taj*; *zitch-zich*. A phonetic segmentation task and a spelling task followed.

Results supported the hypothesis. Phonetic elements symbolized by extra letters were distinguished as separate phonemes almost exclusively by subjects who learned these spellings. Whereas every extra-letter subject included between two and five extra-letter sounds in his or her segmentations, all of the control subjects but two found no extra sounds, the two exceptions finding only one apiece. These findings indicate that the visual forms of words acquired from reading experiences serve to shape learner's conceptualizations of the phoneme segments in those words.

Spellings to Shape Pronunciations

Not only extra phonetic segments but also extra syllabic segments may be acquired when children learn the spellings of words. Evidence for this was collected in a syllable-segmentation task. The words presented were ones whose spellings contain one more syllable than their typical pronunciations. The words are listed in Table 7-2. Of interest was whether knowledge of the spellings of these words might cause subjects to regard them as having the extra syllable in their pronunciations. As evident in Table 7-2, results offered some support. Extra syllable segments were detected more frequently by subjects who knew spellings than by subjects who did not for 10 of the 11 words presented. These results were corroborated in a study by Barton and Hamilton (1980) who found that literate adults segmented multisyllabic words as they were spelled whereas borderline literates omitted the extra syllables in their segmentations. Both studies offer evidence for the formative influence of print over speech, specifically over the pronunciation of words.

Table 7-2. Proportion of Extra Syllables Detected in the Segmentation Task When Words Were Spelled Correctly and Incorrectly

Extra Syllable Words	Proportion of Extra Syllables Detected		
	Correct Spellings*	Incorrect Spellings*	Difference
different	57% (14)	20% (10)	37%
comfortable	100% (7)	35% (17)	65%
decimal	100% (1)	39% (23)	61%
several	38% (8)	44% (16)	-6%
interesting	92% (13)	45% (11)	47%
general	57% (7)	35% (17)	22%
temperature	100% (2)	0% (22)	100%
valuable	50% (4)	0% (20)	50%
camera	50% (12)	25% (12)	25%
miserable	100% (3)	5% (21)	95%
family	56% (23)	0% (1)	56%
Mean	80% (8.5)	25% (15.4)	50%

*Number of spellings given in parentheses (max = 24).

What Happens to Silent Letters?

English word spellings sometimes include letters lacking any referent in sound, usually no more than one letter per spelling. There are various ways that learners might handle this discrepancy between print and pronunciations. One possibility suggested by Kerek's (1976) iconic principle is that silent letters are incorporated into the pronunciations of words. Another possibility is that when spellings are stored in memory, silent letters get tagged as exceptions.

We undertook some experiments to study how silent letters are remembered and whether they differ from pronounced letters. First, we ruled out the possibility that silent letters are incorporated into normal pronunciations of words. In a letter-judgment task, most seventh graders considered pronounced letters to be pronounced and silent letters to lack a referent in speech. Then we conducted five experiments to compare children's memory for silent and pronounced letters in words whose spellings were familiar (Ehri & Wilce, 1982). The task had three parts. First, subjects read a list of the target words. This was to determine which words were already familiar. Next, subjects were told to imagine the spelling of each word in their mind's eye. When they could see it, they were shown a single letter and asked to decide whether or not the letter was present in the spelling. For example, they imagined the word "kind" and were asked "Does it have an N?" Third, they were surprised with a recall task. Each letter was shown and children identified the word they had imagined for that letter. Pairs of words having the same or similar target letters and comparable spellings were selected. In one word, the target letter was pronounced. In the other word, it was silent. Examples are: S–island versus N–insect; W–wrong versus world; T–whistle versus K–freckle. Subjects in the studies were second, third, and fourth graders.

Results of the letter-judgment task revealed that subjects were quite accurate in locating silent and pronounced letters in the words being imagined. Although performance was close to perfect in both cases, children were significantly more accurate in detecting pronounced than silent letters. This indicates that letters are more easily stored in memory when they symbolize sounds. However, it is not the case that silent letters are kept out of memory. They are retained almost as well.

In the letter-judgment task, we measured how long it took for subjects to decide whether the letter was present in the word being imagined. Vocalization latency equipment attached to a slide projector was used to measure reaction time (RT) from the onset of the letter to the subject's response, "yes" or "no." Comparison of RTs to judge silent and pronounced letters for each word pair revealed surprisingly that latencies were shorter for silent letters in the majority of pairs. The pattern held across subjects as well as word pairs and across two experiments involving different sets of words. Superior performance with silent letters was also observed in the recall task where silent letters prompted better recall of the words previously judged than pronounced letters.

From these findings, we conclude that silent letters do have a different status from pronounced letters in memory. Silent letters are harder to remember. However, once learned, they assume greater prominence in orthographic representations than pronounced letters, perhaps because they get tagged as exceptions by the phonetic symbolization process when they enter memory. If this interpretation is correct, it suggests that, contrary to Kerek's (1976) claim, there is not continuing pressure for silent letters to be pronounced. Rather, discrepancies between spellings and speech may be reconciled at the time of acquisition, and this may lay to rest any uncertainty about the pronunciation of silent letters in the minds of learners. Regarding findings of the "pitch/rich" segmentation study mentioned earlier, the reason why extra letters such as the T in "pitch" were carried into pronunciations is that correlates in sound could be found so these letters were analyzed as pronounced rather than silent letters.

To summarize this section, we have proposed that printed language is acquired as a representational system when children learn to read and spell. Various findings indicate that spellings of words are retained in memory by functioning as symbols for sounds, and that, as a result, spellings may influence the way readers conceptualize the sound structure of words and possibly even the way they pronounce words under certain circumstances. In the following sections, we consider additional evidence indicating whether and how acquisition of a visual-spatial alphabetic system affects learners' competencies with spoken language.

Acquisition of Phonological Awareness

The importance of phonemic analytic skills in learning to read has been documented and discussed by a number of researchers (Golinkoff, 1978; Lewkowicz, 1980). Various kinds of phonemic analytic skills have been studied, including the ability to divide words into constituent phonemes, to blend phonemes into recognizable words, and to add or delete phonemes from various parts of blends. Phonemic segmentation is considered to be central in learning to read and spell. However, there is disagreement about how this skill develops and when it should be taught. This controversy was pointed out earlier. Researchers who regard speech as primary and writing as parasitic on speech consider phonemic segmentation to be a prerequisite for learning to read and claim that it should be taught as an oral analytic skill before children are introduced to print (Liberman et al., 1977; Mattingly, 1972). In contrast, we have suggested that the reverse may be more true, that children may learn much about the phonemic structure of words when they learn how to interpret spellings as maps for pronunciations (Ehri, 1979). Evidence confirming phonemic segmentation as a prerequisite is lacking since the distinction between prerequisite and consequent has not been maintained in studies demonstrating its contribution to reading acquisition. Many of the studies are correlational, and it is not clear that the more successful segmenters in the samples of kindergarteners were not already readers rather than prereaders. In segmentation studies showing that experimental groups outperform control groups, the skill has been taught along with other component skills, thus obscuring its specific role in learning to read (Rosner, 1972; Wallach & Wallach, 1976; Williams, 1980). Evidence that segmentation is a consequence of learning to read is also lacking. In our segmentation studies described elsewhere in this chapter, we have shown that learning the spellings of words influences how children divide them into phonetic segments. However, our subjects were older beginning readers, so the data do not clarify whether spellings might be crucial in shaping children's awareness of phonemic segments at the outset when they first begin learning about print.

Research by Elkonin (1973a) and Lewkowicz and Low (1979) indicates that beginners learn to segment better when they are provided with visible models of component sounds in the pronunciations of words. However, their visible models consisted of blank squares or counters rather than alphabet letters. Letters were not examined since it was assumed that letters would only confuse, not help, the child.

A Segmentation Training Study

To settle the question of whether letters help or hinder the acquisition of phoneme segmentation skill, we conducted an experiment in which acquisition with letters was compared to acquisition with nondistinctively marked tokens. (For a fuller description, see Hohn & Ehri, 1983.) We reasoned that since letters provide unique symbols for each sound, subjects taught to segment with letters might be able to use letters as mediators in distinguishing and conceptualizing the separate sounds. To examine this possibility, we selected and screened kindergarteners to be sure that they could neither read nor segment phonemically but that they could identify most of the alphabet letters to be used in training. Triplets with similar vocabulary scores and letter knowledge were formed, and members were randomly assigned, one to each of three treatment groups. In one condition, subjects were taught to segment two- and three-phoneme blends using alphabet letters to represent the sounds (called the "Letter" group). In another condition, subjects were taught to segment in the same way except that they used uniformly marked tokens, each displaying a drawing of an ear, to represent the sound segments (called the "Ear" group). In the third control condition, subjects received no training. Before training began, all three groups were taught letter-sound associations for the eight letters used during training (three vowel letters and five consonant letters whose names contained their sounds). During training, the two experimental groups learned to segment two-phoneme CV and VC blends followed by three-phoneme CVC blends to a criterion of mastery. Acquisition of segmentation skill was then evaluated with a posttest in which subjects segmented both trained and untrained blends with unmarked tokens. Performance with practiced and unpracticed sounds was examined to determine whether segmentation training would transfer to new sounds or whether it would be restricted to the set of sounds practiced during training.

Results are presented in Table 7-3. Statistical tests of performance during training indicated that the two experimental groups did not differ in the time spent in training nor in the number of trials to reach criterion, indicating that neither method was more difficult or time consuming. Analysis of errors during training indicated that Letter subjects were more skillful at breaking blends apart into single phonemes than Ear subjects. Also, incorrect sounds were less likely to intrude in Letter subjects' segmentations than in Ear subjects' segmentations.

Table 7-3. Mean Performances as a Function of Task Variables and Treatment Groups

Pretests	Ear	Letter	Control	Mean	Standard Deviation
Peabody Vocab.	56.9	59.0	58.5	58.1	6.38
Age (months)	67.6	67.6	68.2	67.8	3.94
Training					
Items to Criterion					
per list	25.2	30.6	—	27.9	15.4
Time (minutes)	149	159	—	154.0	56
Posttests					
Phoneme Segmentation (Maximum Score = 6 per cell)					
Training Set	2.88	4.75	1.00	2.88	
Transfer Set	2.50	3.25	.88	2.21	
Difference	+0.38	+1.50	+0.12	+0.67	
Mean	2.69	4.00	0.94	2.54	1.74

Note. There were eight subjects per group.

These results indicate that, during training, letters were helpful in enabling learners to distinguish the correct size and identity of the sound units to be segmented.

Analysis of posttest scores on the segmentation task revealed that both Letter and Ear subjects were able to segment unpracticed (transfer) blends equally well and better than controls, indicating that training with letters and training with uniformly marked tokens were equally effective in teaching general segmentation skill. However, a significant difference favoring the Letter group emerged on the trained sounds. As evident in Table 7-3, Letter subjects were superior to Ear subjects in segmenting sounds they had practiced, with both groups surpassing the control group. This was despite the fact that both experimental groups had been taken to the same criterion during segmentation training. This indicates that letter markers are superior to nondistinctive markers in teaching children to segment sounds symbolized by the letters. Our explanation for the advantage of letters is that they provide learners with a mental symbol system for representing and thinking about specific phonemes and for distinguishing and identifying them in a blend.

As mentioned earlier, the need for distinctive letter markers becomes apparent when one considers how difficult it is in natural speech to distinguish and identify phoneme-size units. The study by Morais et al. (1979) showed that phonetic analytic skills do not develop in the *absence* of experience with print. Very likely the bulk of this knowledge emerges when children begin interacting with print and learning how letters in spellings map sounds in pronunciations. Although beginners may bring some phonemic insights from their experience with speech, orthography helps them stabilize and organize their phonetic knowledge, and it teaches them the full system.

Children's Invented Spellings

Data on children's invented spellings provide more evidence indicating that experience with alphabetic orthography teaches children how to conceptualize the sound structure of words. These findings are especially interesting because they reveal specific ways that naive spellers' phonetic perceptions change as a result of experience with print. Read (1971, 1975), Chomsky (1977), and Beers and Henderson (1977), among others, have analyzed the spellings of prereaders and beginning readers. Also, we have compared the spellings of more and less mature beginning readers.

Comparison of novice and mature spellings reveals that novices do not detect and represent as many phonetic segments as mature spellers. Segments are overlooked not because the children lack knowledge of letter-sound relations but rather because they have not had sufficient experience with conventional print to learn how to divide the sound stream appropriately. When the entire sound of a consonant letter name is detected in a pronunciation, the letter may be selected to stand for both the vowel and the consonant (e.g., YL for "while" and PRD for "pretty"). Correction of these letter-name errors involves learning to attend consistently to phonetic units rather than blends or syllabic units.

One type of phonetic unit proving illusive to novices is the schwa vowel which is commonly omitted in spellings when it occurs before the vocalic consonants /l/ and /r/ and in unstressed syllables (e.g., PESL for "pencil," WOTR for "water," and BLAZS for "blouses"). Interestingly, in learning to detect and represent this phone, some beginning spellers appear to go overboard in their analyses, that is, they find extra segments not symbolized in the conventional system. In spelling /l/ and /r/, beginners have been observed to separate the vocalic phone from the consonantal phone and symbolize each with a separate letter (e.g., FEREND for "friend" and BALAOSIS for "blouses"). (This process of overanalysis is called epenthesis by linguists.) We also saw this happen with some pseudoword spellings written by beginning readers: SHANGCKE (pronounced "shenk") and TRAYULSE ("trels"). These examples illustrate that experience with conventional spellings also teaches children the limits of phonetic analysis, that is, which phones to ignore.

Not only vowels but also consonants are sometimes omitted in novices' spellings. One of the most interesting is the preconsonantal nasal (i.e., the occurrence of /m/ or /n/ between a vowel and a consonant). Examples of children's spellings are BOPE for "bumpy," THEK for "think," and KIDE for "kind." Omission appears to occur because the nasal lacks its own place of articulation. Rather, it combines with the vowel to form a single phonetic segment, a nasalized vowel. This is an instance where conventional orthography actually *misleads* the reader into believing that the vowel and consonant are two separate sounds even though they are one sound phonetically.

To verify that spellings of novices and experts reflect differing conceptualizations of the sound structure of preconsonantal nasals in words, with novices ignoring the nasal and experts regarding it as a separate sound, we compared the phonetic segmentations and spellings of first graders and fourth graders on words containing preconsonantal nasals. Results confirmed this difference. Although first graders represented most of the other sounds in their segmentations and spellings, they omitted the nasal in all but one spelling, and they never segmented it as a separate sound. In contrast, fourth graders included the nasal in all spellings, and varying proportions of subjects marked it as a separate segment in different words: "bank"—81 percent, "beyond"—69 percent, "king"—44 percent. Identification of a separate nasal segment in the final case is interesting since linguists regard /ŋ/ as one phonetic segment. Apparently spellings persuade some learners that there are two sounds.

Another type of shift occurring when novices gain more experience with print is in the classification of sounds. Invented spellings reveal some nonconventional classifications. Affricative sounds are those produced by a stop closure followed immediately by a slow release of the closure. Examples are /č/ in "church" and /ǰ/ in "judge." Affrication turns out to be a salient feature for beginners. In their spellings, they may regard the initial affrication in words such as "truck" and "chicken" as the same sound and symbolize it with the letter H which contains affrication in its name (e.g., HIKN for "chicken" and HEK for "truck"). In contrast, children more knowledgeable about print distinguish between TR and CH sounds. Examples involving /j̆/ are JAGN for "dragon" and KOLAH for "college." Very likely, it is familiarity with orthographic conventions that corrects these deviant analyses. Evidence for this is provided by Barton, Miller, and Macken (1980) who taught preschoolers to segment the first sound in words such as /m/ in "mouse" and /b/ in "bear." Then they examined how these children analyzed the initial consonant cluster /tr/ as in "truck." They found that, whereas nonreaders regarded the cluster as a single sound, children having some reading ability segmented the cluster into two phonemes.

   Another case where novices differ from experienced spellers is in their classification of alveolar flap sounds. In American English dialects when the alveolar flap falls in a stressed syllable between two vowels (e.g., "le<u>tt</u>er" and "mi<u>dd</u>le") the sound produced is closer acoustically to a /d/ than a /t/. However, the sound may be symbolized by either D or T in conventional spellings. Examples of children's misspellings are WADR for "water," DRDE for "dirty," and MITL for "middle." Whereas novices hear the flap as /d/ and spell it D, children more experienced with print recognize that the sound could be /t/ and sometimes use T in their spellings.

   The various shifts in spelling patterns described above are thought to occur as beginners gain experience with print. We collected additional data to determine whether each type correlates with reading and spelling acquisition and distinguishes more from less mature reader/spellers. One inadequacy of invented spelling data is that the words spelled are real. When more mature patterns are exhibited, it is not clear whether this arises because children have acquired knowledge of the general pattern or because they remember those particular word spellings. To tap general knowledge rather than specific word memory, we made up nine pseudowords and gave them to 68 first and second graders to spell. Their knowledge of several patterns was tested, including schwa vowels (G<u>U</u>RB and NACH<u>E</u>R), preconsonantal nasals (SHE<u>N</u>K), affricates (NA<u>CH</u>ER and <u>TR</u>ELS), and alveolar flaps (JU<u>TT</u>Y). The children's ability to read and spell real words was also measured to assess their familiarity with print, and these scores were correlated with target features of the nonsense word spellings.

   Correlations were positive in the case of schwa vowel spellings ($r = 0.57$, 0.62; $p < 0.01$). Better reader/spellers symbolized schwa with a separate letter whereas less mature readers omitted the letter. Also, correlations were positive for preconsonantal nasals ($r = 0.52$, 0.58; $p < 0.01$). Better readers represented the nasal while less mature readers omitted it. Correlations were low and nonsignificant for affricative spellings, primarily because most subjects correctly represented /č/ as CH and /tr/ as TR. Use of immature affricate forms may be limited to children with less print experience than our subjects. Analysis of alveolar flap spellings revealed that most subjects used the letter D which is more accurate acoustically. However, there was some variation according to maturity level: 87 percent of the poorest spellers and 85 percent of the best spellers selected D, whereas only 59 percent of the moderately good spellers used D, the remainder selecting T. This suggests that a shift toward T may emerge temporarily among children once they gain some experience with print. The fact that we used pseudowords never seen before rather than real words may account for the high proportion of D spellings.

Alveolar Flap Judgments

We conducted two additional studies using sound judgment tasks rather than spelling tasks to determine whether experiences with print might shape children's perception of flap sounds in real words. According to the theory developed above, since flaps are ambiguous, spellings should influence how they are perceived in speech, as /d/ or /t/. Children unfamiliar with spellings should analyze flaps acoustically as /d/. Once they acquire some experience with print and learn that flaps are sometimes symbolized by T in spellings, they may shift and perceive them as /t/ in sound. The ambiguity should be resolved for particular words as these spellings are stored in memory and as their letters are interpreted as indicating which sounds are really "there" in pronunciations.

In our first study, first, second, and fourth graders listened to a tape recording of 30 familiar words containing flaps. The words are listed in Table 7-4. In some of the words, flaps were created by adding inflections (i.e., "hottest" and "saddest"). In other words, flaps were present in root forms. Each word was included in a defining sentence. The speaker was careful to articulate a flap when the words and sentences were tape recorded. Children listened, repeated each word, then judged whether they heard a /d/ or a /t/ sound in the middle of the word. They responded by naming the letter (D or T) containing the sound heard. We expected that older subjects' sound judgments would reflect spellings more than younger subjects' since older subjects knew more of the spellings. We expected accuracy to be greater for D-spelled flaps than T-spelled flaps since unfamiliar spellings should be judged acoustically.

Results confirmed these expectations. The proportions of incorrect judgments across age groups and words are reported in Table 7-4. Older subjects' judgments matched spellings more than the judgments of younger subjects. Analysis of the errors of subjects who misjudged at least four words ($N = 35$) revealed a strong bias to say D rather than T, indicating that acoustic cues were being monitored by subjects and used when spellings were unfamiliar. This is one reason why judgments for D-spelled flaps were more accurate than judgments for T-spelled flaps. Also, it indicates that correct /t/ judgments were not random guesses. Five subjects exhibited a bias to say T and four of these were first graders. This provides some evidence that beginners may shift temporarily to exclusive use of T when this becomes learned as a symbol for flaps. Only one of the subjects (a first grader) exhibited no bias favoring D or T in his errors.

Interestingly, flaps spelled with T were not judged more accurately in inflected

Table 7-4. Proportion of Incorrect Responses Across Grades for Each Word

Words	First ($N = 17$)	Second ($N = 20$)	Fourth ($N = 33$)	Mean
Derived (T)				
sitting	35%	20%	11%	
hottest	41%	30%	8%	
smarter	53%	35%	11%	
fatter	59%	25%	8%	
writing	71%	45%	22%	
Mean	51.8%	31.0%	12.0%	31.6%
Derived (D)				
reading	0%	0%	0%	
riding	6%	0%	0%	
harder	6%	0%	0%	
sadder	23%	10%	8%	
maddest	29%	5%	5%	
Mean	12.8%	3.0%	2.6%	6.1%
Nonderived (T)				
cotton	12%	5%	0%	
little	18%	5%	8%	
center	23%	20%	5%	
rattle	41%	20%	32%	
party	47%	15%	11%	
Nonderived (T)				
pretty	47%	10%	5%	
spaghetti	59%	25%	13%	
pattern	65%	50%	24%	
sweater	71%	45%	24%	
letter	76%	40%	8%	
attic	82%	60%	40%	
Mean	49.2%	30.4%	15.4%	31.7%
Nonderived (D)				
wonder	0%	0%	0%	
sudden	12%	5%	3%	
modern	18%	10%	3%	
odor	18%	15%	5%	
meadow	23%	0%	5%	
already	35%	0%	3%	
needle	35%	15%	0%	
ladder	35%	30%	8%	
middle	41%	20%	8%	
Mean	24.1%	10.6%	3.9%	12.9%

words than in nonderived words (see Table 7-4). This was surprising since we expected subjects, young and old, to think of morpheme roots and to recognize the correct underlying sounds in the derived words. The results suggest that lexical derivational sources did not have much impact upon performance in this task. This contrasts with the strong effects of spellings on performance.

Although the above findings are suggestive, they are not conclusive. One weakness is that the evidence is correlational and hence insufficient to support the causal inference that spellings shape learners' conceptualization of sounds. Another weakness is that the sound judgment task may have biased subjects to make their judgments on the basis of spellings rather than sound since they responded by naming the letters containing the sounds they heard in the words.

These weaknesses were corrected with an experiment in which knowledge of spellings was manipulated as an independent variable. Subjects in the experimental group saw and pronounced the spellings of medial-flap words while subjects in the control group practiced only the pronunciations. Then the two groups were given a rhyme-judgment task to see whether experimentals judged rhymes according to spellings more often than controls. In this study, 18 second graders served as the subjects. Scores on word reading and spelling tests were used to form matched pairs, one member assigned randomly to the experimental group, the other to the control group. Experimental subjects were taught to read a set of 16 words containing medial alveolar flaps, half spelled with D, half with T. The words are listed in Table 7-5. Control subjects listened to and repeated the same words. Sentences were provided to clarify word meanings. On the next day, subjects were given a rhyme-judgment task to assess how they conceptualized the flap sounds in the words. In order to minimize the influence of letters on sound judgments, this task was conducted with pictures. First, subjects were taught how to segment the initial syllable of a word so that the flap was at the end of the syllable they pronounced (e.g., "glitter"–"glit"). Note that this procedure has the effect of forcing subjects to pronounce the flap distinctively either as a final voiced /d/ or a voiceless /t/. After learning to segment and pronounce syllables in this way, subjects were taught to decide which of two picture names rhymed with the syllable they had segmented (e.g., which picture, "mitt" or "kid," rhymes with "glit") and to point to the correct picture. After they learned to perform these operations with nontarget words, they were given the target words to segment and judge. The picture choices for each target word are listed in Table 7-5. Note that one picture name rhymes with a pronunciation of the syllable ending in /d/, the other with a pronunciation ending in /t/. It was expected that experimental subjects' selection of rhyming words would reflect medial-flap word spellings more consistently than control subjects' selections. A spelling production test was given at the end to verify that experimental subjects had learned the spellings while control subjects did not know them. Throughout the tasks, the experimenter never pronounced the medial-flap words. Rather, they were always played on a tape recorder along with defining sentences to insure that subjects were not inadvertently cued by the experimenter's pronunciations.

Table 7-5. Target Words and Number of Correct Response Rhymes in Each Training Group (Maximum Correct = 9)

Target Words[a]	Response Rhymes[b]	Groups		Differences
		Experimental (Read)	Control (Pronounce)	
Flap Spelled T:				
Gretel	jet-bed	9	4	+5
meteor	feet-seed	8	3	+5
glitter	mitt-kid	8	5	+3
attic	hat-dad	7	4	+3
notice	boat-road	8	6	+2
cheating	feet-seed	9	7	+2
	Mean	8.2	4.8	+3.4
Flap Spelled D:				
Cadillac	dad-hat	7	3	+4
huddle	mud-nut	8	8	0
pedigree	bed-jet	8	6	+2
modify	rod-pot	8	8	0
shredding	bed-jet	9	9	0
forbidden	kid-mitt	8	6	+2
	Mean	8.0	6.7	+1.3

Note. There were 9 subjects in each group.
[a] Flap-terminal syllable is underlined.
[b] Response which rhymes with flap-terminal syllable is listed first.

Results confirmed predictions. Subjects who learned to read spellings selected significantly more picture names that rhymed with the spellings than subjects who did not see spellings. This shows that, even in a task which is strictly oral with no mention of letters or spellings, the effect of spellings on pronunciations is evident. From values in Table 7-5, it is apparent that performance differences between experimentals and controls were greater for T-spelled flaps than for D-spelled flaps. The reason is that controls were more influenced by acoustic factors and this bias boosted their accuracy on D-spelled words. Analysis of spelling productions verified that subjects who read words spelled the flaps more accurately than subjects who never saw spellings. From these results, we conclude that learning the spellings of words is a causal factor influencing how learners conceptualize the sounds in words when the sounds are ambiguous.

Pronunciations of Words

If spellings alter the way speakers conceptualize the phonetic structure of words, they might also be expected to influence how learners pronounce words when letters in the spellings symbolize somewhat different sounds from those in pronunciations. Evidence for this was found in the syllable-segmentation study described above involving words such as "interesting," "different," and "comfortable." However, our work on silent letters was interpreted to indicate that whether or not this happens depends upon how unpronounced letters are processed when the spellings are stored in memory. Other factors may also influence whether a discrepancy between spellings and sound will cause pronunciations to change in the direction of spellings: how often learners hear other speakers produce the spoken versions of words or their spelling pronunciations; whether or not learners pay close attention to letter-sound relations when they learn the spellings of words.

In our study of children's word spellings, we noticed that younger children sometimes subtly mispronounced words as they spelled them. The deviancy in speech might not have been detected except that the misspelling caused us to listen more carefully. One type of error involved mix-ups between closely articulated lax (short) vowels. Examples are as follows: GIST (guessed), IND (end), GIT (get), MELK (milk), NIKS (necks), LEVD (lived), and CECH (catch). In listening to children pronounce the words, sometimes it was hard to tell which of the vowels was being articulated. The sound produced appeared to be somewhere in between the two categories. Data presented by Lieberman (1980) confirms that in some children's speech there is substantial overlap in the formant frequencies of these vowels. It may be that deviancy or variability in children's vowel pronuncations is reduced when they learn letter symbols for the vowel system and word spellings for particular pronunciations. Having to distinguish among the lax sounds in I, E, and A spellings may strengthen these as discrete categories of sounds. Having to learn vowel letters in the spellings of particular words may clarify their correct or "ideal" pronunciations. Research is needed on this possibility.

There is some evidence that learning the spelling system for vowels influences how people perceive vowel relations and organize them into pairs. Jaeger (1979) conducted a concept-formation experiment in which adults learned to distinguish positive from negative instances of vowel-shift alternations for pairs of words (e.g., positive instances: "profane-profanity" and "deceive-deception"; negative instances: "detain-detention" and "false-fallacy"). Then subjects were given test cases to determine how they were responding on the basis of vowel-shift rules or spelling rules. It was reasoned that if Chomsky's and Halle's phonological system governed performance, subjects should consider "abound-abundant" a positive instance and "presume-presumption" a negative instance. If spelling rules governed performance, however, the opposite judgments should predominate. Results supported the latter. That learning the orthographic system causes subjects to group vowel sounds into short and long pairs was further supported by some of Jaeger's subjects who acknowledged this as the basis for their judgments. Moscowitz (1973) and Templeton (1979) interpret their evidence similarly to suggest that when children learn to read they come to organize information about vowels according to orthographic rather than phonological criteria.

Returning to our spelling study, other subtle mispronunciations of words were also revealed in the spellings of novice readers. Phonemes were deleted: BISACO (bicycle), SOTHING (something), SUCAS (suitcase), and INPORTIN (important). Syllables were deleted: CRANS (crayons), SIGRET (cigarette), AMBLUCE (ambulance), and SPRISE (surprise). The progressive inflection (-ING) was often mispronounced as -EEN or -IN. Since these mispronunciations of words appear to drop out as children learn to read, it may be that print is the instigator of correction.

It is perhaps not surprising to find deviant or variable word pronunciations in children's speech since this is what they often hear. Reddy (1976) and Cole and Jakimik (1978) show how words in connected speech may differ markedly from their pronunciations in isolation. Acoustic and phonetic properties exhibit much greater variability and boundaries between words may be obliterated, for example, "sumore" (some more), "wave-lags" (wave flags), "go 'n see" (go and see), "choc'late" (chocolate), "frien's" (friends), "twen'y" (twenty), "jus great" (just great), and /dIjyu/ (did you). Even in isolation, some words may have alternative forms, a casual pronunciation in which some sounds are reduced or deleted, and a careful pronunciation which includes all the sounds and is truer to spellings (e.g., "chocolate" and "twenty").

Another cause of variable or nonliterate pronunciations of words is dialect. Compared to standard English (SE), Black English (BE) word pronunciations differ markedly from spellings (Burling, 1973). In BE, vowel contrasts are lost in certain consonant contexts (e.g., before /n/ as in "pin" and "pen"). Final TH is pronounced like either /f/ or /v/ (e.g., "wiv" for with and "bof" for both). Consonants are weakened or deleted in many contexts (e.g., "pas" for past or passed, "tol" for told, "hep" for help, and "so" for sore). Kligman, Cronnell, and Verna (1972) collected spellings from Black-dialect-speaking second graders and found several misspellings illustrating the dialect features described above. Groff (1978) looked for the same dialect-based errors in older children (fourth to sixth graders) and found many fewer instances. One explanation for the reduction in errors is that experience with print teaches dialect speakers the spellings of words and also their literary pronunciations. Some correlational evidence for this is offered by Desberg, Elliott, and Marsh (1980) who examined the relationship among reading, spelling, and math achievement scores and dialect radicalism in a group of Black elementary school children. Those who had better command of SE forms were better readers and spellers than those who did not. In contrast, achievement in math was not related to dialect. Although this evidence falls short of indicating a causal relationship, it does suggest that dialect speakers may acquire knowledge of SE word pronunciations by learning to read and spell words.

The impact of print upon pronunciations may not be limited to nonstandard dialect speakers or to prereaders with deviant pronunciations. It may be that the process of learning to read and spell words teaches all readers a literary English dialect reflecting the visible phonology represented in spellings. According to the theory developed above, this literary pronunciation is not necessarily one which renders all the letters pronounced. Rather, it is a pronunciation which reflects how letters in the spelling were analyzed as sound symbols when the spelling was stored in memory. It is held in memory as an "ideal" context-free pronunciation for the word. Its attachment to spellings gives it stability and makes it resistant to further change. Although the pronunciation may not be used in casual speech, it nevertheless tells its possessor how that word is "supposed" to be pronounced.

The possibility that learning to read equips the speaker with literary pronunciations explains a curious phenomenon described by Goodman and Buck (1973) who listened to several BE-speaking children read a text aloud and then retell the story from memory. Whereas the readers showed no dialect miscues in the reading task, they displayed much dialect involvement during their retelling immediately afterwards. The reason why speakers may be able to read text without their spoken dialects is that print activates its own set of pronunciations, those which were established by spellings when they were stored in memory and unitized with pronunciations.

The acquisition of literary pronunciations founded upon and stabilized by spellings stored in memory also explains how writing serves to constrain changes in speech in communities of speakers over time. According to Gelb (1952), English has changed relatively little over the last four or five hundred years as a result of its writing system. This contrasts with dramatic shifts occurring prior to that time. Also it contrasts with rapid linguistic changes evident in modern times among primitive societies which lack a phonetic written language. Gelb points out that some American Indian languages are changing so fast that people of the present generation have difficulty conversing with people three or four generations older.

Bright (1960) presents evidence from a comparison of South Asian Indian dialects to indicate that printed language, by freezing phonology, serves to limit changes occurring in speech over time (Bright & Ramanujan, 1962). He observed that in one geographical area where the Brahmin dialect contrasted with the non-Brahmin dialect in having a written form, phonemic changes over time were less apparent in Brahmin than non-Brahmin speech. However, in another geographical area where neither dialect had a written form, both Brahmin and non-Brahmin speech exhibited phonemic change to an approximately equal degree. It may be that pronunciations are more subject to change in the absence of print since they exist only in the minds and memories of speakers. Print, by casting pronunciations into fixed phonetic forms, gives them permanence outside the minds of speakers and thus frees them from total dependence.

Further evidence for the impact of spellings on pronunciations is provided by Reder (1981) who studied the pronunciations of Vai-speaking adults in Liberia. Vai orthography differs from English orthography in that it is syllabic rather than alphabetic and it includes alternate spellings of words to match variable pronunciations. Furthermore, only some people learn to read and write Vai since literacy is not taught in school and since it has limited use in the culture. Reder compared the pronunciations of words produced by literate and nonliterate Vai adults. The words he studied were polysyllabic words which had variable pronunciations and variable spellings and which were undergoing change in speech (i.e., one form was falling into disuse). For example, the word meaning "moon" could be pronounced either with or without a medial /l/, /kalo/ or /kao/, and written either with or without the /l/ symbolized. Pronunciations with /l/ were much less common in adult speech. Reder examined how frequently medial /l/ forms of the word were produced to determine whether having an internal written representation spelled with /l/ might cause literates to pronounce the /l/ form more often than nonliterates. This is what he found. However, the frequency of this form was not terribly high even among literates. Whereas they included the /l/ form in their writing about half of the time, they included it in speech less than a quarter of the time. These findings indicate that experience with print can influence speech by increasing the frequency of a less common pronunciation, presumably because the two printed forms exist side by side in memory reminding literate speakers of both options. (Equal awareness is suggested by the fact that both spellings were written equally often.) The fact that /l/ forms have not dropped out of Vai speech altogether may be noteworthy. This may be a case where orthography has prevented a less common pronunciation from becoming extinct.

Perception of Lexical Segments

Whereas the spatial model provided by print at the phonological level may actually create in the speaker's mind the idea that speech is comprised of sequences of discrete phonetic units, it is less likely that print implants a completely novel idea about segments at the lexical level since prereaders are familiar with many content words (nouns and adjectives), since syllabic divisions in speech provide some clues to word boundaries, and since words are combined and recombined unconsciously in forming sentences. However, results of several studies show that some types of words are not obvious units of speech to children who are unfamiliar with print. Furthermore, prereaders have substantial difficulty in lexical analytic tasks which require dividing meaningful sentences into words or picking single words out of sentences. In contrast, beginning readers can perform these tasks quite easily (Ehri, 1975, 1976, 1979).

One of the most difficult word types for prereaders to recognize is context-dependent words which evoke little meaning if unaccompanied by other words. Examples are function words such as "might," "could," "from," and past tense forms of irregular verbs such as "gave." Holden and MacGinitie (1972) gave prereaders a sentence-segmentation task and observed that they had difficulty isolating function words from adjacent content words. We found that prereaders had trouble recognizing function words when the words were presented as isolated spoken forms (Ehri, 1979). Since speech does not make children aware that function words are separate units, print may teach this lesson. Learning the printed forms of function words may clarify their pronunciations. Reading function words in meaningful sentences may clarify where the units are positioned and how they operate in speech.

We conducted a study with first graders to compare the effect of two different kinds of reading experiences on subjects' awareness of function words in speech (Ehri & Wilce, 1980a). Half of the subjects practiced reading the function words in meaningful printed sentences. The other half read the words in unorganized lists of words. After reading each list, the latter group heard the words rearranged into the same sentences read by the other group. A word-awareness posttest was given after word training to measure subjects' ability to detect the presence of the function words in spoken sentences played on a tape recorder. The sentences were different from those read or heard during training. The function words received minimal stress and pitch in these sentences. Furthermore, in some of the sentences, the words were "buried" by overlapping sounds from other words flanking them (e.g., "The green frog gave vegetables to the hungry rabbit"). For each sentence, subjects were shown four function words which they had read, they listened to the sentence, and reported whether they heard any of the four words. It was expected that subjects who had read the printed words in sentences would exhibit superior word awareness to list readers. According to our theory, seeing function words in context should ensure that their meanings are stored along with their spellings and pronunciations in lexical memory. Having complete entries for words in memory should make it easier to detect the words in speech. List readers were not expected to acquire complete entries since meanings were not processed when spellings and pronunciations were learned.

Results confirmed our expectations. Subjects who had read the words in context detected significantly more of these words in the spoken sentences than list readers (means = 15.5 versus 13.2 correct out of 20 maximum). Interestingly, burying words in sentences did not make detection any more difficult. This indicates that processing was not conducted on an acoustic or phonetic basis but involved a deeper lexical level of analysis. From these results we conclude that printed-language experiences which integrate spellings with meanings and pronunciations set up word units in the lexicon which are superior for distinguishing these units in speech. Although we showed that the way function words are read influences lexical awareness, we did not test the claim that learning printed symbols for function words is critical for setting these words up as accessible units in the lexicon. This hypothesis awaits investigation.

Once spellings are unitized with the other identities of words in memory, one would expect all of this information to be mobilized when the words are encountered. Even in strictly auditory tasks requiring listeners to process and respond to spoken words, spellings stored in memory should be activated and influence performance. Various studies have indicated that this is true. Seidenberg and Tanenhaus (1979) had adults listen to several target words on a tape recorder and decide whether each rhymed with a cue word. Some of the target words were orthographically similar to the cue word (i.e., *clue–glue*), some were different (i.e., *clue–shoe*). They found that subjects detected orthographically similar rhymes faster than orthographically dissimilar rhymes. In another experiment, they had subjects judge whether two aurally presented words rhymed. Included among the nonrhyming pairs were words with similar spellings (i.e., *bomb–tomb*) and words with different spellings (i.e., *bomb–room*). Here, it took subjects longer to reject words with similar spellings. These results show that the spellings of words are activated in spoken language tasks and influence the perception of rhyme. Similar spellings facilitate detection of positive instances and they interfere with the detection of negative instances.

Jakimik, Cole, and Rudnicky (1980) studied the impact of spellings in a lexical decision task. They had adult subjects listen to a list of spoken words and nonwords and indicate whether each was a real word by pressing a buzzer. The words were arranged so that in some cases successive words shared the same first syllable and spelling (i.e., "barber," "bar"; "napkin," "nap"). In other cases, successive words shared the same first syllable but were spelled differently (i.e., "laundry," "lawn"; "record," "wreck"); Also included were control words which differed in sound and spelling. Other studies of lexical decision performance have shown that, when words are related, the reaction time to judge the second word is faster than when the words are unrelated. Of interest here was whether similarities in spelling and/or sound might enable subjects to recognize the lexical status of the second word faster. Results revealed that reaction times were improved but only when the words had similar spellings, not when they were similar phonologically but different orthographically. Additional experiments were performed which replicated these findings. These results provide further evidence that orthographic representations of words are activated in spoken-language tasks and may even be more influential than phonological representations.

There is another study indicating that orthography may impair the perception of speech when orthographic knowledge creates expectations which conflict with what is actually heard. Valtin (1980) describes this study performed by the German linguist Jung (1977). He constructed pairs of sentences containing verbs which were pronounced identically except that the vowel was long in one case and short in the other. Also, the spellings of the words were distinct, with the difference in vowel length marked orthographically (i.e., *fliegt* versus *flickt*). Pronunciations of verbs were interchanged in some of the sentences recorded on tape. Children with poor and normal spelling skills listened to each sentence and then judged whether its verb contained a long or a short vowel sound. Jung found that the poor spellers perceived the vowel sounds correctly whereas the good spellers judged the sounds according to their knowledge of the words' spellings.

Memory for Words

Evidence presented above indicates that when words are heard, spellings may be activated in listeners' minds. This may influence not only their perception of the words in speech but also their memory for the words. Our study indicating that orthography has mnemonic value has already been described (Ehri & Wilce, 1979). The mnemonic advantage provided by spellings in a word-memory task was demonstrated directly by Sales, Haber, and Cole (1969). They required adults to remember six words displaying vowel variations (i.e., HICK, HECK, HACK, HOOK, HOAK, and HAWK). Subjects either saw or heard the words presented in 48 different sequences. After each, they repeated them back in the order presented. Recall was better when the words were seen than when they were heard, presumably because spellings helped to preserve the words in memory.

Metalinguistic Skills

Another consequence of having a concrete picture of language in one's mind is to facilitate metalinguistic processes. Certainly it is much easier to detach language from its communicative function, treat it as an object, and study its form when one has fixed, visual-spatial symbols to see and manipulate. When speakers can see or imagine the spellings of words, they can better analyze their sound structure. Although most linguists deny that their phonetic or phonemic analyses are influenced by orthography, Skousen (1982) disagrees. Citing evidence that preliterate children analyze speech differently from literate adults, he argues that some of the phonemic analyses of speech proposed by linguists arise solely from orthographic considerations, for example, whether the alveolar flap in words such as "ladder" and "latter" and "pedal" and "petal" is interpreted as /t/ or /d/, the decision to regard the preconsonantal nasal in words such as "sink" and "bump" as a separate phoneme rather than as a nasalized vowel. Additional evidence to support this argument has been presented above.

Metalinguistic skills benefit from print not only when children learn what speech looks like but also when they learn terminology for describing its form. Francis (1973) presents evidence that the meanings children possess for terms such as "word" and "sentence" are dominated by their experiences with print. In explaining what words and sentences are used for, her subjects referred to reading and writing in their replies. Few, if any, mentioned the use of words or sentences in spoken language. When the topic of words in speech was raised, some children claimed that they "thought a pause occurred between all spoken words because there were spaces between words in writing." This comment illustrates how printed language symbols and terminology provide children with the schema for conceptualizing and analyzing the structure of speech.

In a study comparing groups of high- and low-level literate adults, Barton and Hamilton (1980) report that the two groups differed the most in their answers to various analytic questions about language. The high-level literates had available two sources of information to consult. They could analyze the spoken form or the written form and, depending on the situation, they might use either one. In contrast, the low-level group could only respond on the basis of speech.

Concluding Comments

To summarize, it is suggested that written language supplies a visual-spatial model for speech and that, when children learn to read and spell, this model and its symbols are internalized as a representational system in memory. The process of acquiring this system works various changes on spoken language, particularly at the phonetic and lexical levels. Learning spellings as symbols for pronunciations teaches children that words consist of sequences of discrete phonetic units. For sounds which are nonobvious or overlapping or ambiguous in speech, letters clarify whether to and how to conceptualize them as separate sounds. If children's pronunciations are nonstandard and if spellings symbolize standard pronunciations, print may teach children how to say these words. Also, spellings may teach speakers ideal or literary pronunciations for words, particularly if learners speak a nonliterary dialect and if they hear literary pronunciations from other speakers. The experience of reading words in sentences may clarify how speech is divisible into word units and what their distinctive pronunciations and meanings are. The process of associating spellings with pronunciations and meanings may help to establish words as clear, accessible lexical entries in memory. This may be particularly true for context-dependent words whose identities may be harder to distinguish in fluent speech. Once spellings are established in the lexicon, they may influence the perception of spoken words. Spellings may facilitate thinking about similarly pronounced words when the spellings are similar but may inhibit thinking when the spellings are different. Activation of spellings for pronounced words may also improve listeners' memory for the words. Having a visual-spatial model for speech may facilitate performance in various metalinguistic tasks requiring speakers to detach language from its communicative function and inspect its form. Printed language may exert an impact not only upon individual speakers but also upon groups of speakers by fixing their word pronunciations and thus inhibiting change over time. These are some of the effects proposed to result from experiences with printed language. Further research is needed to bolster these claims.

Further research is needed also to investigate how much individuals differ in their susceptibility to the effects of print. From available evidence, it is clear that individuals differ in how they process printed language. Readers vary in their dependence upon context cues for recognizing printed words when they are reading text (Stanovich, 1980). Also, readers vary in the accuracy of their memory for word spellings (Frith, 1978) and in their print-sound recoding skill (Ehri & Wilce, 1983). In our phonetic segmentation study where we observed spellings to affect children's analyses of words such as "pitch" and "rich," we found that subjects differed in how many of the extra segments they detected. Barton and Hamilton (1980) report the same thing. Also, Watson (see Chapter

6) reviews evidence indicating that phonemic analysis skill varies substantially in readers with equivalent reading skill and experience with print (i.e., older, backward readers perform worse than younger, normal readers reading at the same level). This raises the very likely possibility that print has a greater impact upon some readers than upon others, depending upon the way they process print. For example, readers who become very proficient at recoding print into speech and who are good spellers may be most susceptible to the phonological effects of print. Such possibilities await study.

Acknowledgments

Research reported in this chapter was supported by grants from the National Institute of Child Health and Human Development and the National Institute of Education. Gratitude is expressed to Lee Wilce who assisted in the conduct of this work. An earlier version of this chapter, entitled "Effects of Printed Language Acquisition on Speech," was prepared as a chapter for a book edited by Olson, Hildyard, and Torrance, *The Nature and Consequences of Literacy*. The present chapter summarizes some additional evidence on the acquisition of phonological awareness collected after the earlier version was written.

CHAPTER 8

Learning to Attend to Sentence Structure: Links Between Metalinguistic Development and Reading

Ellen Bouchard Ryan and George W. Ledger

Introduction

Metalinguistic activity involves the ability to treat language objectively and to manipulate language structures deliberately. This ability to focus attention on language forms per se becomes possible for children only gradually as their cognitive development proceeds. In the understanding of spoken language, the focus of attention is typically on the meaning of the utterance, and little attention is paid to the particular acoustic forms of the message. Written language, on the other hand, requires analysis and manipulation of language forms in order to extract the meaning from them. According to Vygotsky (1934), Mattingly (1972), and Cazden (1972, 1974), the additional cognitive demands of this analysis and manipulation underlie the observed discrepancy between the adequate speaking and listening abilities of the young, school-age child and his limited ability to deal with written language.

With its dependence on structural analysis, the reading process draws heavily on several metalinguistic abilities (relating to phonological, lexical, and syntactic structure). Adequate phonological skill, which entails accessible knowledge of the sound-spelling correspondences in written text, is required for the child to deal analytically for the first time with segmented phonetic elements. This transition, plus the imperfect letter-to-sound correspondences in alphabetic writing systems and the highly complex pronunciation rules governing the combination of phonetic elements into words, combine to make beginning reading a complex and difficult endeavor for the child (Liberman et al., 1977; Rozin & Gleitman, 1977). Lexical skill, or knowledge of the identity and use of words, must also develop before adequate reading performance is possible. The beginning reader's lack of awareness of the segmentation of words by spaces in text, his difficulty in dealing with context-dependent words (e.g., articles, past-tense verbs, and prepositions), and his initial inability to separate a word from a sentence context all point to the difficulty inherent in initial attempts to analyze written forms (Ehri, 1979). Syntactic skill, involving the ability to utilize sentence information about the order and form of words, also comprises a necessary yet difficult component of beginning reading. The extraction of meaning from written text is an active, constructive process in which knowledge about syntactic and semantic structure must be deliberately utilized.

The present chapter addresses young children's developing ability to focus on sentence structure per se and the role of sentence-processing skill in learning to read effectively. First, the data concerning children's evaluations of sentence acceptability and other structure-based sentence tasks are reviewed. During the years from age five to age eight, children show major performance changes on tasks requiring them to override their natural tendency to deal with sentences in terms of meaning and to attend to the structural features of language. Second, the value of training studies is highlighted for selecting among alternative cognitive interpretations of metalinguistic development. Third, investigations of the links between sentence-processing skill and learning to read are discussed. Although moderate relationships have been clearly established, the direction of causation remains a matter of dispute. The usefulness of strategy training for clarifying this issue and for altering children's orientation to printed sentences is illustrated. Finally, a model of metalinguistic development is described which provides a framework for understanding the cognitive foundations of both reading and grammatical awareness as well as a basis for classifying the diverse array of tasks used to assess grammatical awareness.

Children's Developing Grammatical Awareness

Intuitions regarding the acceptibility, ambiguity, or synonymy of sentences have formed the primary data for many analyses of the structure of language. In recommending emphasis upon judgments rather than speech performance, Chomsky (1957, 1965) assumed that an individual's language knowledge would thus be optimally reflected and that there would be greater homogeneity among the speakers of a language. Although these assumptions may hold for adult informants with the more basic structures, the findings of a host of recent studies appear to indicate that individual and population differences are more extensive on judgmental tasks than in speaking and listening. For example, Gleitman and Gleitman (1970) observed striking differences between graduate students and clerical workers in paraphrasing skill. When given a choice between two meanings for complex noun phrases (e.g., "bird-house black" versus "black bird-house") in a noncommunicative setting, graduate students were much better able to focus on the syntactic structure; and attempts to train the clerical workers to treat these structures in a similar form-oriented fashion were not at all successful. According to Gleitman and Gleitman (1979), the greater variability in judgment performance is due to the higher order of self-consciousness involved. Language judgments are metacognitive activities, in that a prior cognitive process (language performance) must be taken as the object of reflection.

Since adults vary in metalinguistic abilities and since the ability to reflect on one's knowledge and performance represents a high-level cognitive skill, important developmental changes are to be expected in metalinguistic performances. Indeed, even though adults can exhibit greater knowledge of grammar through intuitions than they display in normal production or comprehension, two-year-old language learners clearly cannot. Very young children do, however, exhibit some evidence of monitoring well-formedness (see Clark, 1978): spontaneous and prompted repairs of their own utterances; corrections of the utterances of others; adjustment of own speech to the age, status, and language ability of the listener; and practice or play with language. In the following section, we review the growth in children's ability to reflect upon sentence structure, consider several cognitive explanations for this metalinguistic development, and illustrate the value of strategy training studies for selecting among these alternatives.

Assessing Grammatical Awareness

Normal-vs-Random Word-Order Sequences

Gleitman, Gleitman, and Shipley (1972) describe the rudiments of metalinguistic functioning in three talented 2½ year olds who could make some accurate judgments of "good" or "silly" for well-formed, reversed-word-order, telegraphic, and reversed-word-order telegraphic imperative sentences. All preferred normal word order to reversed, but only the linguistically most advanced child felt that telegraphic sentences were "sillier" than well-formed sentences. Although two of the children were able to correct the sentences judged to be "silly," only a small percentage of the corrections involved syntactic changes without semantic alterations as well.

Noting the bias for semantic corrections in the pilot study of Gleitman et al., deVilliers and deVilliers (1972) examined the judgments of eight two and three year olds. They were asked to judge the acceptability of correct, reversed-word-order (e.g., "Ball the throw") and semantically anomalous (e.g., "Throw the sky") commands and to correct any sentences which they called "wrong." Whereas all the children detected semantic anomaly, only the linguistically most advanced child (on the basis of mean length of utterance) could appropriately judge and correct reversed-word-order imperatives. The occurrence of some correct syntactic categorizations together with inappropriate semantically based corrections indicates that children at that stage may be only partially aware of correct syntactic form. DeVilliers and deVilliers (1974) emphasize the importance of corrections as necessary for distinguishing between the child who knows that a sentence violates the rules, whatever they may be, and the child who knows the particular rule which a sentence violates. This point is underlined in a small study they report in which, despite a high correlation between appropriate categorical judgments and direct word-order corrections, three of twelve four year olds made 80 percent correct judgments with only a negligible number of syntactic corrections.

With regard to school-age children, Bohannon (1975, 1976) has shown that children five to seven years of age tend to exhibit little ability to discriminate between scrambled and normal word-order sentences. In the first study (Bohannon, 1975), the discrimination performance of the first graders was significantly poorer than that of the second and fifth graders. Furthermore, the children in the first grade who did successfully discriminate were found to imitate well-formed sentences at a level equivalent to the fifth graders. In the second experiment (Bohannon, 1976), children in kindergarten, first, and second grade were given the word-order discrimination task as well as imitation and comprehension tasks. The percentage of children who were classified as discriminators changed significantly across grade level from 22 percent in kindergarten to 78 percent in second grade. Superior imitation and comprehension of normal sentences was observed for the discriminators, whereas the corresponding performance for scrambled sentences (where syntactic sensitivity was not useful) did not differentiate between successful and unsuccessful discriminators. Thus, individual differences in the ability to distinguish grammatical from totally deviant sequences (i.e., random word order) still occur among school-age children; and such differences are related to other measures of syntactic skill.

Recognizing Minor Deviations from Normal Sentence Structure

As one might expect on the basis of the Bohannon research with random word order, children's sensitivity to mild syntactic and semantic deviations improves substantially during the school years. Recognition develops earlier for semantic than for syntactic structure since a meaning orientation is helpful only for the former. Yet changes continue to occur during the early school years even in the semantic judgment task. In a study by Howe and Hillman (1973), nursery school children were able to distinguish acceptable from unacceptable sentences quite well when the semantic features of the subject violated the selection restrictions of the verb (e.g., "The apple ate the child"); but they performed at chance level for the less salient conflict between the object and the verb (e.g., "The dog frightened the car"). For children between kindergarten and fourth grade, performance continued to improve with the more striking changes occurring for the more difficult verb-object violations. James and Miller (1973) also noted that semantic violations related to the subject and verb were detected by their five- and seven-year-old subjects much more frequently than the lower-level adjective-noun violations (e.g., "The *happy* pencil rolled off the desk"). Furthermore, seven year olds distinguished more successfully between normal and anomalous sentences than five year olds and were also more proficient in using selection-restriction rules in changing anomalous into normal sentences (e.g., "The *yellow* pencil rolled off the desk") and vice versa.

In order to view young children's deliberate syntactic skill in depth, Beilin (1975) presented sentences with correct (e.g., "A lion jumps once") or incorrect (e.g., "A lion jump once") number agreement between subject and verb in four tasks: imitation, grammaticality judgment, correction to make the sentence sound better, and synonymy judgment. These tasks were administered to 18 children at each of the five ages between three and seven. Imitation of correct number agreement increased substantially between ages three and four while the biggest increase in accurate repetition of the ungrammatical sentences occurred between five and six. Beilin suggests that the latter type of imitation is based on new processes allowing for sentence objectification. For grammaticality judgments and corrections, improvements occurred between ages four and five, five and six, and six and seven, with judgments of ungrammatical sentences moving from 53 percent (chance level) to 87 percent accuracy between six and seven years of age. Synonymy judgments were random except for the oldest group who achieved a very high level of accuracy. Thus, it appears that children experience a notable boost in metalinguistic sophistication between ages five and seven.

A linear improvement has not always been observed. Thus, Kuczaj (1978) presented irregular verb forms in sentence contexts to children aged four, six, and eight years for acceptability judgments. The four year olds distinguished better than chance between grammatical (e.g., "ate") and ungrammatical (e.g., "eated" or "ated") forms, and advances occurred during each interval for some forms. However, a curvilinear relationship between age and judgment was obtained for ungrammatical forms of the type "ated," which the middle group accepted to a much greater extent than either the younger or older children. These patterns paralleled the age changes in spontaneous productions of overgeneralizations. Hence, it appears that young children can display their syntactic knowledge at a level closer to their productive competence when they can focus on the acceptability of a single word than when attention to sentence structure is also required.

In our first two metalinguistic studies (Scholl & Ryan, 1975, 1980), five- and seven-year-old children were asked to classify grammatically primitive (e.g., "Where the car go?") and well-formed (e.g., "Where does the car go?") sentences by pointing to the photograph of the likely speaker (either a young child or an adult). These discriminations had to be achieved on the basis of form alone, without any support from the meaning. Discrimination between primitive and well-formed sentences (negatives and interrogatives) for children in the first experiment was far from the all-or-none judgments expected of adults, even though the older children did perform better than the younger ones. Thus, more explicit instructions, a feedback condition, a repetition task, and a group of nine year olds were included in our second study. Even though judgment accuracy was near perfect for the oldest group and discrimination was better in the other groups than in the earlier study, room for improvement was evident for both five- and seven-year-old children. All three groups did quite well on the repetition tasks, even though repetition of well-formed sentences was better than ungrammatical sentences for all. Moderate correlations between judgment and repetition performance were obtained for five and seven year olds. Surprisingly, systematic feedback regarding judgment accuracy had no effect upon subsequent judgment performance.

A replication of the first study, conducted by Flahive and Carrell (1978), showed that four year olds did not discriminate successfully between primitive and well-formed sentences. This lack of structural sensitivity in preschoolers was also apparent in a study by Leonard, Bolders, and Curtis (1977). The attempt to examine the relationship between preschool children's actual usage of grammatical features and their judgments of these features in a formal task was thwarted by children's failure to attend to form. On first glance, these data seem to be contradicted by the almost error-free acquisition by preschoolers of discrimination among case grammar categories (e.g., actor, object, and locative) reported by Braine and Wells (1978). However, such case categories are semantically based and could be distinguished in the paradigm employed with little or no specific attention to structure.

Acceptability and synonymy judgments have also been obtained from young, school-age children for syntactic structures with which they may not yet be proficient in terms of production and comprehension. With these more complex structures, the distinction between comprehension and metalinguistic control cannot be sharply made. By manipulating the existence or lack of contextual support, investigators have been able to show that children often rely on semantic strategies to interpret the meanings of structures such as passive sentences (Bever, 1970; Powers & Gowie, 1977) and temporal connectives (French & Brown, 1977). Thus, tests of comprehension which allow the children to use semantic probability cues underestimate the age at which they fully understand a syntactic structure. Emerson (1979) obtained both comprehension measures and acceptability judgments for compound sentences with "because" from children ages six, eight, and ten years. The judgment and comprehension data showed significant improvements between each of the age groups, and even the ten year olds did not exhibit good comprehension in the absence of contextual support. Hence, the weaker judgment and correction performance of the younger children seems to be attributable to less ability to focus on structure at both the level of comprehension and of reflection. Similarly, the more accurate judgments and corrections of six and seven year olds over five year olds for compound sentences with "if" (Emerson, 1980) cannot be attributed solely to a growing sense of awareness nor solely to enhanced knowledge about the specific language rules involved. Beilin's (1975) finding that above-chance synonymy judgments for active-passive sentences only appear for children at least seven years old could be linked to improvement in syntactic comprehension as well as to the new analytic understanding of formal relationships between sentences which he proposed.

In order to examine the development of syntactic skill further, we (Ryan & Ledger, 1979b) tested approximately 100 kindergarten, first, and second graders with judgment, repetition, and correction tasks for parallel sets of sentences with five types of syntactic deviations (wh-questions, negatives, articles, irregular verbs, and separable verbs). Instead of the child-adult classifications employed by Scholl and Ryan (1975, 1980), judgments of right and wrong were elicited since the correction task depends on that distinction. In contrast to previous investigations (deVilliers & deVilliers, 1972; Gleitman et al., 1972), the judgment of grammaticality was separated from the correction of ungrammatical sentences.

The earlier procedure differentially reinforced the two judgment responses, requiring an additional response (i.e., correction) for "wrong" judgments. Significant improvements across grade levels were observed for judgments (for grammatical sentences only), corrections, and repetitions. Also, significant partial correlations (age controlled) were obtained between reading performance and the above three task scores. However, judgment of ungrammatical sentences yielded anomalous findings, correlating negatively with other measures and deteriorating across grade level. Thus, it seems that even among second-grade children, judgments are not yet adultlike in accuracy nor in the underlying decision process.

With regard to the anomalous patterns for judgments of ungrammatical sentences, several related findings are available and some possible interpretations can be offered. In a follow-up study, judgment of grammatical and ungrammatical sentences for the kindergarten participants again correlated negatively; however, the judgments of ungrammatical sentences did not correlate negatively with any other measures. The corresponding correlation in a study with first graders (Ryan & Ledger, 1979a) was again negative, but was not significant. Accurate corrections correlated with overall judgment accuracy in both studies and also correlated most systematically with other measures of syntactic sensitivity obtained in the first-grade study. The correlational patterns obtained for judgments suggest that children may correctly judge grammatical stimuli on the basis of a global sense of well-formedness, whereas their reaction to ungrammatical strings is less systematic. At the very least, we must conclude that the basis of simple dichotomous judgments is obscure for children younger than eight years and that they ought to be accompanied either by corrections or explanations. Presumably, older children making predominantly accurate discriminations process all stimuli in a similar, consistent fashion; but young children may not yet have integrated their response modes.

In a related investigation with 100 children from four to eight years of age, Hakes (1980) elicited judgments of "good" versus "silly" and explanations for grammatical and ungrammatical sentences, which were similar to the ones used as stimuli in our research. Consistent with earlier studies and our data, the youngest children (ages four and five) seemed to base their judgments primarily on meaning rather than structure, whereas substantially better performance was observed for the older children. Furthermore, the development of judgments of acceptability closely paralleled the maturity of synonymy judgments (a more meaning-based sentence task). However, in contrast to our data, the difference between younger and older children was more evident for the deviant sentences, with younger children accepting more deviations as normal than the older children. The Hakes (1980) finding fits with intuition if one views development in terms of a linguistically based progression from attention to major deviations to more subtle distinctions. On the other hand, if the child is progressing gradually from attention to meaning and contextual aspects of the stimuli to structural features, the relation between structure and accuracy might be less predictable.

Clearly, further research is needed to determine the conditions under which the judgment process depends on the well-formedness of the stimuli. A procedure developed by Bialystok (1979) for high-school learners of French may prove particularly useful for explaining the anomalous negative correlations obtained in some of our child judgment studies as well as the contrast between our judgment data and those obtained by Hakes. In her experiment, judgments for French sentences were elicited under six conditions, differing factorially in time of response (immediate versus 15-sec delay) and detail of required response (correct-incorrect versus form class of error versus specific rule violation). Interactions with sentence grammaticality indicated that the correct items were accurately identified best in the simplest condition (dichotomous decision only) and in the immediate response condition. However, the judgment of incorrect items benefitted from the delay. Using Bialystok's distinction between implicit and explicit knowledge, it seems that correct sentences are best judged with an implicit sense of good form whereas formal analysis and additional time allow the learner to discover more errors in incorrect sentences. Thus, as in our studies, the optimal judgment process appears to differ according to the correctness of the stimulus sentence. Second, judgments in a right-wrong context (as in Ryan & Ledger, 1979b) seem to be based on a more superficial, intuitive reaction to the stimulus than similar judgments when corrections and explanations are also required (as in Hakes, 1980). Finally, the former condition enhances judgments of grammatical sentences while inhibiting optimal performance for incorrect sentences. A study of child sentence judgments under the same six conditions employed by Bialystok would be most useful for elucidating the development of judgment processes.

The sentence judgment and correction skills of language-disordered children between the ages of five and seven were compared to those of linguistically normal peers by Liles, Shulman, and Bartlett (1977). The two groups with equivalent vocabulary levels did not differ in their recognition of semantic anomaly; but the language-disordered children did exhibit a deficit in the recognition of word order and syntactic deviations. Their corrections were substantially inferior in all three categories, but they were particularly inadequate for syntactic errors. Hence, the syntactic difficulties observed in the spontaneous speaking/listening of language-disordered children were reflected in their intuitions of grammaticality.

The fact that substantial development occurs in sentence-judgment performance during, and even beyond, the elementary-school years is highlighted by several studies with older children. Within an ambitious examination of the cognition and metacognitive correlates of reading ability among third and sixth graders, Forrest-Pressley (1983) obtained a strong effect in a combined judgment and correction score for grade. Third-grade children averaged 66 percent accuracy; while sixth graders achieved only a level of 88 percent accuracy with sentences no more difficult than those typically used in our studies with much younger children. In an attempt to examine whether twelve-year-old children are sensitive to degrees of grammaticality as outlined within transformational grammar, Moore (1975) obtained grammaticality judgments that neither corresponded to adult judgments nor to linguistic predictions. Judgments of ambiguity have revealed substantial growth in structural sensitivity, especially in terms of syntax, across all the elementary-school years (Hirsh-Pasek, Gleitman & Gleitman, 1978; Keil, 1980; Kessel, 1970; Shultz & Pilon, 1973). Furthermore, these studies (all cross sectional) also indicate that individual differences within age groups are greater than the age-related differences. Longitudinal research which follows the development within individual children across several years is definitely needed to establish whether such differences in verbal talent are stable characteristics.

Cognitive Interpretations and the Need for Training Studies

Metalinguistic development concerns the gradual shift of attention from meaning to structure in tasks requiring deliberate control over language forms. The essential feature of the beginnings of linguistic awareness seems to be flexibility of strategy—the ability to decenter, to shift one's focus from the most salient attribute of a message (its meaning and contextual setting) to structure (the ordinarily transparent vehicle by which meaning is conveyed). Both Piagetian and information-processing views of cognitive development predict that this metalinguistic attention shift would not be observable until the age of five and that a growth spurt would occur between the ages of five and eight. From the first perspective, a child must be in the concrete operational period before he can decenter and coordinate two stimulus dimensions simultaneously. From the second perspective, the developmental memory literature (see Flavell, 1977) suggests that the disinclination of children younger than eight years to behave strategically and actively also imposes limitations upon their metalinguistic performance.

Development of Cognitive Operations

The significant growth in children's ability to separate sentence structure from sentence meaning observed among children between the ages of four and eight years is clearly in line with the notion that metalinguistic skills develop along with other aspects of concrete operations. Beilin (1975) argued that concrete operational thought might be a prerequisite for the ability to make appropriate metalinguistic judgments about sentences. Indeed, he found a significant relationship between synonymy judgments for active-passive sentences and his measures of concrete operations. However, the existence of a number of nonoperational children experiencing success on the judgment tasks disconfirmed the strict prerequisite hypothesis. Using similar reasoning, Hakes (1980) hypothesized that the emergence of adultlike linguistic intuitions was the reflection in the child's language of the achievement of concrete operational thought. He argued that the shift from judgments based on meaning and understandability (i.e., language in context) to judgments based on how the meaning is conveyed (i.e., language in isolation) depends on concrete operations, and even that structural awareness is actually an additional concrete operation. A nonparametric scalogram analysis for the sample of 100 children revealed that a single underlying factor could account quite satisfactorily for performance on three metalinguistic tasks (sentence acceptability judgments, synonymy judgments, and phonological awareness) and conservation. In a brief summary, Flahive and Carrell (1978) also report a substantial partial correlation (age controlled) between conservation and a sentence-transformation task among children ages five to nine. These studies are important first steps in exploring the relationships between concrete operations and linguistic judgments. Similar arguments have been made and supportive data obtained for the operational character of phonological and lexical awareness (see Ferreiro, 1977; Lundberg, 1978; Papandropoulou & Sinclair, 1974).

Growth in Strategic Activity

It is possible that young school children's tendency to react to language forms passively, rather than analytically, is a part of their more general cognitive failure to apply known facts and strategies deliberately in situations where such application is not specifically cued. The tendency to behave strategically in memory tasks as well as the ability to verbalize instructions about the memory process (i.e., metamemory) have been shown to increase dramatically between the ages of five and eight years (see Belmont & Butterfield, 1977; Flavell, 1977). In many memory studies, younger children have been led to perform similarly to older children by relatively brief training in the utilization of simple active strategies.

Since the achievement of deliberate control over one's memory strategies may well involve similar cognitive processes to those required by metalinguistic tasks, it seemed worthwhile to examine the impact on young children's sentence judgments of strategy training. We had observed (Scholl & Ryan, 1975) that the kindergarten children appeared to be making their judgments impulsively, with shorter reaction times than the older children. Also, the anomalous findings obtained by Ryan and Ledger (1979b) suggested that judgment performance may have been hampered by an inappropriate task orientation which might be easily altered. Relatedly, support for the modifiability of judgment accuracy is offered by Brodzinsky (1977). Whereas impulsive children exhibited inferior spontaneous comprehension of verbal jokes (with phonological, lexical, or syntactic ambiguities) in comparison with reflective children, both groups performed equivalently under a prompted condition in which all children were made to "stop and think."

Hence, we (Ryan & Ledger, 1979b) designed a study to test the hypothesis that strategy training would enhance children's metalinguistic performance by guiding them in applying their syntactic knowledge. Kindergarten children scoring below the 67th percentile on a sentence judgment and correction screening test were selected to participate. The control and training groups were equated on pretest judgment and correction scores as well as age, sex, and nonverbal intelligence. While the control group (N = 14) participated only in pretesting and posttesting sessions, the training group (N = 14) received six brief sessions in which they were taught a "stop and think" strategy to improve their judgments (i.e., pause, whisper target sentence, attempt to correct it, and make the judgment on the basis of whether a correction can be made). During training a sequence of stimuli graded in difficulty was employed: article-noun pairs (e.g., "the boy" versus "boy the"), two-word sentences, three-word phrases, three- and four-word sentences, and finally, six- to seven-word sentences containing subtle syntactic errors. Feedback on use of the "repeat-and-correct" strategy, judgment accuracy, and adequacy of corrections was given throughout the sessions. Statistical analyses for group (training versus control) and grammaticality (normal versus ungrammatical) revealed no significant effects although the experimental group did perform slightly better on the posttest than the control group. Similarly, no group differences were observed for the correction task nor for a generalization task (repetition of structured and unstructured strings of nonsense words).

Even with systematic feedback, prompting to use the strategy, and very simple training stimuli, the kindergarteners were unable to perform at near-perfect levels during training. As shown by Huttenlocher (1964), several children even experienced difficulty in reversing two-word phrases (i.e., they repeated "boy the" as "the boy the"). Poor performance of kindergarten children on tests of sensitivity to grammatical structure seems not to be readily modifiable with the procedures we employed. These negative results support and extend earlier findings that accuracy feedback alone did not enhance sentence judgments (Bohannon, 1975; Scholl & Ryan, 1980). Perhaps more promising training programs could be developed, but this study demonstrates clearly that kindergarten children are not easily brought to shift their attention to sentence structure. The lack of training effects here stands in marked contrast to the dramatic effects obtained in our semantic integration training studies to be discussed later in this chapter. In those projects, kindergarteners uniformly and readily responded correctly to instructions to insert articles appropriately into a meaningful word sequence. Hence, it seems that young children can profit more from brief training in the use of a meaning-oriented sentence strategy than from similar training in a structure-oriented strategy.

If the brief instructions and feedback on the use of a "repeat and correct" strategy had succeeded in enabling the kindergarteners to display their knowledge of grammaticality more adequately, then we would have had evidence that a remediable passive (or inappropriate) task orientation was limiting young children's metalinguistic performance. Similar training studies with slightly older children may prove more effective, as it may well be that the average and below-average kindergarten children in our project had not yet reached a stage where they could benefit from such training. Another strategy-based prediction worthy of future experimental tests is that children who spontaneously employ organizational or rehearsal strategies in simple memory tasks exhibit greater sentence awareness than age-matched children who do not.

Relationships Between Sentence-Processing Skill and Early Reading

Vygotsky (1934) argued that the abstract nature of the reading task required deliberate attention to sentence structure beyond the spontaneous use of structure typical in normal speaking and listening. This distinction between deliberate control over language structure and spontaneous linguistic skills has proven useful in characterizing cognitive advances during the early school years in abilities to deal with language stimuli (Mattingly, 1972). A special aspect of the abstractness of the reading task, highlighted by Olson (1977) and Donaldson (1978), is the need to interpret the meaning of sentences without the aid of naturalistic context. As Olson has pointed out, reading and writing provide the primary opportunities for dealing directly with sentence meaning itself rather than with a sentence merely as one of a variety of available cues to the speaker's intended meaning. Thus, the lag of deliberate control behind spontaneous knowledge as well as the lack of immediately relevant context during reading interfere with the beginning reader's ability to apply his syntactic and semantic knowledge to text. The large amount of attention devoted to decoding by novices further detracts from the limited potential for utilization of phrase and sentence structure.

Downing's (1979) cognitive clarity model of learning to read makes the critical distinction between function and feature and shows how a clear understanding of both aspects is important in learning to read. According to the model, children typically approach beginning reading instruction in a state of cognitive confusion about the purposes and technical features of language in general, and of reading in particular. Hence, a child will progress slowly in learning to read if he is not aware that the goal of reading is the extraction of meaning and that decoding and word recognition are means to that goal. Furthermore, metalinguistic knowledge of language features (e.g., phonemes, syllables, sound-symbol correspondences, words, and sentence structure) and flexibility in the application of that knowledge are also necessary for obtaining meaning from print. (See also Downing's chapter in this book).

Reading for comprehension is an activity which requires the appropriate interweaving of top-down, conceptually driven processes and of bottom-up, text-driven processes (see Rumelhart, 1977a). In order to become a fluent reader, the child must become adept in the deliberate application of knowledge about technical language features in the service of specific meaning-oriented purposes. Thus, the children exhibiting advanced metalinguistic knowledge concerning sentence structure (as in the studies described in the previous section) can be expected to make faster progress in learning to read.

Assessing Links Between Sentence-Processing Skill and Reading

In terms of judgments, corrections, and repetitions of orally presented sentences, several correlational investigations have yielded relevant data. Within our laboratory, we have examined the proposed relationship between syntactic sensitivity and reading in six experiments. Judgments were found to be highly correlated with reading readiness for kindergarteners in the research of Scholl and Ryan (1980), while repetition of grammatical and ungrammatical sequences was related to reading for second graders. Judgments for grammatical sentences, repetitions, and corrections were significantly related to reading in the first study reported by Ryan and Ledger (1979b), whereas only the first measure was related to word recognition in the second (kindergarten) study. With first graders (Ryan & Ledger, 1979a), corrections correlated with reading scores even though neither corrections nor judgments discriminated between skilled and less skilled readers as defined by a median split. In two other Notre Dame projects, a combined sentence judgment and correction score was found to differentiate between gifted and average second graders (Peck & Borkowski, 1980); and a judgment score (based on selection of the grammatical sentence from a pair of almost identical sentences) significantly correlated with reading ability for both first- and third-grade children (Borkowski, Kurtz & Reid, 1980).

For our most recent project (Willows & Ryan, 1983), we administered intelligence, reading, and sentence-processing measures to children in first, second, and third grades. Partial correlations controlling for nonverbal reasoning, vocabulary, and digit span revealed significant correlations with reading for sentence judgment, sentence correction, repetition, oral cloze, and word substitution. This last task, an informal measure of parts-of-speech knowledge, was designed for this study. Children were required to substitute a specific word for its appropriate counterpart in a target sentence, thereby creating a different meaningful sentence. Factor analysis identified three factors: reading (including reading cloze), grammatical sensitivity, and word substitution. The oral-cloze task loaded equally on the first two factors. The low correlations between word substitution and the more traditional grammatical measures suggest further research would be profitable with this new task and the related matching words subtest of the Elementary Modern Language Aptitude test (Carroll & Sapon, 1959).

Bohannon (1979) has reported two relevant studies concerning the relationship of his word-order discrimination task and reading. In the first, reading measures for six year olds were moderately correlated with successful discrimination between scrambled and normal sentences. In the second project, children between five and seven years of age were given the word-order discrimination task, and a measure of reading achievement was taken a year later. A median split on sentence discrimination yielded a reading-level advantage of more than one year for the discriminators.

Several other studies have examined linguistic intuitions among good and poor readers in the middle grades. Forrest-Pressley (1983), in the project mentioned earlier, found that third-grade good readers achieved combined judgment and correction scores equivalent to poor and average readers in the sixth grade. Furthermore, third-grade poor readers achieved a disparate level of only 58 percent accuracy while sixth-grade good readers achieved a near-perfect level of 94 percent. Fourth-grade good and poor readers were compared by Killey and Willows (1980) on a task requiring the children to listen to taped passages and to identify which sentences contained an error, pinpoint the word(s) involving an error, and correct the error. Although the groups did not differ on the global classification of sentences, the poor readers did exhibit less sensitivity to linguistic information (syntactic as well as semantic) in pinpointing and correcting the errors. Relatedly, superior readers in grades one through five were better able than less adequate readers to explain the surface structure and underlying structure ambiguities in verbal jokes (Hirsh-Pasek et al., 1978). Menyuk (1981) reports two projects (one with fourth graders only and the other with fourth graders as well as older students and adults) in which structural paraphrase, ambiguity detection, and grammatical judgments/corrections correlated with reading performance at each age level. It appears, then, that relatively poor performance on tests of linguistic intuitions is characteristic of less skilled readers throughout the elementary school years.

In addition to grammaticality intuitions, a number of other tasks have been employed to examine the links between reading and sensitivity to sentence structure. For example, Vogel (1974) compared dyslexic second-grade boys with normal reading peers on nine measures of oral grammar and obtained significant differences in favor of the normal readers for seven measures (including oral cloze, sentence repetition, spontaneous speech production, and knowledge of morphology). Similarly, Hook and Johnson (1978) found inferior morphological knowledge for reading-disabled ten year olds.

Skilled and less skilled readers from first and second grades were compared in two studies. Ryan, McNamara, and Kenney (1977) found that the better readers outperformed the poorer readers on word discrimination (distinguishing single words from nonwords and two-word phrases), sentence comparison (identifying the particular word which occurred in only one of two presentations of a sentence), and a multigrammatical function word task (generating a sentence with the second meaning for a word such as *fly*). In a recently completed master's thesis with skilled and less skilled readers in first and second grade, Garson (1980) assessed the relationship between beginning reading ability and the utilization of syntactic and semantic knowledge on three tasks: written-word segmentation, sentence anagram, and oral repetition. All tasks included normal and anomalous sentences; and the segmentation and repetition tasks included random word lists as well. The older children and the more skilled readers were predicted to exhibit greater ability to take advantage of the syntactic structure, even in the meaningless anomalous stimuli. Analyses of the anagram and segmentation data revealed that the second graders performed better than the first graders and the skilled readers performed better than the less skilled readers. Grade by reader interactions indicated that the differences between skilled and less skilled readers were more pronounced for the younger group. A similar pattern resulted for the repetition task, but the effects for grade and reader only approached significance. The predicted ordering of sentence types (normal > anomalous > random) occurred for all tasks, but the groups did not show differential sensitivity to sentence context (i.e., no interactions occurred with grade or reading ability). Hence, even at the beginning of learning to read, the use of sentence context is important. Even though these tasks differentiated between skilled and less skilled readers, the accuracy data did not allow the interpretation that differential sensitivity to sentence contexts was the underlying reason. Yet, additional analyses of the utilization of sentence structure in the sentence anagram task revealed a substantial advantage for the more skilled and older children.

In a related study, Willows and Ryan (1981) assessed the extent to which grammatical information guided the reading performance of matched pairs of skilled and less skilled readers in the fourth, fifth, and sixth grades. The two tasks (reading of geometrically transformed text and cloze performance) were selected to reduce the availability of graphemic information and to encourage reliance on contextual cues. While both groups made considerable use of contextual information in the two tasks, the skilled readers made proportionately greater use of both syntactic and semantic information, even when only the prior context was considered. Furthermore, no change occurred across grade levels, indicating a stable difference between the two types of readers in their sensitivity to contextual cues. These studies of differences between skilled versus less skilled readers have shown, then, that poorer readers typically rely less on contextual cues while reading and are less likely to use the active, organizing strategies necessary for good reading comprehension.

Our final correlational findings (Ryan & Ledger, 1982) address the links between sensitivity to sentence structure (as assessed by an oral cloze task) and reading, other metalinguistic acts, and other cognitive abilities for 60 kindergarten and first-grade children. For the group as a whole, oral cloze performance correlated significantly with Gates-MacGinitie word recognition, writing own name, oral word blending, word discrimination, listening comprehension, concept of reading, and nonverbal reasoning. When children in the two grades were considered separately, it was found that overall the correlations tended to be much stronger for the older group. For each grade, the ability to complete an orally presented sentence by filling in the appropriate content or function word was related to the traditional reading readiness measures (word recognition and word blending) and to the children's concept of reading.

In sum, clear differences in oral sentence processing skill have been demonstrated between better and poorer readers. However, the underlying causes for the correlations between metalinguistic behavior and reading performance have not been identified. Since the major changes in metalinguistic performance occur during the readiness and beginning reading years, many researchers (see Mattingly, 1972; Rozin & Gleitman, 1977) have suggested that certain levels of linguistic awareness (in particular, at the level of sound structure) are prerequisite for learning to read. On the other hand, the fact that reading requires the deliberate application of sound, word, and sentence knowledge suggests that learning to read (and to write) would be a major factor in promoting linguistic awareness (see Ehri, 1979). For example, the separation by spaces of function words (e.g., *the, is,* and *of*) signals their distinct syntactic identities with a physical cue, for which no parallel exists in the spoken language. Similarly, the written language highlights the sentence unit with capital letters and periods. Undoubtedly, both of these views have some validity in that a bidirectional causal relationship holds between metalinguistic ability and reading. Although our emphasis has been upon the ways in which oral language sets the stage for reading, future investigations must seek to describe how and to what extent metalinguistic and reading levels influence each other.

Traditional group comparisons or correlational analyses (even with a metalinguistic measure in one year followed by a reading measure the following year) cannot yield information regarding the direction of causation. However, cross-lagged panel correlations based on measures of both skills at two points in time and training studies can help to elucidate this important question. The effects on linguistic awareness of reading could best be examined by training at the preschool level, before the children are introduced to reading in school. The prerequisite hypothesis can be assessed best in terms of the effects on reading acquisition of training on particular aspects of linguistic awareness. Although the bidirectional relationship hypothesis has not been directly tested in the realm of syntactic measures, Goldstein (1976) did find some evidence for this plausible conjecture for phonological ability. Phonological awareness predicted which four year olds would learn to read in a special preschool reading program, and the children taught to read subsequently exhibited greater phonological awareness.

Training Sentence Strategies for Reading

Studies in which sentence strategies are trained can be very useful in the examination of individual differences in reading comprehension. From a theoretical standpoint, the success of brief strategy training or of instructional manipulations for young and poor readers can assist in the identification of performance deficits which are not related to problems of competence. From a more practical point of view, such success in altering sentence orientations while reading can lead to substantial improvement in comprehension. These two kinds of implications of strategy training are illustrated in this section.

Several studies have shown that children five years of age and older can readily learn pictograph or logograph symbols for individual words, but that youngsters five and six years old as well as older poor readers do not tend spontaneously to integrate a sequence of them into a meaningful whole (Denner, 1970; Farnham-Diggory, 1967; Ferguson, 1975; Keeton, 1977). In order to examine spontaneous integration as well as the effects of semantic integration training for prereaders and young readers, we developed a pictograph sentence reading and recall task (Ledger & Ryan, 1982). The main goal was to examine the extent to which initial failure to integrate a meaningful sequence of pictographs was due to a modifiable lack of strategic orientation. The strategy taught was composed of two components: to read the pictures as a sentence and then to act out the meaning of the sentence with available toys. Kindergarteners initially treated the pictograph sequences as if they were random sequences, but three 15-min training sessions led to tripling of sentence recall levels in an uninstructed posttest two weeks after training. In a second study (Ryan, Ledger & Robine, in press), similar effects were found for both kindergarten and first-grade children. The inclusion of a condition in which children were taught only the sentence component demonstrated that children of both ages could profit from the sentence orientation but that the deep processing required by the toy enactions yielded substantial additional improvement. Analyses of concurrent measures of strategy use during posttests and generalization indicated that, whereas both strategies were readily acquired, only the sentence strategy was transferred to slightly different task situations.

The impracticality of toy enaction as a reading strategy and the fact that it did not generalize led us to consider an imagery strategy as a candidate for training in the pictograph sentence context. Instructions to picture the meanings of what is being read in one's mind have been somewhat effective for children over the age of eight years (Lesgold, McCormick & Golinkoff, 1975; Levin, 1973; Pressley, 1976), especially when the child is instructed to imagine the meanings sentence by sentence. Even though imagery instructions have been effective for enhancing the listening comprehension of children as young as five years under some circumstances, the increased information-processing requirements involved in "reading" pictograph sequences might well preclude any effects. On the other hand, the pictographs themselves provide the pictorial base from which to form an interactive image representing the sentence meaning. In two studies with kindergarten children (Ledger & Ryan, 1981; Ryan, Ledger & Weed, 1983), an imagery strategy has been shown to be almost as effective as actual actions for recall of pictograph sentences in the original task. Moreover, as predicted, the imagery strategy has shown some evidence of being more generalizable to similar pictograph tasks.

In our research, we have focused on demonstrating that sentence processing while "reading" pictographic messages has a substantial strategic component. Hence, we have not examined the practical impact of long-term training in sentence processes. However, our studies and the imagery studies with older readers (Levin, 1973; Lesgold et al., 1975) do suggest that such a training program might be very useful for children experiencing difficulty in the transition from decoding to comprehension. Moreover, the pictograph sentence paradigm provides an ideal setting for teaching the sentence processing subskills of reading before children can decode (see Woodcock, Clark & Davies, 1969).

Moving from simple meaning-seeking strategies, we should also consider sentence strategies which focus directly upon structure. For example, Weaver (1979) taught third-grade children to use such a strategy to solve sentence anagrams. The children learned an elaborate sequence of wh-questions to ask themselves as they searched for the sentence verb, its subject and object, and various modifiers. Although such a strategy could not be explicitly used while reading because of time constraints, it was expected that experience with this structural strategy would attune the children to structural cues in their reading. Indeed, instruction in the strategy not only facilitated sentence anagram performance but also yielded some improvement in reading comprehension. A recent replication (White, Pascarella & Pflaum, 1981) and experience in our own laboratory indicates that poor readers can only benefit from such a structure-based strategy if they have sufficient analyzed knowledge about sentences. Relatedly, Short and Ryan (in press) obtained quite striking changes in story comprehension by poor readers with brief training in the use of a story-grammar strategy. In this case, the fact that training raised poor readers' comprehension to the level of skilled readers argues persuasively against earlier interpretations which attributed good-poor-reader differences in use of story grammar to differences in the underlying knowledge (Dickenson & Weaver, 1979).

A Framework for Investigating Metalinguistic Development

Two Underlying Dimensions

During the past 10 to 15 years, much research concerning metalinguistic develop-ment as well as its links with beginning reading has been reported. Now that we have a general idea of what develops, we need to seek explanations for these developments and for the interrelationships among metalinguistic ability, reading, and general cognitive growth. With this important goal in mind, Bialystok and Ryan (in press) have presented a metacognitive model which explains the rela-tionships among three domains of language use (conversation, reading/writing, and metalinguistics) in terms of two underlying dimensions. Whereas most defi-nitions of metalinguistic ability conflate analyzed knowledge with control over that knowledge (e.g., Vygotsky, 1934), some definitions emphasize primarily the first aspect (e.g., Gleitman & Gleitman, 1979) or the second (e.g., Cazden, 1974). The model proposed is based on the separation of these into two potentially or-thogonal dimensions: (1) cognitive control and (2) analyzed linguistic knowledge.

The *cognitive control dimension* represents an executive function which selects and coordinates the language information appropriate to task solution. Hence, this dimension resembles the Piagetian notion of decentering, the ability to ignore a salient attribute to coordinate multiple cues relevant to problem solu-tion. The more salient the irrelevant cues and the more abstract or complex the relevant cues, the greater the burden on cognitive control. Tasks in which the natural direction of attention toward meaning is appropriate and sufficient usu-ally require little control; those which involve attention to form and coordina-tion of formal and meaning information typically require moderate amounts of control; and finally, those which involve strict attention to language structure despite the compelling salience of meaning require high levels of control.

The *analyzed linguistic knowledge dimension* refers to the extent to which the form-meaning relationships known by a language user have become objective and explicit. A natural consequence of such analysis of language knowledge is an increase in one's consciousness of that knowledge and hence one's ability to express it. However, intermediate levels of analysis, where clear verbalization is not yet possible, are frequently sufficient. The necessity to rely on more expli-cit language knowledge tends to correlate with the lack of supporting context. In the most extreme cases, the young, language-learning child can often interpret speech addressed to him mainly through the rich accompanying context, whereas working out the rules underlying use of a particular grammatical structure is an activity completely outside any linguistic or extralinguistic context.

Relationships Among Domains of Language Use

If one considers the language skills of the child between ages four and eight years, the prototypical requirements involved in three language-use domains can be differentiated, even though some overlap does occur. First, speaking and listening competently in conversational settings involves little cognitive control (i.e., attention is naturally and appropriately directed toward meaning) and little analyzed knowledge (i.e., context is rich). Second, reading and writing activities involve more cognitive control (i.e., to attend to language forms and to coordinate this information with evolving text meanings) as well as more analyzed knowledge of sounds, words, and sentence structures. Third, performing metalinguistic tasks such as correcting ungrammatical sentences depends upon a relatively high degree of cognitive control (i.e., to attend to the structure for its own sake) and a relatively high degree of analyzed knowledge of language structure. Hence, the relationship between linguistic awareness and reading is now viewed in terms of their overlapping demands upon the two underlying dimensions.

Within our theoretical framework, then, children are considered to show marked growth during this age period along both of the underlying dimensions, and reading-metalinguistic relationships are interpreted in terms of these underlying changes. In particular, the process of learning to read can be segmented into three metalinguistic stages: (1) analyzed knowledge of the functions of print (i.e., to convey meaning); (2) analyzed knowledge of the features of print and the control to relate them to units of oral language; and (3) the control to coordinate forms and meanings to achieve the goal of extracting meaning from text. While stage 2 involves both analyzed knowledge and cognitive control, stage 1 is primarily related to analyzed knowledge and stage 3 is primarily a matter of cognitive control. The acquisition of skill in stage 2 tends to be the major hurdle for beginning readers and hence is the focus of early reading instruction. However, some children flounder because they lack knowledge of the functions of print. Moreover, the passage from stage 2 to stage 3 is not automatic for many children. This last transition depends especially upon analyzed knowledge and control over sentence (and discourse) structures.

Classifying Tests of Grammatical Awareness

One of the important consequences of the present framework for future metalinguistic research is the provision of a principled basis for task selection and task development. Although the diverse collection of grammatical awareness tasks which have been employed in the literature cannot be precisely mapped onto our two-dimensional space, three major clusters can be identified. Moreover, the burden upon cognitive control or analyzed linguistic knowledge can be increased independently by separate task manipulations.

The first cluster, *sentence evaluation and explanation,* is composed of a set of tasks which emphasize analyzed knowledge of sentence structure. Sentence judgment and correction, which have been the main tasks in grammatical awareness research, require the same amount of cognitive control (to focus on grammatical form) but differ in degree of analyzed knowledge required. In fact, one can rank the following series of tasks in terms of moderate to very high demands upon analyzed knowledge: judgment, error location, correction, explanation, statement of rule, and statement of a system of interrelated rules. Also in this cluster are school tasks in which children are asked to define terms such as sentence or verb, to identify the sentence subject and predicate, to state language rules (e.g., the rule differentiating the use of bring and take), and to diagram sentences.

The judgment and correction tasks can be modified so that much greater cognitive control is necessary. Presentation of true and false sentences for grammaticality judgments would require respondents to ignore the compelling meaning dimension while attending to the presence or absence of grammatical mistakes. Similarly, the correction of grammatical mistakes requires much more control when false, ungrammatical sentences (each of which can simply be made either true or grammatical) are involved. With the original two tasks and these modifications, one can examine the relative difficulty of the knowledge and control dimensions in a 2 × 2 design (see Ryan & Bialystok, 1983). In addition, we can interpret the time and level of analysis manipulations introduced into the grammatical judgment paradigm by Bialystok (1979) within this framework. That is, the time manipulation alters the opportunity to exercise control while the level manipulation alters the explicitness of the knowledge required.

The second cluster, *sentence repetition and substitution,* consists of a set of tasks which stress cognitive control. Although ordinary sentence repetition does not require much control or analyzed knowledge because reliance upon sentence meaning can be very helpful, variations of the task do add to the burden, especially for control. For example, repetition of structured nonsense (Weinstein & Rabinovitch, 1971) requires attention to structure in the absence of supporting meaning, and verbatim repetition of grammatically deviant sentences (Beilin, 1975; Menyuk, 1969; Scholl & Ryan, 1980) necessitates suppression of the natural tendency to normalize. Also demanding much control is Ben-Zeev's (1977) word substitution task in which a given word must be substituted for a target word in a sentence, thereby violating both syntactic and semantic rules. Bialystok and Ryan (in press) have created several other tasks by contrasting this one, which requires no analyzed knowledge, with the matching words subtest of the EMLAT (Carroll & Sapon, 1959), which involves only analyzed knowledge of how words function within sentences. Control is manipulated in terms of whether the requested substitution yields an anomalous sentence while knowledge is manipulated in terms of whether the word to be replaced must be selected by the respondent or is identified by the experimenter.

Finally, the third cluster of grammatical tasks, referred to as *sentence completion,* are those which stress the coordination of both forms and meanings. Tasks which tap only a moderate amount of attention to form and subsequent coordination include the cloze task and the written segmentation task (in which respondents segment a sentence printed without spaces into words). Sentence anagrams, in which a randomly ordered list of words is to be reordered into a meaningful sentence, represent the more challenging task type. As Bialystok and Ryan (in press) propose, the analyzed knowledge burden of the anagram task can be lessened by providing cues concerning the grammatical structure of the target sentence, and the appropriate use of control can be supported with hints about systematic ways of solving the task.

Future Research

With this theoretical framework to guide future research, it should be possible to make considerable progress during the next decade in clarifying the nature of metalinguistic development and its links to reading and to other aspects of cognitive development. In particular, research hypotheses can now be more specific concerning the kinds of growth to be expected over particular age intervals and the kinds of differences to be expected between skilled and less skilled readers at specific ages. Moreover, grammatical tasks can be selected and devised in order to assess particular aapects of cognitive/linguistic ability. For example, interrelationships among tests of grammatical, sound, and word awareness can be expected to be strongest when they tap cognitive control and analyzed knowledge in approximately the same manner. Furthermore, it can be expected that metalinguistic tasks which basically assess coordination of form and meaning would relate most strongly to the emergence of beginning reading comprehension.

Summary

Development of Sentence Awareness

After reviewing the empirical evidence concerning the improvement in children's awareness of sentence structure, we can conclude that important changes occur between the ages of four and eight years of age and that children's sense of grammaticality continues to progress toward adult levels throughout elementary school. Substantial improvements in sentence judgments, sentence corrections, and repetitions of normal and ungrammatical sentences occur between kindergarten and second grade. Furthermore, individual differences on these metalinguistic tasks intercorrelate, suggesting that most of these performances flow from a common underlying sense of sentence acceptability and ability to access that knowledge deliberately. In addition, sensitivity to sentence structure was found to be moderately related to attention to language form on lexical and phonological tasks. Our two-dimensional model provides a framework within which to select and devise specific tasks to test specific hypotheses concerning meta-linguistic development.

In agreement with Hakes (1980, 1982) and Lundberg (1978), we contend that the observed age changes reflect a major cognitive achievement, the ability to shift attention from the meaning of sentences and their associated context to the linguistic forms themselves. Indeed, some correlational data suggest that the transition toward concrete operations (also involving decentering) is related to growth in metalinguistic ability. Yet, we have also proposed that another poten-tially orthogonal dimension, analyzed linguistic knowledge, is important to con-sider. Initially, progress toward attending to language form and the beginnings of analysis of one's language knowledge tend to occur together. However, as child-ren continue to mature past the stage of concrete operations, these two dimen-sions can be expected to become more and more distinguishable. Within an information-processing paradigm, it has been proposed that poor performance on metalinguistic tasks may derive from the young child's general tendency to behave passively, rather than strategically, on decontextualized school-type tasks. Although our single attempt to train kindergarten children to employ a correc-tion strategy to enhance their sentence judgments was not at all successful, it may be that more advanced children would have been able to profit from such training and also that metalinguistic tasks involving coordination of form and meaning will be more amenable to training. Finally, further investigations of the links between the development of grammatical awareness, operational thought, and strategic behaviors need to be undertaken.

Links Between Sentence Processing and Early Reading

Within a metalinguistic framework, reading is viewed as requiring an understand-ing of the meaning functions of print; deliberate attention to the phonological, lexical, and sentence structures of language; and coordination of this featural information with the goal of extracting meaning. Consequently, those children with relatively good metalinguistic ability would be expected to learn to read more quickly and attain higher levels of success in reading than children strug-gling to deal deliberately with decontextualized language in oral tasks. Numerous studies have demonstrated the predicted correlational links between individual differences in beginning reading and oral sentence processing skill as well as lexi-cal and phonological awareness. The fact that performance on a variety of sen-tence-processing tasks (judgment, correction, repetition, oral cloze, and picto-graph sentence recall) was found to be related to beginning-reading achievement supports a model of reading that stresses comprehension processes from the very onset of reading readiness instruction. Future research must address the causal bases for these interrelationships by separating the dimensions of cognitive con-trol and analyzed linguistic knowledge as much as possible. It has been argued that strategy training studies are especially valuable in distinguishing beween performance differences due to capacity or basic knowledge differences and those due to readily modifiable strategic orientation. In our own research within the pictograph sentence paradigm, we have identified some aspects of early reading comprehension strategies that even prereaders can acquire readily.

CHAPTER 9

Theory and Practice in Learning to Read

Douglas Pidgeon

Some Theoretical Considerations

The aim of this chapter is to examine, with reference to a recently completed study (Pidgeon, 1981), the relationship between certain theoretical aspects of the learning-to-read process and classroom practices. Teachers concerned with the task of teaching beginners are not always cognizant of the theory underlying their activities, and indeed some have such little interest they question that it has any relevance at all. The intention, therefore, is first, to take a brief look at different theoretical views; second, to examine the actual practices found in a sample of reception classes; and, finally, to relate the two, outlining the model prepared for the study mentioned. It should be made clear at the outset that the concern here is with the process and practice of learning to read and not with the actual exercise of reading once the elements of the task have been mastered. Indeed, the study dealt not only with the early stages of reading but also with what are usually described as prereading activities, it being appreciated that the two are very closely linked.

As will be noted later, the process of learning to read and write bears a similarity to that of learning spoken language. It must be noted, however, that the latter requires no instruction at all—learning to achieve mastery of oral communication, albeit often to only a limited extent, is somehow acquired by the growing child with no direct help—excluding the odd grammatical correction by parents or teachers. On the other hand, seemingly the reverse applies to learning to read and write. Here, with a few exceptions from mostly gifted children, the vast majority do need specific instruction. Under normal circumstances, reading and writing are not just acquired as children mature.

Since the acquisition of spoken language does not require any instruction, the study of its theoretical basis might appear to have little value; yet it is the very complexity of the grammatical structure of language and the fact that young children can apparently work this out for themselves that has led to extensive theoretical speculation. At the same time, one has only to compare the divergent theories of Skinner (1957) and Chomsky (1968), whose names and views are quite unknown to all but a few parents, to illustrate how unnecessary theories are in language acquisition.

Is it necessary, therefore, to have a theoretical framework when it comes to learning to read and write? Does the fact that some kind of instruction is required make all the difference? It might be argued that it all depends upon what is meant by "theory." On the one hand, there are the ideas which over the centuries various people have had about how reading should be taught, and, on the other, as with language acquisition, there are basic theories about how children actually learn, which, of course, may well influence the practical approach taken to initial instruction. Until possibly within the past 20 or so years there is no doubt that the former dominated the reading field. From the 16th century on, nearly every method and approach known today had been suggested at one time or another—alphabetic, whole word, whole sentences, phonic, etc. (Fries, 1963). It was not until the present century, however, that studies of their relative merits were undertaken, with the question of starting with a meaning emphasis approach (whole word) or a code emphasis approach (phonics) given prominence. Chall (1967), reviewing this work, drew two conclusions that bear on this present account: first, that a code emphasis method is one that "views beginning reading as essentially different from mature reading" and, second, that it "produces better results." Chall, however, had serious misgivings about research on beginning reading, expressing the view that too much of it was trying to prove that "one ill-defined method was better than another ill-defined method." She advocated more coordinated studies to give "definite answers"—presumably to produce better defined methods—but she gave no indication as to how these might be formulated. Barton and Wilder (1964), on the other hand, in their extensive review of reading researches over the period from 1930 to 1960, while also condemning them as being mostly of "poor quality and non-cumulative," did make a plea for the development of a systematic model which "would lead us toward a theory of learning to read." They also noted that the "isolation of the reading field has recently been threatened by invasions from experts in other fields," and mentioned notably linguistics and experimental psychology.

Jenkinson (1969) endorsed the move away from the pragmatic approach to solving reading problems and set out the sources of knowledge required for theory building which included the encompassing of other disciplines. In seeking to explain the failure to evolve theories of reading, she quoted Wiseman's (1966) reminder that the distinguishing mark of a scientific theory was not that it could be proved true but that it must be susceptible to disproof. This is a very important point which has tended to be overlooked in the attempt to evolve theories of both learning to read and language acquisition. The problem as it relates to learning to read will be referred to later, but it can be illustrated by comparing Chomskyan and Skinnerian views of verbal behavior. Chomsky attacks Skinner's hypotheses (Chomsky, 1959, 1971), yet as McLeish and Martin (1975) demonstrate, the latter being behavioral in nature can be put to experimental test, while Chomsky, in order to explain the child's acquisition of grammar, demands the need for an innate linguistic competence and a language-acquisition device, both of which are mentalistic and therefore not readily susceptible to disproof.

It would appear that it is the need for instruction that distinguishes learning to read from acquiring a language, and yet, at the same time, it has been the emphasis on trying to prove which method of instruction is superior which has been the greatest stumbling block to focusing attention on what are now seen as important theoretical issues. Theories of learning to read, like theories of language development, are essentially concerned with the nature of the process: that is to say, some operation is performed by the central nervous system on a given input to produce a particular output and the theory attempts to explain the nature of this operation not only for the furtherance of knowledge but also to serve, at least in the case of reading, as a guide for appropriate practical activities permitting maximum efficiency in the accomplishment of the task. Basically, in reading the input is graphic and the output is meaning, which is internalized but which can be externalized as speech or some other action. It may be noted that, with language development, the only apparent difference is that the input is aural instead of graphic—a fact which appeared to influence the early theoretical ideas. The first incursions of linguists into reading stressed the similarity of the comprehension aspects of both listening and reading, while accepting the different ways in which the physical stimuli—sound and sight—impinged on the central nervous system. The process of learning to read, as Fries put it, was simply "the process of transfer from the auditory signs for language signals which the child has already learned to the new visual signs for the same signals" (Fries, 1963). Somewhat earlier, Bloomfield had made the same point (Bloomfield, 1942), insisting that, in beginning reading, learning the alphabetic code was a necessary prerequisite to discovering meaning. In accordance with his theoretical ideas, he required that a start should be made with regularly spelled words so that the general relationship between letters and sounds should first be discovered.

The practical outcome of this early linguistic influence, which stressed starting with phonics, was to add fuel to those opposing the widely accepted whole-word, meaning-emphasis approach (Flesch, 1955; Daniels, & Diack, 1956), and led to an increase in studies comparing the two approaches and, in the course of time, to further debate on the theoretical issues involved. Danks (1980) discussed in some detail the fundamental question of whether the process of comprehension was the same or different for spoken and written language, noting, however, that the strict separation between decoding on the one hand and comprehension on the other may be a "fiction that ultimately cannot be maintained" since the decoding processes for listening and reading being necessarily different may influence the comprehension stages. Danks, stressing the many methodological issues and factors involved, provided no clear answer to the question, but did relate how the answer would influence classroom practices. "If there are no fundamental differences between listening and reading, then a general language experience approach including both listening and reading would be sufficient. If, however, differences between listening and reading appear in one or another comprehension task, then training on that task would be appropriate" (Danks, 1980), whether the task was concerned with vocabulary, syntax, the organization of ideas, or some other aspect.

It is interesting to note that Danks, writing after a decade or so of more theoretically oriented studies in the Chomskyian tradition, does not draw the same conclusion as would Bloomfield and Fries, namely, that if the comprehension of listening and reading were the same, then the practical outcome must be an emphasis on decoding. He did, however, note that educational psychologists as diverse as Thorndike (R.L.) and Goodman espouse the unitary view. The former, analyzing data from several modern empirical studies, produced evidence to support the thesis first advanced by Thorndike (E.L.) (1917) that beyond the first stage of decoding, reading consisted largely of a "thinking" or reasoning component only (Thorndike, 1974). For Goodman, the essence of reading was seeking meaning, and he too maintained that the same process was involved in arriving at meaning whether from an aural or visual input. For him, however, meaning did not exist in the surface structure of language arrived at by precise decoding, but was constructed from past experiences of language. Comprehension was not, therefore, a passive, automatic process as Bloomfield and Fries declared, but was an active process, and learners possessed an innate competence which enabled them to discover the deeper meaning through the use of grammatical and semantic mechanisms (Goodman, 1966, 1970). The influences of Chomsky's theories are clearly seen here and their application to reading has also been expounded by Smith (1971, 1973), who contended that the comprehension of reading, since it depended upon an innate knowledge of "linguistic universals," did not have to be taught but was acquired in the same way as spoken language was acquired. It followed from this that reading instruction per se was unnecessary and Smith, in fact, argued that the function of the teacher was not so much to teach reading as to help the child learn, and reading programs often merely stood in their way— "children learn to read by reading" (Smith, 1973).

The implications of this theoretical approach meant that teaching children *about* reading was quite unnecessary. The essence of reading was a search for the deeper meaning, and learning facts extracted from the surface structure, such as phonic associations, only made the task harder. As Downing (1979) queried, however, even if there was backing for some of the theoretical ideas of Goodman and Smith, did these inevitably give the complete explanation of the reading process and did it follow that what was true for fluent readers necessarily applied to those learning to read? Other writers, notably Eeds-Kniep (1979) and Stott (1981), have attacked the intrusion of psycholinguists, particularly Goodman and Smith, into the field of reading mainly because of their contempt for phonics teaching about which they were "poorly informed" (Stott, 1981).

What is important for this account is the influence of these psycholinguistic theories on classroom teaching. Not merely is phonics disparaged but the approach to reading requires a complete revision. The idea of immersing children in written language much as they are when learning to understand spoken language was put forward in a technique called "assisted reading" by Hoskisson (1975b) and was severely criticized by Groff (1979). This technique does not appear to have been widely employed, but, without a doubt, the psycholinguistic invasion has had some effect. Certainly, in Great Britain, starting directly with phonics has gone out of fashion and meaning emphasis approaches of one kind or another for beginning reading appear to have predominated throughout the 1970s.

A further theoretical advance, stemming from the much earlier work of Vygotsky, was developed in the early 1970s which postulated that listening and reading were not parallel processes. Vygotsky distinguished two stages of knowledge acquisition; in the first, a concept evolved spontaneously and unconsciously, and, in the second, it slowly became conscious. Reading was more difficult to acquire than speaking and listening, since it was only a way of representing speech—a spontaneous activity—and, therefore, involved the second stage (Vygotsky, 1934). Mattingly put this a different way, suggesting that acquiring an understanding of spoken language was a primary linguistic activity involving language-acquisition mechanisms, many of which were not accessible to immediate awareness, while reading was a secondary activity, which, in contrast, did necessitate an awareness of certain aspects of language behavior, which he called "linguistic awareness." He contrasted the facility of oral-language learning and the apparent difficulty of learning to read by regarding reading as a "deliberately acquired language-based skill, dependent upon the speaker's awareness of certain aspects of primary linguistic activity" (Mattingly, 1972).

This notion of linguistic awareness, sometimes called metalinguistic knowledge, was elaborated by other writers. Cazden described it as the "ability to make language forms opaque and attend to them in and for themselves," and she noted that it was "less easily and less universally acquired than the language performance of speaking and listening" (Cazden, 1974). Ehri put forward a slightly different idea and drew the distinction between implicit knowledge and metalinguistic awareness, describing the former as that which "governs the child's ability to process and comprehend speech or print," noting that it "emerges earlier and is quite separate from metalinguistic awareness which entails the ability to focus upon, think about, or make judgments about the structures comprising language" (Ehri, 1978). Mattingly himself later described linguistic awareness, which he considered to be prerequisite for learning to read as "the ability of a speaker-hearer to bring to bear rather deliberately the grammatical and, in particular, the phonological knowledge he does have in the course of reading" (Mattingly, 1979a). The proposition that awareness of the grammatical rules underlying language may indeed be prerequisite for the attainment of reading proficiency was stressed by Ryan. For her, metalinguistic knowledge or linguistic awareness "involves the ability to focus attention upon the form of language in and of itself, rather than merely as the vehicle by which meaning is conveyed," and defined it as part "of the general cognitive ability to utilize knowledge deliberately and consciously" (Ryan, 1980).

These theoretical advances still postulated language-acquisition mechanisms which were actively engaged both in learning to speak and in learning to read.

But they were no longer the same and, while the former was largely unconscious and therefore mentalistic, the latter involved a conscious knowledge of at least some of the rules of language. Mattingly maintained that possession of this metalinguistic knowledge varied considerably from child to child. For some children—the linguistically aware—their language-acquisition mechanisms continued to function beyond the point necessary for simply processing spoken sentences, and, hence, when it came to reading, the phonological segmentation and the morphological structure of words were intuitively obvious so that the orthography seemed reasonable and no direct obstacles prevented their learning to read. But there were others whose language-acquisition mechanisms ceased operating once they had passed the period of learning to talk and they then atrophied, and the principles by which the orthography transcribed words seemed quite mystifying so that learning to read presented problems. In such cases, the task of the teacher was essentially "to rekindle this awareness by getting the language acquisition machinery started again" (Mattingly, 1979b).

The linguistic awareness theory has been taken one step further by Downing who has drawn on the notion of cognitive confusion advanced by Vernon as a prime cause of reading failure and has expounded what he has called the "cognitive clarity theory" of reading (Downing, 1979). Following an extensive study of reading failure, Vernon had concluded that "the fundamental trouble appears to be a failure in development of [the] reasoning process" (Vernon, 1957, p. 48). She described the cognitively confused child as being "hopelessly uncertain and confused as to why certain successions of printed letters should correspond to certain phonetic sounds in words." Downing perceived the necessity of removing this and other confusions, and he summarized his theory in eight postulates which, put very briefly, claimed that children approach the task of reading instruction "in a normal state of cognitive confusion about the purposes and technical features of language" and that "the learning to read process consists in the rediscovery of (a) the functions and (b) the coding rules of the writing system" (Downing, 1979; see also his chapter in this book).

The ideas inherent in the concept of linguistic awareness were in essence still psycholinguistic, although they now contained the need for learners of reading to have conscious views about the nature of the process involved—in direct contrast to the earlier notions of Smith and Goodman. Vernon and Downing, however, were more concerned with the practical aspects of learning to read, and thus Downing's contribution was to reformulate the ideas of linguistic awareness reached by psycholinguistic analysis into a more practically based theory more readily susceptible to disproof.

Existing Practices in Reception Classes

The purpose of this brief account of theoretical developments over the past 20 years or so has been to set the stage for a description of the position adopted in a recently completed study concerned with the development and evaluation of a new reading program. The idea for the study came from a relatively informal survey of just under 120 reception classes in some 50 or so infant schools scattered throughout Great Britain. A judgment sample only was employed which, although it included schools from a broad cross section of socioeconomic backgrounds, was deliberately biased towards those from the lower end of the scale. The visits, mostly carried out over the period 1973-1975, were made by the author who spent some time observing the reading instruction in each class containing children who had just started school, and discussing with the teachers the reasons for the general approach adopted and the specific activities which they employed, dwelling particularly on the theoretical knowledge they possessed of the learning-to-read process. Most of the schools had three intakes each year, in September, January, and after Easter, so that generally one visit was sufficient to cover progress over most of the first year. In about a quarter of the schools, however, two visits were made spread over one year. It should be pointed out that, although the official age of starting school in Britain is five years, a great many schools employed the practice of accepting "rising fives" (children who would only become five some time during the ensuing school year) and thus the age range of the observed pupils on starting school was from four and a half to five and a quarter years.

It is probably no more than a statement of the obvious to say that there appeared to be almost as many variations in teacher practices as there were teachers. What was more relevant, however, was the wide range in the teacher's knowledge and understanding of the processes they were attempting to teach. At the same time, it did seem possible to make certain generalizations. Many of the teachers, for example, seemed appreciative of the fact that prereading activities of one kind or another were necessary before some children should actually be given reading instruction. There was considerable diversity, however, about what prereading activities were important and where these ended and reading itself began. Many teachers used one or more of the many commercial packages available, covering mostly visual and/or aural discrimination, but very few indeed sought specifically to ensure that all their pupils had acquired an adequate degree of understanding of what the nature of the reading process comprised. While many teachers had a fairly clear idea in their minds what their prereading program was, by no means all of these, and very few of those who had no explicitly planned program of prereading work, could give any theoretical justification for the activities they pursued. Certainly, the line between prereading and reading was very blurred and confused (cf. Standish, 1959). Except in a few instances where a child's linguistic handicap appeared so great that taking any steps toward teaching them to read seemed quite pointless, most teachers, whatever else they did, moved on to teaching reading as quickly as possible. For the most part, this occurred during the first term, sometimes after only a week or two of schooling, although some exceptions were noted. Two teachers, for example, in different schools made a deliberate point of not making any formal moves towards teaching reading until the second term, even if, as appeared to be the case, a number of children in the class were more than ready to make a start. In both these instances, however, the prereading activities undertaken included learning phonic associations.

For the most part, formal instruction began with the development of a small sight vocabulary by means of printed notices, flash cards, or simple, amply illustrated books. A fairly common practice was for the teacher, claiming to be using the language experience approach, to write out a short phrase or sentence to be copied under a child's drawing—usually with no previous instruction on how to form the letters. Progression to a specific reading scheme or combination of schemes was made as soon as the teacher felt a child was able to cope; and coping too often meant simply being able to learn the associations between the graphic representations of words and their sound equivalents. Progress invariably was measured by the amount that pupils could remember; that is, they were judged by how far they had moved through the reading scheme, although a great many instances were found where pupils were basically learning words by heart. Phonics, or the associations between specific letters and the sounds they represented, were fairly universally taught as a group exercise and children were encouraged to try "sounding out" words which they met in their reading as soon as possible. Unfortunately, many pupils failed in this task, sometimes because they were unable to remember a particular association, but more often because they had no clear idea of what they were meant to be doing—it was one thing to learn parrot-fashion the association between, say, the symbol "i" and the sound /i/, but quite another to understand the generalized concept that a series of marks on paper represented or signaled a specific set of sounds. And, without this understanding, the whole exercise of "word building" from a row of marks to a combination of sounds that conveyed meaning was both a mystifying and a meaningless task.

The variation among teachers cannot be emphasized enough. There were some who were fully aware of the implications of the task that confronted the child struggling to say what the teacher wanted, while looking at a row of meaningless marks on paper, but it would be true to say that a great many did not. This survey revealed some interesting facts about the classroom practices of teachers of reading and also about the knowledge, or lack of it, that they possessed to justify their activities. It is intended to report the full results in some detail elsewhere, but some of the more important points are listed below with a few appropriate comments:

(1) *Too many assumptions were made about the prior knowledge thought, often erroneously, to be possessed by pupils.* This applied especially to the "technical" language used in teaching reading; words such as "sound," "word," "top," "middle," etc. (Downing, 1970; Downing & Oliver, 1973-1974), but also to many other reading prerequisites such as left-to-right directionality, the orientation of letters, the segmentation of sounds into words, and even the purpose of reading, and the fact that a relationship existed between spoken and written language (Reid, 1966; Downing, 1971-1972; Francis, 1973, 1977).

(2) *The learning load was far too great for many pupils.* This was observed both

in the instructions given by teachers and in published reading materials. An analysis carried out by Malt (1977) demonstrated that, on average, the first book alone of eight popular reading schemes introduced the pupil to 26 words and 20 letters and employed 27 phonemes and 34 graphemes. This abundance of learning appeared quite acceptable for beginners who had been well prepared by earlier language experiences, but it led only to confusion for those who had little awareness of what it was they were meant to be doing.

(3) *There was an almost universal tendency for teachers to be solely concerned with teaching children to read with very little attention paid to ensuring that they learned how to read.* There is a subtle difference implied here that is related to the theoretical aspects considered earlier. Learning *how* to read involves an understanding of the nature of the process or possessing "linguistic awareness." Learning *to* read could simply mean learning the associations between marks on paper and a particular sound or a row of marks and a given meaning. In practical terms, learning *to* read is a short-term activity with the learner ending up possibly permanently on the infamous plateau so well known by teachers of reading.

(4) *The reading ability of pupils was mainly judged by the progress made through a reading scheme.* There were slight variations, of course, especially where some teachers utilized more than one set of reading books; but, almost universally, the criterion of progress was how much individual pupils could read. While there may seem nothing at fault with this procedure and, beyond a certain level of reading ability, it provided a reasonable assessment, it had what appeared to be unacceptable consequences. It meant that many teachers felt under pressure either from parents or colleagues to "get their pupils reading," and this often led to a greater emphasis on associative learning by heart at the expense of the development of any clear understanding of how the reading process really worked. In other words, the pupils were pushed into performing a task which was called "reading" when they still had little idea of the actual nature of the task, including remaining unaware of the connection between what they were doing and understanding and speaking their own language. Diack made the same point when he said "The trial of the teacher's patience comes when she is trying to get the child to understand what the letters are there for. Children will appear to be able to read long before they are actually able to do so in the full sense of the word" (Diack, 1965).

(5) *Few teachers made any systematic check with beginners concerning their existing knowledge of the reading process but were inclined to rely on intuition about when and how to begin actual instruction.* Moreover, with many teachers, a certain ridigity was observed in the methods and techniques employed. For example, a particular teaching approach, such as building up a small sight vocabulary before introducing phonics, was adhered to for nearly all pupils irrespective of the wide differences in previous language experiences, motivation, and learning abilities which existed among them.

(6) *The time spent actually learning was, on average, a small proportion of the time devoted to instruction.* There was great variation here clearly related both to the skill of the teacher concerned and to the linguistic backgrounds of the pupils. Teachers who were unable to organize their reading instruction well and/or had to deal with large classes of pupils from poor or "different" linguistic backgrounds—the very ones who had the most problems in learning to read—invariably produced the poorest performance. In such classes, the average "efficiency" of instruction was often little more than 10 percent. That is to say, of the time the teacher devoted to trying to teach reading, the average pupil was concentrating only for a little over 10 percent of that time on learning some even remotely related aspect of the task. This crude efficiency index increased to around 60 to 70 percent in some middle-class area schools with well-organized teachers, thus emphasising the importance of both motivation and well-organized instruction. In relation to this last point, Southgate, deprecating the "swamp them with books and they will soon learn to read attitude," states "children will fail to learn to read in infant classes unless a good deal of guidance and instruction is undertaken by the teacher. There are some children who would neither be 'motivated' nor 'ready' by the time they were eight or nine or ten, if someone did not do something about it" (Southgate, 1972, p. 39).

(7) *For many pupils the incentive to learn to read was low, leading to frequent lapses of concentration.* This also applied far more to pupils from poor linguistic environments. The difficulty in concentrating seemed to arise mostly from two sources—the fact that for many pupils the learning tasks they met in reading were alien to what their previous backgrounds had either provided or had led them to expect, and also to the fact that in many cases the pupils did not understand what they were being asked to do and hence failed to see how or why they should do it. Although, in general, teachers were very well aware of the importance of their pupils' feelings, learning to read was regarded almost exclusively as a straightforward cognitive task, and sometimes insufficient attention was paid to its affective aspects. The fact that failure could be a *cause* of failure was not always appreciated in the early stages and pupils were often expected to continue, sometimes for weeks and even months on end, with tasks (such as plodding through a reading book) which they did not understand and on which they had little hope of success. There seemed to be an unbroken vicious circle of inability to concentrate, leading to failure which promoted a negative self-image reinforcing the inability to concentrate.

(8) *Initial instruction almost universally involved associative learning.* Whatever deeper meaning might be extracted from a line of print by a mature reader, the initial instruction given by the vast majority of teachers consisted of getting pupils to learn that two things were related. Using a whole-word approach, pupils were expected to associate the graphic display of a row of marks on paper with spoken sounds and thus with meaning, while, with phonics, they were taught to associate a particular mark with a specific lan-

guage sound. This kind of learning could be effected—and very often was—with the learner in complete ignorance of its purpose. Moreover, in the whole-word approach, the associations were between an unknown (the graphic representation) and supposed knowns (the phonological and semantic information presumably stored in the pupil's internal lexicon or memory). In phonics, the association was between a graphic representation and a language sound—both unknown for beginners, since, although they might use a particular phoneme attached to others in speech, they invariably had no concept of it existing as an entity in its own right. Of course, given sufficient time and enough repetition, even the slowest learner would acquire a given set of associations, conveying the impression that progress was being made in mastering the reading process. Inevitably, however, the law of diminishing returns would begin to operate and, unless some additional knowledge or understanding was acquired, pupils tended to slow up and even stop their learning—clearly one of the major factors which held them on the infamous plateau (cf. Diack, 1965).

(9) *Teaching was invariably from sight to sound.* Whether a whole-word or a phonics approach was employed, pupils were presented with a graphic representation and then given the spoken word or sound to be associated with it. At first sight this may seem the natural order of progression since reading is basically a visual activity—at least the input is visual. But knowledge of the spoken language precedes any activity directed at learning to read. Nearly all children, with the possible exception of the congenitally deaf and possibly some non-native speakers who form special cases, learn first to speak their language (however crudely) before learning to read. In other words, meaning is conveyed through the sounds of a language initially, and it would appear to make sense in learning to read as in any other task to move from the known to the unknown, that is, from sound to sight, and not try to operate the other way round (Pidgeon, 1979).

One general point which emerged from this survey was that few of the reception class teachers had very much knowledge of the theories advanced for how children learn to read or understood themselves much about the nature of the learning process involved in the task. The practices they adopted appeared to have been largely based on a combination of what was thought to be appropriate and bits of gleaned knowledge. The same approach was used with each new intake with occasional modifications arising from using new published materials noticed in advertisements or at publishers' exhibitions or recommended by colleagues. None of this is meant to imply that this group of teachers were not fully dedicated to their jobs or did not have the best interests of their pupils at heart.

Nevertheless, despite these generalizations about how teachers decided on their approach, theories about reading had clearly filtered through over time and had exerted an influence. For example, the alphabetic method was rarely seen in any form, and only a small proportion of teachers plunged in at the beginning with direct phonic instruction. In various degrees, it was the ideas of Goodman and Smith that were most in evidence even though few teachers had even heard of their names or appreciated that what they were doing was based largely on their theories. The notion of linguistic awareness, however, was quite unknown, which was not really surprising, since, at the time of the survey, the general implications which arose from it were far from clear.

The main object of the survey had been to review, in general, the relationship between known theory and current teaching practices and, in particular, to attempt to build a theoretical framework from which a new practical program could be constructed. The overall impressions gained from the survey, including the points listed above, were therefore carefully considered, resulting in the emergence of four general principles. These were:

(1) There should be a definite sequence in the learning order of basic concepts such that no pupil should be asked to learn a concept which subsumed knowledge of earlier concepts which were possibly not fully understood. This implied not only the introduction of an overall structure in the teaching of reading but that pupils should always have an understanding or awareness of what they were being asked to learn.
(2) Some form of continuous feedback to the teacher providing information on the progress being made in the learning sequence of individual pupils was essential. The first principle could not be properly employed unless a constant check was carried out on the developing awareness of each individual pupil.
(3) The early learning activities should possess a meaningful structure designed to provide short periods of inherently motivated learning and the avoidance of fragmented imposed instruction which often failed to contribute any significant incentive. To bring this about, the instructional materials must be such as to arouse and sustain interest, thus helping to promote a positive self-image.
(4) Some understanding of the structure of spoken language was an essential prerequisite of learning how to read. This, in the practical context of the classroom at a level that was meaningful for young beginners, appeared the appropriate way to ensure the development of linguistic awareness.

Although derived solely from observations of the task of learning to read, it will be noted that the first three of these principles are all found within the general concept of mastery learning (Carroll, 1963; Bloom, 1968, 1971a; Block 1971). The aim of mastery learning is to get all, or nearly all, pupils to a specified criterion level in a particular learning task (in this case, learning to read and write) by providing the optimum time and conditions required by each individual learner to achieve mastery of that task. In practice, this means developing a structured learning sequence—usually a sequential set of objectives to be achieved—and providing appropriate learning activities for each objective together with feedback information indicating whether or not the essential principles have been grasped. The practical application of the model also takes full account of the affective side of learning (Bloom, 1971b, 1976). The important aspect of mastery learning is, of course, the structured learning sequence, and it was to determine what this should contain and how it should be sequenced that resort was made to the theoretical issues discussed earlier and to the fourth principle listed above. The general position adopted considered the acquisition of a task such as reading to consist essentially of three practical stages. These were: (1) developing an awareness of the nature of the reading process; (2) mastering the essential features of reading and writing; and (3) developing the techniques of fluency so that performance becomes automatic. Although arrived at independently, these stages are not dissimilar to those advanced by Fitts and Posner (1967) relating to all human skill learning. They postulated first the *cognitive* phase in which the learner becomes familar with the relevant features and nature of the task; second, the *mastering* phase in which the skill is practised until mastered; and third, the *automaticity* phase where further practice is undertaken until the skill is performed without conscious effort.

Unfortunately, the impression gained from the survey was that, with reading, little conscious effort appeared to have been made by teachers to distinguish among the three stages stated above, with the result that not only were many children plunged straight into the practical task of basic learning (stage 2) before they had even the faintest idea of what reading was all about (Reid, 1966; Downing, 1970), but, all too frequently, they were expected to acquire the fluency techniques of stage 3 (e.g., developing a sight vocabulary) when they should have been concentrating on the more important stage 1 aspects of discovering the "how" and "what" of reading. In a sense, such children were placed in the position of trying to decipher a code when they not only had no idea of the key, but they did not even know that it was a code or even what a code was. It should be made clear that these comments do not apply to all children. Those from linguistic backgrounds who are surrounded by books and are read to constantly from a very early age do, as Mattingly puts it, become linguistically aware or, as he states, come "to know the phonology of the language so that the morphophonemic representations of words in [their] personal lexicon(s) match the transcriptions of the orthography" (Mattingly, 1979a, and in Chapter 2 of this book).

Relating Theory and Practice

In developing the theoretical basis for the study to be carried out, a start was made from the proposition that, in learning how to read, confusion might occur if a pupil was asked to learn a given task which itself assumed the possession of some prior knowledge not yet in fact acquired, and a series of sequential steps were developed, using an approach similar to that put forward by Gagné, namely, to work backwards from the end product of the learning and, at each point, asking "What does the learner need to know immediately prior to learning this?" (Gagné, 1965). The major steps have been outlined in an earlier paper (Pidgeon, 1976), but an example might help to illustrate the general proposition. As the survey demonstrated, the vast majority of reception class teachers tended to introduce reading through some variant of the whole-word approach, generally building up an initial sight word vocabulary. Now, while it is certainly true that an association between the graphic form of a word like "dog" and the concept of a dog (or of a cat if the teacher, playing the devil instead of God, were so to misinform the pupil) can be formed by a child, it is clear that, for such an exercise to provide any insight into the reading process, it must be assumed that the child (1) knows the meaning of the spoken word concerned (and the meaning of other "technical" words which the teacher might use in illiciting the required response); (2) has developed, using Piaget's term, the symbolic function, or, in other words, knows that a series of marks on paper are just that—a series of marks on paper—and at the same time can also indicate other different things, e.g., the sounds of a word, and the representation of the concept it implies; (3) is aware of the conventions used when extracting information from the visual presentation of words, e.g., left-to-right eye movements, orientation with respect to letters, and the segmentation of words and letters; (4) has some knowledge of the sound structure of the language, that is, has some awareness that language consists of basic sound units and how spoken words are constructed from them; and (5) knows the particular relationships between the specific graphemes and phonemes in a given word.

Taking the practical view of the reception class teacher, it would seem that the initial building up of a sight word vocabulary might not be as useful for all children as is generally supposed—a view in line with the analysis made by Groff that it is a "wasteful and ineffectual practice" (Groff, 1974, p. 577) and supported by Chall's review of research evidence comparing code emphasis and meaning emphasis approaches to beginning reading (Chall, 1967). Of course, teachers introduce phonic analysis sooner or later, but what may be questioned is whether it is too late for some children or, indeed, whether the right approach is being used at all. Figure 9-1 is an attempt to represent diagrammatically the usual approaches by which a child is expected to extract meaning from the printed word. Route 1, on the left, illustrates the whole-word approach; the child is presented with a printed word and, in some way, is informed of its spoken form with which meaning is associated. With repetition and practice, the direct association is committed to memory so that eventually the meaning comes directly from the visual presentation and becomes available through route 3 on the far right. Associative learning of this kind could, of course, be applied to a purely logographic script and takes no advantage of the phonemic properties of one that is alphabetic or syllabic. The argument adopted here is that, without employing such phonological knowledge as is available in an orthography, heavy restrictions must necessarily be placed on the learner. Psycholinguistic research into the question of whether any phonetic recoding is necessary in order to obtain access to meaning when reading still shows considerable diversity of views (Barron, 1978; Feldmann, 1978) and appears to have little relevance for the beginner in the classroom. In practical terms, however, utilizing the phonetic properties of an alphabetic script involves (1) recognition of the component parts of a printed word, (2) learning the specific grapheme/sound relationships,

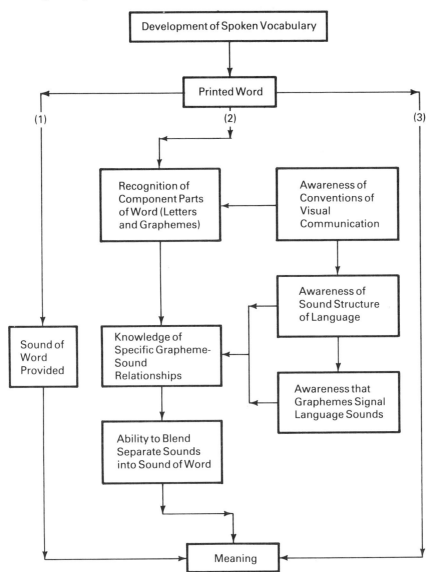

Figure 9-1. Routes to reading comprehension by usual look-say and phonics teaching methods.

and (3) combining the separate sounds into the total sound of the word with which meaning is associated. This phonic analysis is illustrated in route 2 of Figure 9-1, and the problem for many teachers is that graphemes can be associated with sounds by rote learning, that is, without any appreciation of the precise relationship between spoken and written forms of the language, which to be understood completely involves conceptual learning. Even the first step of recognizing words and their component parts, which has usually been regarded as a perceptual task, has been demonstrated in a study by Downing, Ayers, and Schaefer (1978) to be primarily conceptual. The view taken here is that to be able to employ phonemic principles with understanding requires the kind of prior awareness detailed in the boxes to the right of Figure 9-1; without this, there will be a tendency for pupils to revert mostly to the simpler technique of relying on building up a sight vocabulary alone, thus removing the possibility of their ever being able to master the process of fluent reading.

It should be made clear, of course, that by no means all children experience the difficulties outlined here. Many develop an awareness of the basic constituents of the reading process before they even get to school, largely, it is suggested, because they will have been read to extensively in their early preschool years and much attention will have been paid to their language development. This is one explanation of Mattingly's statement that only some children become linguistically aware by continuing to learn the "grammar" of their language (i.e., the rules which explain its nature) to the point which enables them to learn how to read. It is argued that far too many children who fail to achieve any fluency in reading do so because they lack the kind of awareness through which they can acquire a resonable grasp of how language sounds are structured to form words and how this structure can be represented on paper. They are forced, therefore, to depend upon a minimum sight vocabulary and a few inadequate and not understood phonic associations. Hence, they remain on the infamous plateau where reading becomes a chore and the expectation of failure itself helps prevent any further progress. Gleitman and Rozin discuss this situation and state that, "Despite phonics instruction, some children seem to have acquired merely a sight vocabulary of some whole words after tedious years of schooling . . . . But this is exactly what we would expect as the outcome of learning a logographic system; there is a slow accretion of items and, in the absence of over-riding motivation, a diminishing return as the number of items increases" (Gleitman & Rozin, 1973, p. 478).

It follows from what has been said so far that the first stage in teaching beginners to read must be to take them through a series of steps designed to ensure that they possess an awareness of what the process consists in *before* any attempt is made to introduce the elements of reading per se—that is, before word/meaning or grapheme/sound associations are presented for learning. An attempt to illustrate the alternative pathways which this entails for the beginner is given in Figure 9-2. Special emphasis is placed on acquiring an awareness of language sounds and how words are constructed from them. It would seem clear that, for many children, how their spoken language works is as much a mystery as how reading works. Downing has pointed out that most children learn to become skilled in speaking without knowing how they do it, but he adds that, in developing literacy skills, a child "has to become aware of his own language behavior if he is to understand how written language operates" (Downing, 1977). The same point is made by Elkonin, who, after discussing the part played by sound in the reading process, concludes "Hence, it follows inevitably that the first problem to be

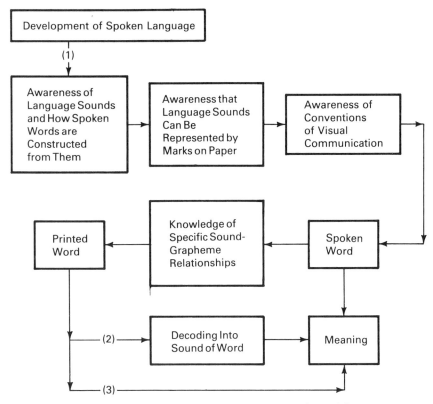

Figure 9-2. Route to reading comprehension through planned language awareness.

solved before teaching literacy is to reveal to the child the sound structure of spoken words; not only the basic sound units of language, but also how language in constructed from them" (Elkonin, 1973a, p. 553). In Figure 2, the equivalent of route 1 in Figure 9-1—the initial acquisition of a sight vocabulary—is missing entirely and is replaced by row 1, comprising a series of steps through which children gain an awareness of what reading is all about. Row, 2, which follows only after this linguistic awareness has been successfully achieved, reads from right to left and involves the learning of specific sound/grapheme relationships, always presented, however, in the context of meaningful words. Route 2 in row 3 illustrates how unknown words are subsequently decoded and is equivalent to route 2 in Figure 9-1, except that now the learner has a clear awareness of how the process works before the exercise is attempted. Route 3, of course, provides direct access to meaning from the graphic presentation and is now clearly seen as a fluency technique. In fact, for mature readers, as for beginners after a short period of practice, routes 2 and 3 operate together. Reference can now be made to the three stages of task acquisition as applied to learning to read, and which form the basis of the new program mentioned earlier. Stage 1 is designed to ensure an awareness of what the reading process consists, allowing the pupil to gain an insight into how it works. Stage 2 is concerned with mastering the essential features of decoding and encoding, that is, learning the basic require- ments of reading and writing; and stage 3, which overlaps with stage 2, deals with the development of the techniques of fluency.

Great importance is attached in the program to stage 1, since, regretfully, as has been pointed out, for many children, in the rush to get them reading, part or all of it is missed out (point 4 from the preliminary survey). This does not matter a great deal for those whose language background has provided them with the necessary awareness, but it is vital for those who are plunged into stage 2 with very little, if any, idea of what they are meant to be doing. Stage 1 has four steps which arise logically from a consideration of both the theoretical and practical points discussed earlier and indicates what pupils need to know *before* learning to read, that is, before attempting direct associative learning. They need (1) to possess the ability to understand *and respond* to spoken language, which includes using minimal motor skills (e.g., making meaningful marks—*not* letters— on paper), following simple oral instructions and acquiring a basic oral vocabu- lary including some of the technical language of reading instruction; (2) to be able to appreciate the significance of visual communication particularly in rela- tion to order, which includes knowledge of the symbolic function and the left- to-right and top-to-bottom conventions; (3) to have a general awareness of the sound structure and segmentation of spoken language, that is, to be aware that there are basic units of sound in the language and to understand how these are combined to form spoken words with which meaning is associated; and (4) to know that speech sounds can be represented by marks or symbols on paper and that groups of such symbols signal both the sounds and meaning of words.

An understanding of these steps of stage 1, particularly the last two, means that reading instruction can start with a known spoken word and then proceed to the learning of the graphemes which represent its constituent sounds on paper—learning that is now meaningful and not just associative. Of course, for the process to appear logical and not confusing for the learner, the many appar- ently illogicalities of English spelling must initially be avoided or circumvented by the use of a few diacritical marks or a regular initial teaching medium (such as i.t.a.). Such means were employed in the learning materials for the reading pro- gram developed from the work described here.

Stage 2 consists of instruction in both reading and writing, except that the procedure adopted departs from the practices normally employed, since (1) the general approach is from sound to sight and not, as is more usual, from sight to sound; (2) the rate of introduction of new ideas and concepts can be controlled by the teacher to suit the working pace of each individual, and (3) the awareness of how the reading process works, developed in stage 1, permits certain strategies to be followed which would otherwise lead only to cognitive confusion. A num- ber of other novel features are also included within the whole program arising from the fact that it is based on the principles of mastery learning.

Little needs to be said about stage 3, since the techniques involved are introduced as early as possible in stage 2, with practice provided at every opportunity. One of the first techniques developed is the building up of a sight vocabulary, and a second is the growing use of both syntactical and semantic information obtained from what is being read, which permits meaning to be gleaned from context without the apparent intervention of sound. While the importance of stage 3 cannot be overstressed, its development does not form a major role in the present program. Rather, it is stage 1 and its implications for stage 2 which are given particular emphasis.

For the classroom teacher, the implications of the latest changes in the underlying theory of how children learn to read are really quite drastic, although at first this may not be apparent. Early theories suggested that, since the same passive mechanisms operated for comprehending both spoken and written language, all that was necessary was to turn the visual code into an auditory one, and the emphasis was placed squarely on beginning with phonics. Later, however, linguistic influences suggested that this was far too simple an explanation. Meaning, which was the important issue, lay not in the surface structure of the spoken or written word, but existed in an underlying deep structure and, as in learning spoken language, children possessed an innate competence for extracting the meaning from print. It followed from this that children could learn to read by being immersed in written language as they are in speech, although care must be taken that instruction did not get in the way. In practice, few reading teachers went as far as this, but it meant that the stress was largely centered on meaning emphasis approaches stressing whole-word or whole-sentence learning. The mechanics of reading to determine the surface structure, although not entirely neglected, were, however, relegated to second place.

A subsequent theoretical advance, however, suggested that learning a spoken language was a primary activity dependent upon implicit knowledge, while learning to read was secondary and required as a prerequisite a metalinguistic and conscious knowledge of how language was structured. If this knowledge was missing, instruction in reading could be meaningless, leaving the learner in a state of cognitive confusion. In practice, this change implied that, before the task of teaching reading per se was undertaken, a check was needed on the extent to which beginners were aware of the nature of the reading process, and, if deficiencies were noted, particularly in the conscious awareness of language structure over and above that required for merely interpreting and using spoken words, then action was needed to remedy these.

An informal survey of reception classes in infant schools in Britain revealed that teachers were only indirectly influenced by theoretical work, but the classroom reading activities observed suggested the need for an approach with young beginners that accorded very closely with current theoretical thinking. The model, produced as a result of the survey, was used to develop, with an appropriate set of learning materials, a reading program for young children. A subsequent evaluation of this program produced supportive evidence to accord with predictions and, thus, did not suggest any disproof of the theoretical stance adopted (Pidgeon, 1981).

CHAPTER 10

Learning to Read and Skilled Reading: Multiple Systems Interacting Within and Between the Reader and the Text

Harry Singer

Introduction

The substrata factor theory of reading explains that reading consists of multiple systems, including perceptual, linguistic, cognitive, motivational, affective, and physiological, that interact within the reader and between the reader and the text. An obvious implication is that any theory of reading that is based on only one of these systems must be inadequate. In this chapter it will be shown that this implication applies to Mattingly's chapter's linguistically based speculations on the process of reading and learning to read. Further, it will be shown that, in contrast, more adequate explanations have come from other linguists (Gleitman & Rozin, 1977) and from some psychologists (Gough, 1976; Adams & Collins, 1977). Furthermore, it will be shown how multiple systems within the reader must be drawn upon in learning to read by providing a brief description of instructional processes and strategies in teaching reading. Finally, two of Mattingly's claims will be tested. First, although phonological processes may be bypassed when deaf subjects learn to read, hearing subjects only seem to bypass them as they develop automaticity in reading. Second, there seems to be no evidence to support the hypothesis that there is a critical age for learning to read. Our conclusion is that, although teachers may draw upon linguistic processes in teaching reading, to date there is no valid evidence that linguistic awareness is necessary for learning to read.

Reading is a multidimensional process consisting of three major systems: perceptual, linguistic, and cognitive. These systems are undergirded by physical and physiological substrata and directed by the reader's purposes, values, attitudes, and perspectives (Holmes & Singer, 1964; Anderson, Spiro & Montague, 1977; Singer, 1983). Readers mobilize or allocate their attention consciously or unconsciously to utilization of these systems according to their momentary purposes and the demands of the task (Singer, 1976b). As they do so, they show an amazing ability to integrate these systems and their supporting and directing processes in remarkably flexible ways into a smoothly flowing performance.

But when we try to understand how readers interact with texts to perform this most remarkable species-specific feat, we tend to fractionate it into its constituent systems and subsystems. We then run the risk of overemphasizing one or another of these systems or subsystems. If this overemphasis remained in the realms of theory and research, we would not be concerned. But when a popular theory or a research finding, whether in the field of perception, linguistics, cognition, or even sociology, the major fields which study reading, has implications for teaching reading, it often gets translated uncritically into educational practice (Athey, 1976).

However, an implication from a particular field when applied to practice can have beneficial or detrimental effects. Sometimes it can have both. For example, during the 1960s, linguists explained that grammar, particularly syntactic competence, played a powerful role in the performance of reading. They adduced evidence to show that readers, even beginning readers, use their syntactic and semantic competence to anticipate or select words or word-meaning choices that are appropriate to the context. They also taught teachers to differentiate between competence (what an individual knows) and performance (what an individual does). Accordingly, teachers began to differentially diagnose responses in reading. For example, substitutions and repetitions in oral reading were no longer necessarily "errors" in reading, as many oral-reading tests had indicated, but they became meaningful responses or attempts of the reader to get or clarify meaning. These contributions were beneficial. But, unfortunately, teachers were also led to believe that dialect constituted a "barrier" to comprehension (Goodman, 1965). After some four years of trying to remove this barrier, teachers were told that it was not a barrier after all (Goodman & Buck, 1973). A review of research on the relationship between dialect and reading indicated that at no time had this barrier hypothesis been supported by empirical evidence (Singer & Lucas, 1975). Perhaps as a consequence of emphasis on dialect as a barrier to comprehension, researchers began to find that some teachers were subjectively misclassifying children, assigning more dialectally different children to remedial instruction than would have been referred had their test scores been used as the basis for referral. We should learn from these experiences that it is necessary to validate theoretical and research implications before putting them into practice in the classroom. For these reasons our reaction to Mattingly's speculative chapter will be to first briefly summarize his explanation of reading and then point out some of its shortcomings.

Summary of Mattingly's Position

Mattingly categorizes reading into two modes: analytic and impressionistic. In analytic reading, readers understand a written sentence when they identify its written words as corresponding to specific items in their lexicons and make a grammatical analysis of the sentence. (Of course, Mattingly is cognizant that understanding a written sentence is not just the sum of its corresponding lexical entries, but instead, one part of a sentence affects the interpretation of another part of the sentence.) In the second mode, readers guess the meaning of the text by getting semantic associations from familiar orthographic patterns (text-based data) and use prior knowledge (reader-based resources) to get the gist of the text. Thus, meaning is a function of the interaction between text-based data and reader-based resources.

Mattingly then draws on two types of evidence: (1) A linguistic analysis which shows that lexical items can be transcribed morphemically. But in an alphabetic system they are transcribed morphophonemically in order to represent morphemes consistently. (2) An information-processing analysis which indicates that a preliminary representation of information is phonetically recoded and stored in short-term memory where it can affect interpretation of later parts of the sentence. Then, when the reader has achieved a semantic representation of a sentence, it is stored in long-term memory.

Critique of Mattingly's Position

Although Mattingly may not have intended his information-processing analysis to be comprehensive, we nevertheless have to recognize that it is an oversimplified version. Gough's (1976) linear model of what may occur in one second of processing a sentence provides a more detailed and more comprehensive analysis. But even Gough's model does not adequately represent the complexities of the reading process. A more complex model depicts the process of reading as involving an interaction among perceptual, linguistic, cognitive, and executive processes within the reader and between the reader and the text (Adams & Collins, 1977).

An additional limitation of Mattingly's position is that his analysis and his analogies to speaking and listening are all at the sentence level. But comprehending a passage is not simply a summation of understanding the individual sentences in a passage. Current theories of reading comprehension go beyond the sentence level to include chunking of information across sentences (Bransford & Franks, 1971); use of story grammar for assimilating stories (Rumelhart, 1977b); mobilization of schemata for constructive and reconstructive processes, and for storage of information in long-term memory and retrieval (Anderson et al., 1977); and the operation of inference and interpretative processes on propositions stored in long-term memory (Kintsch & van Dijk, 1978; Lee, 1979).

Mattingly's analysis of how an individual acquires the ability to read is similar to transformation-generative grammar's exlanation for language acquisition: Information carried by the orthography is somehow able to trigger the "innate, highly specialized and inaccessible" mechanisms used in generating a spoken sentence. Yet Mattingly recognizes that in reading "the boundary between grammatical knowledge and performance is crossed." Performance includes extralinguistic factors, such as purposes and goals; perception, which involves formation of feature detectors and unitizing mechanisms for print; and other component processes which also have to be learned, such as rules for identifying words and for using syntactic and semantic constraints on lexical choices, and integration of these processes with already developed language abilities and prior knowledge (Tzeng & Singer, 1981).

However, all of these processes cannot be taught at once. A teaching strategy must be adopted that reduces the initial instructional tasks to a level where the beginner can be successful and can progress a step at a time towards the more complex levels that characterize a skilled reader. The skilled reader appears to have bypassed phonological processing and encoding, and may operate at a level that has been called "lexical reading" (Chomsky, 1970), but skilled reading may be merely so efficient in these intermediary processes that it only appears to be "lexical reading." To understand how this teaching strategy can be realized, we will briefly review what can occur in beginning reading instruction.

Beginning Reading Instruction

The age for beginning reading instruction varies within a close range around the world. For example, reading instruction begins at age five in Scotland, six in the United States, and seven in Sweden (Downing, 1973). At the age range of five to seven, children universally have acquired sophisticated control over the syntax of their language, a vocabulary of about 5000 words, which is about the size of vocabulary used in everyday oral language, and a phonological system that is adequate for generating word pronunciations and appropriate intonations. Hence, beginning readers can orally express their needs and interests to peer group members and to adults.

However, in order to read, the primary-grade student still needs to acquire (1) word identification processes or rules for relating print to phonological processes, or as Mattingly would prefer, to the lexicon and (2) integration of his or her language abilities and prior knowledge with rules for identifying printed words. In learning to identify printed words, the student must learn to discriminate the distinctive features of letters, learn graphomorphophonemic correspondences, and acquire ability to blend the correspondences into a unified whole-word response (Gibson, 1965; Samuels, 1976b). Since letters vary, the beginning reader also has to learn to generalize the same phonological response to perceptually different letters, such as upper- and lower-case forms, and to give different phonological responses to the same letter when it is in different environments, such as /k/ in *cat, cot*, and *cup*, but /s/ in *city, celery*, and *cycle* (Singer, 1976a). In short, the beginning reader will have to learn to use the principles of conjunctive, disjunctive, and relational concepts in acquiring responses to letters. For example, a vertical line with another line across it conjointly determine the letter *t* which is pronounced /t/. However, if it is in an environment followed by a *ch*, it is assimilated to the sound /č/, and if it precedes *ion*, it is pronounced /š/. Moreover, some words have morpheme boundaries which have to be observed prior to use of graphophonemic responses, such as *hatcheck girl*. Then some words have vowel shifts that are not realized in spelling, as in *nation*

and *nationality*, but are in pronunciation. To further complicate reading acquisition, some letters have become silent over the last 500 years, such as *k* in
knight, and some letters that are silent in one form of the word are pronounceable in another form, as *sign* and *signal*. Also some words have been borrowed
but have kept their spelling intact, such as *beau*. Thus, learning to read draws
heavily upon perceptual and cognitive processes, as well as syntactic, morphemic, and phonological processes. Also, beginning readers must acquire and
use knowledge about English orthography.

   In addition, the beginning reader must learn the metalanguage of reading
instruction, which includes the names of letters, characteristics of prose passages,
such as paragraphs, sentences, words, and punctuation, plus other linguistic
features of text, such as the direction of print (Downing & Oliver, 1973-1974).
Since some properties of oral language, such as intonation and pauses in word
and phrase boundaries, are not present in print, the beginning reader will also
have to learn to infer appropriate pronunciations from constructed meanings and
from punctuation cues, such as the exclamation mark.

   Fortunately, the beginning reader in almost all reading programs learns to
read by using materials that drastically reduce the complexity of the task in the
initial stages of reading. Indeed, most beginning students in the United States use
basal readers which draw upon a vocabulary that is usually familiar to children.
The basal readers contain stories in which the characters engage in activities that
are almost always within the realm of children's common cultural experience
and knowledge. Hence, beginning readers are likely to have the necessary semantic hierarchies and knowledge structures for comprehending text (Anderson et
al., 1977). They are also likely to be familiar with the roles, plans, sequences of
events, and goals that occur in beginning reading materials (Schank & Abelson,
1977). Moreover, they can use already acquired knowledge of simple story grammar to assimilate most basal reader stories (Mandler & Johnson, 1977; Rumelhart, 1977b).

   To ensure a necessary background for reading, teachers engage children in
such prereading activities as discussion of characters and events in the story and
have them relate the discussion to their own experiences. Teachers also preteach
"new words" in isolation and in context, that is, they teach the words that their
students previously have not had to identify in print. As children gain skill in
word identification, the teacher provides fewer cues and requires students more
and more to use their own cues to word identification; thus, the teacher develops independence in word identification (Singer, 1968). When students do read
stories under teacher guidance, they are learning to transfer or apply their newly
learned word identifications and their language abilities and knowledge in
reading-connected discourse. Hence, it is not surprising to find that, as early as
the end of the first grade, some children have already learned to use their syntactic, semantic, graphophonemic, and graphomorphophonemic abilities in
identifying printed words (Weber, 1970b).

   Basal readers also exercise frequency control over words, that is, they introduce new words gradually and then have them occur repeatedly and cumulatively

in successive stories. Consequently, beginning readers can capitalize upon previously acquired rules for word identification and upon a growing repertoire of words that they gradually learn to recognize automatically (LaBerge & Samuels, 1974). Thus, with only a few new words per story for learning additional symbol-sound rules, with other words that can be identified from previously learned rules, and with a growing repertoire of automatically recognizable words, students can read increasingly longer and more complicated stories. These instructional characteristics of basal readers, which have now been in use in the United States ever since the 1850s when McGuffey introduced his eclectic readers, represent the result of successive revisions. Use of this cumulatively improved teaching tool, with its highly informative teacher's manual, explains, in part, why reading instruction in the United States has been so successful.

But basals are not the sole reading materials in the school. Most classrooms also have a small library of children's literature and expository materials. Furthermore, many elementary schools even have their own central libraries. Hence, children can apply their developing skill to reading fiction and nonfiction of their choice and within their levels of reading ability.

However, adequacy of materials alone is not enough. We know that all children in the normal range of intelligence can learn to read (Singer, 1977a); but to do so, they have to have adequate time to learn and appropriate conditions of instruction (Bloom, 1971a; Carroll, 1963), such as properly paced introduction of new words and skills (Barr, 1973-1974). If students have teachers and school programs that meet these conditions, they are likely to master the acquisition phase of reading (Singer, 1977a, 1978). Of course, they will achieve this instructional goal at varying periods of time; some will achieve mastery as early as second or third grade, but others may not do so until sixth or seventh grade. At that time, individual differences in the initial phase of reading will have decreased. Concomitantly, individual differences in the second, but overlapping phase of reading instruction, the ability to gain information or learn from text, will have increased and will become increasingly wider as a group of students progresses through school (Singer, 1977a). Hence, teachers throughout the grades must learn to teach students whose general reading achievement varies greatly (Singer & Donlan, 1980).

If conditions of schooling are not adequate, then some children will begin to suffer from a cumulative gap between the new words or subskills that have been introduced and those that they have learned, as shown in Figure 10-1. Hence, they gradually become disabled readers. Consequently, when reading specialists try to determine why students are having difficulties in learning to read, they should not only examine *students* to determine whether they have "learning disabilities," but also *teachers* to discover whether they have "teaching disabilities," and *schools* to find out whether administration and organization of the reading program facilitates or interferes with conditions for a consistent, cumulative, and coherent instructional program (Singer & Beasley, 1970; Singer, 1977b; Singer, 1982).

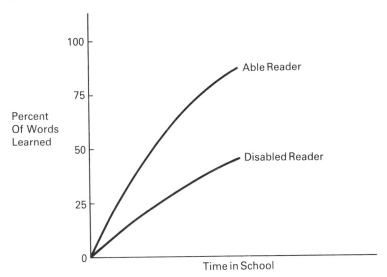

Figure 10-1. Hypothetical curves to show cumulative gap between able and disabled readers in words learned over time in school.

With this brief overview of conditions for learning to read and for teaching reading, we have a basis for examining some of the issues in beginning reading instruction raised by Mattingly's chapter.

Linguistic Awareness

Since first graders can adequately pronounce some 5000 words in their oral language and since these are the words they will have in their initial reading instruction, we know that their lexical knowledge is adequate for learning to read. But Mattingly (1979a,b, pp. 1-2) speculates that beginning readers also have to have "linguistic awareness," which he defines as "specially cultivated metalinguistic access to certain aspects of the grammatic structures of sentences." Mattingly also asserts that the linguistically aware child is intuitively cognizant of phonological segmentation and the morphological structure of words, despite discrepencies between orthographic transcriptions and the child's morphophonemic forms. In other words, the linguistically aware child can recognize the difference in sounds between words pairs as "sip" and "slip" and can distinguish the three phonemes in such a word as "bag." Furthermore, the correlation between phonological segmentation ability and reading achievement in beginning readers is high (Liberman et al., 1974). But correlation is not causation. A critical review of the evidence does not permit any conclusion to be drawn yet as to whether linguistic awareness is causally related to reading acquisition (Ehri, 1979, Singer, 1981).

Although Mattingly draws solely upon linguistics to explain the process of reading, he argues against the view that the reader converts orthography (or letter patterns) to sounds but agrees that *somehow* the lexical character of orthographies gives the reader word identities, which elicit the phonological, syntactic, and semantic information in the reader's lexical entries. Although Mattingly's explanation may apply to skilled readers (Banks, Oka & Shugarman, 1981), he may be overgeneralizing in attributing to beginning readers a direct pathway from print to lexicon.

At least we have no evidence that beginning readers of an alphabetic orthography bypass phonological processing in the pathway from print to lexicon. We do know that beginning readers apparently tend not to perceive whole words because in their eye-movement behavior during reading they have an average of two fixations per word, which contrasts with skilled readers' single fixation per one and one-quarter words (Buswell, 1922).

If the author of this chapter is allowed to speculate, he would hypothesize that current phonic methods teach children to learn implicit rules for relating print to their phonological processes, which, in turn, give them access to the semantic and syntactic properties of lexical items stored in long-term memory (Reed, 1965). However, as the children become skilled readers, they may achieve automaticity in processing highly familiar words (LaBerge & Samuels, 1974) by bypassing phonological processes, responding to printed words as though they were ideographs, and processing them by going directly from print to their lexicon. Then they might still phonetically recode the words for storage in long-term memory, just as Chinese readers do in reading their characters (Tzeng & Hung, 1981).

Mattingly does not attempt to explain how to teach children to identify printed words. Consequently, we cannot evaluate whether and how his concept of linguistic awareness is applicable to reading acquisition. But Gleitman and Rozin (1977) do provide a plausible explanation. Their explanation involves not only linguistic but also perceptual and cognitive processes in interaction with linguistic processes and abilities. In short, their explanation is consistent with the hypothesis that multiple systems are involved in reading acquisition (Katz, 1980; Katz & Singer, 1982, 1983). They postulate that to learn to read, students have to have access to their phonological processes, which are usually at an unconscious level of operation, *but can be made conscious through instruction*. They then explain that phones, the perceived surface sounds of language, cannot be segmented physically because the acoustic properties or formants of any letter in a syllable are spread across the entire syllable. For example, an audio tape cannot be snipped apart to separate the three phonemes in the word "bag." But phonemes, which are the distinctive sounds in a language, can be *abstracted cognitively* and, through this cognitive process, can be perceived as separate sounds. However, this process of abstraction requires information from both phonetic and morphemic levels. That is, there is not a one-to-one correspondence between letters and systematic phonemes. For example, such homophonic words as *know* and *no*, *pain* and *pane*, and *in* and *inn* are orthographic representations which are partly phonographic and partly morphemic. Consequently, alphabetic letters and letter sequences are related to language in a complex way. Gleitman and Rozin conclude that "the relations between cognitive-perceptual alphabetic units (phonemes and morphophonemes) and perceived sounds of speech (syllables and phones) are very abstract; [they are] mediated through a set of covert rules which are essentially closed to consciousness" (p. 50). Moreover, the child's "difficulty in understanding the phonological basis of alphabetic orthography (the child's insufficient access to the segmental nature of his or another's speech) is the major cognitive barrier to initial progress in reading." But this barrier can apparently be removed through instruction. In short, what the child can learn to do is "cognitively separate the sequential phone from its embedding in real speech." Gleitman and Rozin do not say so, but the child has to have attained certain cognitive capacities in order to abstract phonemes. These capacities include decentration and class inclusiveness that the Piagetians categorize as part of the concrete stage in the development of logical thinking (Elkind, Larson & Van Doorminck, 1965).

Following their explanation of word-identification processes, Gleitman and Rozin point out that a writing system, such as Chinese logographs, which tracks meaning is cognitively easier but puts a greater burden on memory because of the larger number of logographs that have to be acquired, while an alphabetic script is cognitively more difficult because it is more abstract but places a lower load on memory and gives the reader access to new words.

Then, to help children develop linguistic access to the segmental nature of speech and to facilitate acquisition of reading ability, Rozin and Gleitman (1977) proposed that beginning readers follow a sequence that would repeat the evolution of written language and would progress from the cognitively easy to the cognitively difficult. Using this rationale, they had children learn to read by following a curriculum which started with semasiography (picture writing), went on to logographs, next to a rebus (a transitional stage from meaning to sound-based units), and finally to syllables and the alphabet (abstract phonemic units). After trying this curriculum out in Philadelphia for one year, they found that their students had not achieved any more than a matched control group on the California Achievement subtests of word recognition, letter recognition, and initial and final phonemes. Nevertheless, Rozin's and Gleitman's analysis that the syllable is a basic unit from which readers can perceive and abstract phones and relate them to alphabetic characters is very persuasive. However, the syllabic unit may work for the Japanese language because it has relatively few syllables, some 40 to 60 (Henderson, 1982), but the syllable is not an efficient unit for the English language which has over 1000 syllables. Learning each of these syllables would place a cognitive burden upon English-speaking students that would be like the one that Chinese students have in learning logographs, but the English-speakers would not have the compensating payoff in meaning that the Chinese get. Indeed, Chinese characters appear to be optimally related to the Chinese oral language, which virtually entirely consists of monosyllabic morphemes (Wang, 1981).

Rozin and Gleitman explain that they did not observe any significant gain in their experimental group because the method used for teaching their control group was very similar to their experimental treatment. Indeed, the Rozin and Gleitman steps for teaching children to abstract and form graphophonemic and graphomorphemic correspondences are quite like the prescriptions given in textbooks for teaching teachers how to teach reading. (For example, see Heilman, 1977; Ruddell, 1974.)

Reading Without Phonological Processes

Is it necessary for students to learn to read by relating print to phonological processes? Or can individuals bypass their phonological processes and learn to read through visual perception and abstraction alone? Research on reading acquisition in deaf children leads us to pose this question and provides some evidence for answering it. For example, Gibson, Shurcliff, and Yonas (1970) found that congenitally deaf children were also able to learn and transfer to novel words what were assumed to be graphophonemic relationships. At least, the deaf children performed like the hearing children who were taught to form graphophonemic relationships and were then tested on another set of novel words. The deaf children could have been successful at the task by visually apprehending internal orthographic patterns and utilizing knowledge of transitional letter probabilities, such as a T can be followed by an H or an A but never by a B or a P within the same morpheme boundary. (This qualification excludes such words as *hatband* and *hatpin*.) Even subjects with normal hearing can learn orthographic sequences that have no phonological basis (Reber, 1967). Conrad and Rush (1965) found that most deaf subjects do use a visual code. They asked both hearing and deaf subjects to remember and recall a sequence of five consonants drawn from a set of nine (B, F, K, P, R, T, V, S, and Y). The intrusion errors of the hearing children were predominantly phonologically similar letters, such as B and T, while the errors of the deaf, especially the profoundly deaf, were intrusions of graphically similar letters, such as B with R and P. They also confused some letters, such as T and V, that are confuseable letters in finger spelling (Locke, 1970), which suggests that some deaf children could have been using kinesthetic cues in learning and recalling letters. Subsequently, Conrad (1970) found that a group of profoundly deaf, but bright, children who had been trained in speech and oral communication fell into two distinct groups: a speech-based and a shape-based coding group. It appears that at least some individuals can bypass phonological processing in learning to read. This bypass may be a function of how they are taught. Possibly even hearing students could be taught an alternate route from print to their lexicon, bypassing phonological processing as skilled readers do. For example, Buswell (1945) found that a nonoral method of teaching reading to elementary students over grades one to six resulted in significantly fewer lip movements and greater speed of reading. But this method of instruction, although intriguing, has not been tested under controlled conditions and evaluated with current sophisticated information-processing designs and instruments.

However, hearing subjects who do use the alphabetical principle in learning to read may use phonological processes initially, but can apparently bypass them when they have developed automaticity in word identification. This bypassing occurs when they are reading material that is not difficult for them and when they are engaged in a semantic task (Tzeng & Hung, 1981). Apparently most readers do use phonemic recoding in beginning reading (Liberman et al., 1977). But some children have difficulty in dividing words they hear into their constituent phonemes and relating them to print. They could benefit from a logographic system (Rozin, Poritsky & Sotsky, 1971) or perhaps from a method of teaching English orthography as though it were not alphabetic.

Even though a logographic writing system can map onto a lexicon directly from the beginning of instruction, the Chinese nevertheless use a syllabary, "pinyin" (literally "spell-sound"), as a transitional orthography accompanying the regular characters as a key to their pronunciation. The Japanese start reading instruction with a syllabary ("kana") but substitute characters ("kanji") as children gradually acquire them because meaning can be communicated with fewer kanji than kana; consequently, reading can then be more rapid (Downing, 1973).

Hence, even logographic writing systems use phonetic recoding as an intermediary process and as a transitional system for teaching reading. The shift to communicating through characters (Chinese) and kanji (Japanese) may be analogous to a change in perception of English words as readers develop automaticity for them. They may perceive them as ideographs, and this perception may enable the readers to bypass phonological processing and map the printed words directly on their lexicon. Thus, even when phonological processing can be bypassed by a logographic writing system, a transitional system that capitalizes on phonological processing as an intermediary stage is helpful in teaching reading. But even if the stimulus is alphabetic, not logographic, readers bypass phonological processing as soon as they can, but of course can revert to it when necessary. Hence, phonological processing may not be necessary for teaching beginning reading, but it greatly facilitates learning to read alphabetic and nonalphabetic writing systems.

Critical Age for Learning to Read

Is there a critical age for learning to read, as Mattingly seems to suggest? Although children may need some cognitive and linguistic prerequisites, there is no evidence that we know of where delay in reading instruction beyond a critical age had interfered with subsequent reading acquisition. For example, the Laubach method has successfully taught thousands of adults to read (Laubach & Laubach, 1960).

Perhaps what Mattingly is referring to as a critical age can be better understood by distinguishing between (a) learning to read and (b) using reading as a means of gaining information or learning from text (Singer, 1977a; Singer & Donlan, 1980). Early mastery of the reading-acquisition phase of reading development by some individuals means that they will have a longer period of time to read and learn from text. Hence, they can build up their knowledge base, their conceptual frameworks, and their semantic system, which are necessary for interacting with and comprehending texts (Anderson et al., 1977). Consequently, when these early readers are assessed on comprehension tests, they are likely to perform at a level closer to their expected level of attainment and at a higher level than those students who are otherwise comparable but learned to read at a later age. It takes time to read and build up a knowledge repertoire; although some later readers can quickly master the acquisition process and even develop automaticity for most words, there is no shortcut to acquisition of knowledge gleaned from texts.

Linguistic Awareness and Reading Instruction

Whatever Mattingly means by linguistic awareness, to his credit, he disclaims any intention of using the concept of linguistic awareness as a prerequisite or as reading readiness instruction prior to initiating reading instruction. Instead, he merely suggests that children be given opportunities to communicate and to play with their oral language as a means of enhancing their linguistic awareness. But even though he characterizes children as linguists who hypothesize and test for new words to add to their lexicons, we would hope he would also disclaim any attempt to try to teach children to do what professional linguists cannot do yet, at least not completely, that is, explicate the unconscious and automatic mechanisms and processes that they employ for producing speech and try to consciously use them in learning to read or in understanding written sentences.

If teachers were to try to teach linguistic awareness as such, their reading instruction might produce similar detrimental results as occurred when teachers in the 1900s tried to teach children phonetics (Smith, 1965) and in the 1960s, at least in California, when they tried to teach elementary children the metalanguage and rules for phrase structure and transformational-generative grammar as a means of learning to compose sentences. Of course, we can make use of the transformational-generative grammar as Ruddell (1976) has done in his research (1968), in his model of the reading process, and in his textbook on teaching reading (Ruddell, 1974), which he followed up with his basal reader series (Ruddell, 1977).

We should be aware of the pitfalls of past practices in applying new theoretical insights; specifically, we should avoid teaching children linguistic metalanguage and linguistic analyses in reading instruction; we should not assume that children can think or can learn to think like professional linguists or even better than professional linguists can now think about spoken or written language.

Indeed, Mattingly's metaphor that linguistic awareness is like a muscle that atrophies without practice conjures up in my mind teachers having children do linguistic muscle-strengthening activities or something called linguistic awareness exercises. Such activities, depending on how linguistic awareness is taught, not only interfere with children's acquisition of reading ability but also with their ability to speak. In fact, any motor act that has developed to the stage where it operates successfully at an unconscious level suffers when performers try to recapture the conscious conditions under which they originally learned it.

Perhaps Mattingly would avoid the pitfalls we perceive in making children linguistically aware. Perhaps he only means to raise their phonological processes to a conscious level in the process of instruction as Gleitman and Rozin advocate so that children can relate them to segments of print, without requiring children to specify rules for the relationship. Perhaps he would have teachers teach phonics as they currently do which only involves acquisition of *implicit* rules for relating sounds to printed words.

However, we do not know whether Mattingly's concept of linguistic awareness is identical to Gleitman's and Rozin's explanation that phonological processes are raised to a level of consciousness *during instruction* in reading acquisition. Although the Gleitman-Rozin hypothesis is plausible, it has not yet been experimentally confirmed. A related, but alternate, hypothesis in the precondition for reading is not phonological awareness, but *the capacity for becoming aware*. Hence, the appropriate prediction would not be whether the child exhibits phonological skill but whether the child can *learn* phonetic segmentation (Henderson, 1982).

We also do not know how Mattingly would develop linguistic awareness for reading acquisition. He mentions having children play with language *prior* to reading instruction and draws an analogy to a muscle when he talks about *exercising* linguistic awareness. Therefore, he implies that linguistic awareness is something that could and should be developed in isolation, not as a byproduct of instruction in reading acquisition. If that is his concept of when and how to develop linguistic awareness, then he is relying on it to transfer to reading. But the general evidence on transfer of training is that such transfer is tenuous; it would be better to develop linguistic awareness in the specific situation in which it is to be used, as Gleitman and Rozin would do in teaching reading.

These reservations are not about the concept of linguistic awareness, but how and when it should be developed, whether it is necessary for reading acquisition, and, if so, the degree of awareness that is necessary. As has already been said, it may be unnecessary and perhaps even harmful to require *maximum* linguistic awareness, that is, not only ability to use but also to explicate rules for relating sounds to print. We do not want to return to *phonetic* instruction in reading acquisition when children had to learn and state the rules for sounding out printed words. The author of this chapter would, however, subscribe to *phonics* instruction which is merely learning to relate sounds to print, that is, the ability to segment and relate phonemes, combination of phonemes, syllables, and words to corresponding segments of print. Research has adduced evidence to support phonics instruction (Samuels, 1976b, Singer, 1976a). Unfortunately, Mattingly does not address any of these issues. Hence, it is not known whether Mattingly claims linguistic awareness is a cause, concomitant, or a consequence of learning to read.

At the very least, before we ask teachers to add exercises in linguistic awareness to their reading-readiness repertoire, we should require Mattingly and others who advocate this vague concept of linguistic awareness to formulate it more precisely, to state their arguments in the form of testable hypotheses, to specify what teachers should do to make students linguistically aware, and to adduce evidence that linguistic awareness is causally related to reading acquisition. Perhaps we will then find that teachers are already developing linguistic awareness as a byproduct of reading instruction, and if not in reading, then in spelling instruction (Ehri, 1980b).

Until we have valid evidence of the role of linguistic awareness in reading acquisition, we should not make it a specific instructional component. However, we can add other linguistic components to instructional procedures that Mattingly refers to and that are valid as Ruddell (1968, 1974, 1976, 1977) has demonstrated. We can help children identify morphemes in their orthography and teach them to relate their phonological, semantic, and syntactic processes to printed words. We can also provide them with linguistic information on printed words so that they can form hypotheses, test them, and corroborate new entries in their lexicon. Also, we can help them relate these entries to morphemic and morphophonemic patterns in their orthography and even to parse sentences into phrase structures without trying to make them aware of their linguistic mechanisms and processes. We know that individuals can and do learn without awareness. Moreover, the rules for relating sound to print are so complex that, if we tried to make them aware of all the rules, it would be an exceedingly difficult task. We know that it takes some 2000 rules to program a computer to read aloud (Cushman, 1980), but perhaps children can learn to read with fewer rules, possibly with as few as 200 rules (Gough, 1982). If students had to learn to read with awareness of all the rules for having a computer read aloud or even only those necessary for humans to read aloud, the rate of instructional progress in teaching reading would probably be much less than the current rate. Nevertheless, we can and do capitalize on linguistic and other systems interacting within and between the reader and the text to help students learn to read and achieve what Huey (1908) called this most remarkable species-specific accomplishment.

CHAPTER 11

The Development of Metalinguistic Abilities in Children Learning to Read and Write

Renate Valtin

Terminology

In the preceding chapters different aspects of linguistic awareness regarding both functions and features of oral and written language have been referred to. In the existing literature the term "language awareness" is used with a still broader meaning. As Sinclair (1981) notes, this term "seems to include all the capacities and activities concerning language and language judgment which are not themselves a part of (or very closely tied to) production and comprehension processes. Any reflections, ideas, knowledge, or explicit formulation of underlying principles, rules, etc., concerning language structure, functions, or the rules for its use have been classified under the label 'linguistic awareness' or 'metalinguistic activities'" (pp. 44-45). This already very broad picture of abilities becomes even more incoherent when further metalinguistic abilities related to written language are included under this topic. According to Weaver and Shonkoff (1979), linguistic awareness "includes knowing what reading is; knowing conventions of print . . . ; and knowing the concepts of a letter, word, sentence, or a story" (p. 30).

The term "cognitive clarity" as proposed by Downing seems better suited to designate this latter type of knowledge connected with written language because it refers not only to (meta)linguistic but also to (meta)cognitive elements. Knowing the conventions of print, moreover, may be related to different cognitive levels. Young children's knowledge of print and recognition of literacy behavior may be more figurative knowledge in the sense of Piaget (recognizing relevant physical–visual or behavioral–cues) without the conceptual knowledge of a literate adult. The term "cognitive clarity" clearly relates to operative knowledge.

This research topic has become incoherent not only because of the broad range of different processes studied but also because of the different assumptions and methods used in the studies. As Sinclair (1981) states, "Most of the experimental work carried out has concentrated on one or the other aspect of language as it is understood by present day linguists and psychologists. Our present knowledge and theories about language are far from complete . . . . Moreover, there is no reason to suppose that the child's thinking (or the naive adult's for that matter) is either concerned with the same matters as linguists, or that it follows the historical development of linguistics" (p. 45).

It would be of little help in clarifying the vague concept of language awareness merely to offer a new definition for it in the absence of a theoretical framework that allows one to specify aspects of language awareness and to establish a developmental sequence. Therefore, in this chapter and in Chapter 12 an attempt is made to analyze various phenomena of language awareness and to offer a theoretical framework that allows a differentiation between various forms of awareness as well as a suggestion as to their developmental order and their relation to the learning of written language. This model claims that, under existing circumstances, the child acquires the linguistic notions of word and phoneme mainly through instruction in reading and writing (spelling) in school. Some research studies are presented that show how the preschool child's everyday concept of a word undergoes a transformation under the influence of reading instruction in school, in the sense that a multiple-core concept of a word is narrowed to a concept that is mainly oriented to visual strategies.

The Concept of Language Awareness

Clark's Taxonomy of Metalinguistic Abilities

Clark (1978) has reviewed literature on phenomena of language awareness from naturalistic observations, interviews, and experimental studies and has classified different types of awareness. The following list contains those different types in their order of emergence and can be interpreted as a preliminary and tentative taxonomy of instances of language awareness:

(1) Monitoring one's ongoing utterances
(2) Checking the result of an utterance
(3) Testing for reality
(4) Deliberately trying to learn
(5) Predicting the consequences of using inflections, words, phrases, or sentences
(6) Reflecting on the product of an utterance (Clark, 1978, p. 34).

Clark characterizes the different forms as follows:

> The first is the ability to *monitor* one's own ongoing utterances. This activity is a prerequisite for spontaneous repairs, practice, and adjustments of one's speech style to different listeners. Another skill is the ability to *check* the result of one's utterance. Even very young children check to see if the listener has understood, and if not, try again. Rather later, they start to comment explicitly on their own utterances and on

those of others. They also correct others. Another skill is *reality testing*: children check on whether a particular word or phrase has "worked" in the sense of getting the listener to understand what they were saying. . . . A fourth metalinguistic skill is that underlying deliberate attempts to *learn* language. Children practice not only sounds and sentence structures but also the speech style characteristic of different roles. The last two skills . . . seem to emerge rather later than the others. In *predicting* the consequences of using particular forms, children use language or make judgments about it out of context. They supply the appropriate inflections to indicate plural, past tense, or diminutive; they judge utterances as appropriate to particular settings or speakers; and they correct sentences that are "wrong." Finally, in *reflecting* on the product of an utterance, children may be doing something that is never called for in other forms of metacognition. With language, it is possible to reflect on language structure independent of its actual use. Children identify specific linguistic units—anything from a sound up to a sentence; they provide definitions of words; they construct puns and riddles, and exploit other forms of verbal humour, and they explain why some sentences are possible and how they could or should be interpreted (p. 35).

Clark points to the difference between implicit and explicit knowledge and states that children "show implicit knowledge of different linguistic units—words, syllables, and phonetic segments—long before they can reflect on those units explicitly" (p. 36). Clark's claim that her phenomena of language awareness correspond to Vygotsky's (1934) second stage of the acquisition of knowledge ("the gradual increase in active, conscious control over knowledge already acquired," Clark, 1978, p. 36) is not persuasive, however. Hakes (1980) has pointed to important differences between earlier and later forms of metalinguistic performances. While earlier forms seem to arise spontaneously in the ongoing course of conversing, older children are not only able to discriminate spontaneously the properties of utterances but they are also capable of deliberately reflecting on speech. Early spontaneous metalinguistic performances of young children are quite rare, while older children and adults *can* make them more frequently upon request. A third difference is that young children show less variety in the aspects of language on which they comment. Hakes concludes "that the change in children's metalinguistic abilities is a change in the systematicity and variety of their performances and in the extent to which they can engage in such performances deliberately" (p. 107).

The Analysis by Andresen and Januschek

A more systematic analysis of Clark's types of language awareness has been provided by Andresen (1982) and Andresen and Januschek (1983). They make the criticism that Clark sees all phenomena of language awareness (from the first spontaneous repairings of one's speech by the two year olds to the explicit comments on the grammaticality of sentences by older children) as conceptually undifferentiated and as a manifestation of the conscious manipulation of lan-

guage. Above all, they complain that Clark did not consider the motives and
intentions of the children for specific metalinguistic activities and that she did
not mention whether or how the early types of metalinguistic activities change
in the course of development. These critics of Clark's work postulate that a
theory of language awareness must be connected to a general theory of action.
Based on an action-theory-oriented model of language acquisition, they propose
various criteria for the systematic description of metalinguistic activities:

(1) Motives and intentions of the speaker (e.g., explaining misunderstandings
    and language learning)
(2) Mode of action to which the metalinguistic performance refers (role play,
    rhyming, and telling puns)
(3) Nature of the action (to comment, to correct, and to formulate a rule)
(4) Object (sounds, syllables, syntactic relations, and consequence of a speech
    act)
(5) Degree of explicitness (implicit, explicitly referred to but without adequate
    terminology, and explicit verbal statements)
(6) Spontaneity (spontaneous, response in an interview or an experiment, with
    and without preparation and training)
(7) Chronology

Andresen (1982) and Andresen and Januschek (1983) state that the metalinguistic activities in the Clark taxonomy seem to appear as a direct function of
age development and not as a function of the social context in which these
activities are embedded. Clark also seems to ignore the effect of schooling.

Using the criteria of the intentions of the speaker, the authors propose the
following differentiation of the spontaneous metalinguistic activities described
by Clark:

(a) Most of the types of awareness in children up to the age of five occur in the
    ongoing course of conversing as spontaneous repairs of speech which are
    motivated by the concern to be better understandable. Children seem to be
    aware not of their speech but of the failure of their attempts to make their
    speech intelligible to others. Out of the context of speech children are not
    able to identify those parts of their speech that they correct spontaneously.
(b) Young children are able in role-play to comment on the appropriateness of
    verbal behavior with regard to different roles (for instance, the appropriate
    voice). They are also able to choose a more polite form of an utterance. But
    this cannot be regarded as a manifestation of reflection on isolated elements
    of speech. Children judge the behavior of persons and view those language
    aspects (for instance, the "right voice") as social indicators or characteristic
    attributes of specific persons. Focus of attention is on verbal behavior as
    part of an action and part of a play situation that requires the following of
    specific social norms. Children of the age of about three to five are able to
    speak about language in a given context, but are not able to abstract from
    this context and make language an explicit object of thinking.

(c) These two forms of metalinguistic performances that are embedded in a specific communicatory activity or where speech can be viewed as part of an action are conceptually differentiated from other forms—as practicing and playing with language—that are not directed by the concrete speech situation. Andresen gives as an example the utterance of a young girl of about five years of age whose father was a merchant (Diplom-Kaufmann): " '*Di*plomkaufmann' must be wrong. It must be '*Der*plomkaufmann.' " This girl has reflected on regular properties of the language, detected that in German the article "der" refers to the masculine gender and "die" to the feminine gender, but she had wrongly identified the first syllable of the morpheme "Diplom" as an article.

In creative manipulation of language (variation of language forms, rhyming, creation of nonsense words, riddles and puns), language is made an object that is operated upon. The children act on language features, on phonetic or morphological forms, or attributes of the semantic structure. Andresen emphasizes that these types of language awareness are implicit, not yet explicit. She refers to the findings of a study by Hirsh-Pasek, Gleitman, and Gleitman (1978) who investigated children's ability to explain jokes that turn on various kinds of language ambiguity. In their sample of children aged six to twelve, they observed three developmental stages. Some six year olds were able to understand the meaning of each phrase, but not the essence of the joke, as the following example shows:

Joke   : "Did you ever stand on a pet?"
             "Stand on a pet? I should say not!"
             "I have: on a carpet!"
Subject: Well, he said have you ever stamped on a pet—I never would.
Exp.   : You wouldn't? Well, what kind of pet did he stand on?
Subject: He said I haven't stepped on a rug.
Exp.   : Did he say *rug*?
Subject: Uh-huh.
Exp.   : I don't think I heard that.
Subject: I think I did (Hirsh-Pasek et al., 1978, p. 126).

In the next stage the children understand the ambiguity, that is, they are able to detect potential alternative interpretations, but are not able to explain them. Only later in development is the child able to make judgments and give explicit explanations. The detection of ambiguity was clearly related to the linguistic variables that were manipulated in the stimulus material. Ambiguities that depended on underlying representations and were closer to the level of meaning were easier to judge than ambiguities that turned on superficial representations and referred to superficial syntactic and phonological factors.

In order to explain these discrepancies between perceiving language (linguistic abilities) and judging language (metalinguistic abilities), Hirsh-Pasek et al. refer to two possible interpretations. The first is that "the brain makes available to consciousness only some of the computations relevant to language perception," and especially the more meaningful levels of representation (p. 127). The second interpretation—but one which does not seem to exclude the first one in the view of the present author—is that metalinguistic performance is "a single example of a more general 'metacognitive' organization in humans. That is, a variety of cognitive processes seem themselves to be the objects of higher-order cognitive processes in the same domain" (Hirsh-Pasek et al., 1978, p. 128). Andresen, in her analysis of Clark's phenomena of language awareness, tries to relate those results with the findings of Hirsh-Pasek et al. (1978) while using as a general framework the theory of the Russian psycholinguist A. A. Leontev.

A Theoretical Framework Provided by Leontev

Leontev (1973, 1975) differentiates between a "model of language ability" referring to psycholinguistic processes and representing the implicit knowledge and a "model of language" that represents explicit knowledge as it is interpreted by linguists.

Leontev (1981) describes the speech process as follows:

> Man does not immediately begin with speech, with the choice and combination of sounds, words, and constructs. As in every purposive activity, there has to be a plan (or intention, or programme) for any future utterance. Such a programme is generally of a visual nature; the content of the utterance emerges as it were in the mind's eye of the speaker in the form of a picture, schema, etc. This programme is retained in the conscious mind (operative memory) until it is no longer necessary, i.e. until we have said what we wanted and passed on to the next utterance. The speech process consists in the translation of the programme into a strict linguistic form, which in the mother tongue is a more or less automatic procedure (we are not considering the written language, to be distinguished from oral language above all by the deliberate and conscious character of the choice and combination of its components) (p. 26).

His model of generating verbal utterances is organized in different levels. In normal, everyday speech the "directrix" (Leitlinie) has the focus of attention. Speakers are aware of their intentions and the meaning of the utterance, but usually they are not aware of the linguistic means by which the internal pogram is realized. In certain cases, however, for instance, when communication fails, it is possible that lower levels of language can be raised to awareness. This process is labeled by Leontev as "actual awareness," where certain segments of speech (usually not words but predicative phrases or semantic units) are focused upon and the attention switches momentarily from the intention of an utterance to the linguistic means. This actual awareness is spontaneous and momentary. Andresen points to the fact that Hake's description of the early forms of metalinguistic performances in young children corresponds to Leontev's "actual awareness." Leontev states that this process where certain language segments are momentarily in the focus of attention occurs spontaneously and does not need to be learned in the sense that it requires new capabilities (Leontev, 1975, p. 263).

This actual awareness is to be differentiated from "real consciousness" or "conscious awareness." Leontev characterizes this process in terms of Vygotsky's concepts of deliberate mastery, generalization, systematicity, and control. Conscious awareness requires not only objectivation of language (the ability to make language an object of thinking) but also explicitness: in order to identify linguistic units, the subject must have acquired the concept of this unit.

According to Vygotsky and Leontev, conscious awareness is related to formal instruction in school, instruction in grammar, reading and writing, and in foreign languages. We will return to this point.

Andresen uses Leontev's model to explain the observed differences in difficulty between syllable and phoneme segmentation in children. As many studies have demonstrated (e.g., Liberman et al., 1974), it is easier to segment into syllables. Andresen points out that the segmentation of syllables need not be connected with the conscious identification of this unit, since syllables are a basic element of the motor program of speech. Children are able to segment words into syllables by clapping or tapping, thus transferring the speech rhythm into other rhythmic movements without being aware of the unit syllable. The apparent difficulty children have with phonemic segmentation may result from the fact that the phoneme does not belong to the "model of language ability" but to the linguistic model of language. Therefore, as Andresen states, it must be learned consciously and can only be isolated by an act of conscious awareness. The attention paid to formal aspects of language is very difficult for children to achieve because they usually (as adults in everyday speech do) focus on the meaning aspect. Focusing on syllables is easier than on phonemes because syllables are a speech motor unit. Leontev (1975) states that children are able to isolate and to make an object of actual awareness only those segments that are operative units of speech or psycholinguistic units (for instance, semantic units or syntagmas when segmenting a sentence). Children are also able—upon request —to segment into syllables and, as Leontev calls it, into "initials" (consonants which are identified with the whole syllable). Without specific instruction children are not able to isolate vowels in syllables. The findings presented by Zhurova (1973) demonstrate that children aged three to five have extreme difficulties in isolating the first sound of a word, even when words were used in which the breakdown of their structure did not change what was specific to the pronunciation of the individual sound. Thus "conscious awareness" refers both to psycholinguistic units that children are able to segment spontaneously and to linguistic units which they have to learn and which have a new psychological quality (deliberate mastery and control). When Clark (1978) speculates that children analyze language into explicit units rather late because they "have first to learn a vocabulary for talking about how they use language" (p. 37), she neglects this important cognitive step.

A Developmental Sequence of Language Awareness

Leontev's model of generating verbal utterances is a descriptive model of how a (skilled) speaker transforms his or her internal program of a future utterance into an actual utterance. It is not a model for explaining the development of speech. But, as Andresen has suggested, it may serve to differentiate three developmental stages of language awareness levels "from the dimly conscious or preconscious speech monitoring which underlies self-correction to the concentrated, analytic work of the linguist," as Slobin (1978, p. 45) describes it.

These stages can be characterized as:

(1) *Unconscious awareness or automatic use of language.* Early forms of metalinguistic activity are embedded in a communicative situation and serve to establish an effective communication. The child is not aware of his or her speech but can become aware that speech acts fail.

(2) *Actual awareness.* Spontaneous creative manipulation of language and answers in interviews and experiments show that children become increasingly able to abstract the language from the action and the meaning context and to think about some of the properties of the form of language. Their knowledge of language units is still implicit, however, and related to psycholinguistic units of speech.

It is suggested that children's creative manipulations of language forms are not a manifestation of a reflective usage of language or of a new "ability," but a manifestation of the child's mastery of language rules. Elkonin (1971), in reviewing some Russian studies on this topic, comes to the same conclusion:

> Children's independent word forming, or so-called word creativity, has been cited by some authors (Chukovsky and Gvozdev) as primary evidence of the presence of special linguistic sensitivity in a preschool child. It appears to us that word creativity is not something exclusive, but must be viewed as a manifestation, a symptom of the child's *mastery* of linguistic reality. At the base of children's word creativity lie principally the same rules that are found at the base of mastery of the inflected system of language. Realistically, the child performs not a lesser nor a less active task during mastering, for example, of such an abstract and entirely formal category as gender, having a very important significance in Russian language (p. 139).

Elkonin continues in his argument that not only was the assumption of a special linguistic sensitivity a great obstacle that "blocked the paths for clarification of objective rules for the language mastery process," but also the so-called "glass theory" proposed by Luria was "not a lesser hindrance" (p. 139). Elkonin points out that from the beginning "the word is perceived by the child first of all in terms of its material, acoustic aspect" (p. 140) and "that the most important condition in mastering the grammatical structure of language is the formation of orientation toward the sound of the word" (p. 150). These arguments cannot be denied, but they are no argument against Luria, who had emphasized that the child had difficulties in isolating the word from its meaning, an aspect that relates to conscious awareness.[1]

The finding that kindergarten children are not able to judge whether two words are rhymes or have great difficulties in producing rhymes on request (Calfee, Chapman & Venezky, 1972) is in line with this argument that in verbal play rhyme producing is not a conscious activity (as, for instance, Levelt, Sinclair & Jarvella, 1978, p. 2 claim). Rather, it may be viewed as an incidence of "actual awareness."

(3) *Conscious awareness.* The child has acquired the ability to deliberately focus on and manipulate linguistic units. This sort of knowledge is explicit and has a new psychological quality. Many authors suggest that this explicit, deliberate, and systematic knowledge is dependent on formal instruction or at least some miminal form of instruction, and that the acquisition of reading and writing helps to enhance this conscious awareness (Vygotsky, 1934; Leontev, 1973; Andresen, 1982, 1983; Andresen & Januschek, 1983; Januschek, Paprotté & Rohde, 1979a,b; Paprotté, 1979).

Later on in this chapter we will investigate how certain forms of spontaneous awareness of language features may change under the influence of school instruction.

Our definition of conscious language awareness is at variance with Mattingly's concept of linguistic awareness as outlined in his chapter in this book. Mattingly describes linguistic awareness as a "specially cultivated metalinguistic consciousness of certain aspects of primary linguistic activity," but states that it is "not a matter of consciousness, but of access. This access is probably largely unconscious, but the degree of consciousness is not very relevant" and "access, to repeat, does not imply consciousness."

Chomsky (1979), in her reaction to Mattingly's paper, has formulated a critique which expresses the ideas of the present author:

> In discussing linguistic awareness, Mattingly deals with the notion of *access* to implicit grammatical knowledge as the major factor in linguistic awareness. Linguistic awareness, he maintains, is not a matter of consciousness of grammatical knowledge itself, but of access to grammatical knowledge. Fine. This view, basic to any formulation of linguistic awareness, asserts that speakers can reflect on language and make judgments about, say, the grammaticality of a sentence, without being able to state the grammatical rules that operate to produce it. Speakers recognize when the rules are broken, although they have no consciousness of the rules themselves. . . .
>
> I agree with Mattingly that *access* to grammatical rules is a necessary component of linguistic awareness, but I disagree with his view that *consciousness* is not relevant to linguistic awareness. Because he rules out consciousness as a component of linguistic awareness, he is able to attribute the highest degree of awareness to very young children: 'during the period of active language acquisition, grammatical knowledge would be

highly accessible . . . after language acquisition has ceased to be a major preoccupation, grammatical knowledge should tend to become less accessible.' I think that, on the contrary, linguistic awareness does entail consciousness on the part of the speaker, resulting from the ability to reflect on language and view it objectively. This ability develops slowly in children, and it is far easier to raise linguistic consciousness in adults than it is in young children (pp. 1-2).

A similar position is held by Gleitman (1979), who wrote as a comment on Mattingly's claim: "Intuition giving is a manifestation of the sophisticated individual's ability to manipulate and report upon his knowledge; as such, it has a relatively conscious and aware component" (p. 3). "Despite the obvious fact, then, that adults have more intuitions than young children, Mattingly goes on to argue that access is present in early childhood, but rather withers away in later life" (p. 1). She concludes that Mattingly's "identification of the language-acquisition state with the state of linguistic awareness (having intuitions, etc.) is untenable" (p. 3).

Language Awareness and Cognitive Factors

The relationship between language awareness and other cognitive factors, and hence to cognitive development in general, is far from being clear, as a short overview of different positions outlined in the literature on this topic reveals. While most authors would agree in defining language awareness as language abilities (production and perception) plus some sort of additional cognitive or reasoning abilities, Mattingly, in his chapter, proposes another solution. Because he has a specific assumption about the language-learning process, he does not need an additional cognitive ability in order to explain metalinguistic performance. On the basis of Chomsky's theory that all human beings have a specific innate linguistic capacity, Mattingly claims that the child's "general theory of language is innately given." Taking this conception of the infant language learner as a linguist, he suggests that, during the period of active language acquisition, grammatical knowledge is highly accessible to the child that thus possesses linguistic awareness. That this access to one's grammatical knowledge *does* require consciousness and thus additional cognitive abilities has already been pointed out. The other claim, that knowledge of language is innately given, also seems to be untenable. Bruner (1978), among others, has emphasized that, prior to language learning, the child already has acquired knowledge about the world and that "prelinguistic concepts provide guides for the learning of forms of utterance that relate to them and refer to them" (p. 245).

A conceptual analysis of metalinguistic activities, however, may result in the differentiation of three aspects that are closely related:

(1) The ability to focus attention on language forms per se, or the ability to treat language objectively and freeing it "from its embeddedness in events" (Donaldson, 1978, p. 89)

(2) The acquisition of concepts of oral and written language, such as phoneme, word, sentence
(3) The ability to make deliberate utilization of phonological, syntactic and semantic structures of language.

With regard to the first aspect, it is plausible to relate it to cognitive developmental stages. Lundberg (1978) calls the central aspect of linguistic awareness an attention shift from content to form and points to a relationship with Piaget's concept of decentration. Donaldson (1978) suggests that such language activities involve "the temporary suspension of overt action and a turning of attention inwards upon mental acts" (p. 78), which is a late developmental achievement.

Hakes (1980) hypothesizes that metalinguistic activities and concrete operational cognitive abilities both have a common developmental basis: the "increasing capacity to engage in controlled cognitive processing and, in particular, an increasing ability to stand back from a situation mentally and reflect upon it" (p. 100). In his empirical investigation which attempted to support this claim, Hakes ignored the influence of school, however.

The concepts of oral and written language which the child acquires cannot be viewed as a natural consequence of cognitive development in general but must be seen as a function of the specific demands a society makes on its members. Heeschen (1978), for instance, in investigating the metalinguistic vocabulary of the Eipo, neolithic horticulturists in New Guinea, observed that their vocabulary and judgment focused on appropriateness and content, not on the form of an utterance. In Western cultures the concepts of linguistic units which children and adults acquire are mainly a result of literacy. All the linguistic features which are represented in a given orthography must become objects of awareness to the user of this orthography. This view implies that individuals from cultures with different scripts (logographic, syllabic, or alphabetic) will differ in their awareness of various linguistic units, e.g., illiterate adults in Portugal failed a phonemic test (Morais, Cary, Alegria & Bertelson, 1979). Likewise, individuals with poor and normal reading and spelling abilities differ in their linguistic awareness (for a review of studies see Valtin, 1980, 1981, 1982b).

In this context it is interesting to note that written language has an influence not only on the phonological and lexical level of spoken language but also—as Hjelmquist (1981) has demonstrated—on the adult's conception of spoken discourse. When asked to judge utterances or paraphrases from a real conversation or when confronted with transcriptions of verbal discourse, people usually "do not seem to be aware of the specific characteristics of spontaneously generated verbal discourse such as repetitions, incomplete sentences, interruptions, stutterings, etc." (p. 69). Their judgments are oriented to norms of written communication.

Turning to the second and third aspects in the above list, some authors regard them as identical. Vygotsky (1934) states that "control of a function is the counterpart of one's consciousness of it" (p. 90) and points out that all higher intellectual functions are characterized by reflective awareness and deliberate control. Donaldson (1978) proposes that consciousness of an operation is necessary for its control but differentiates the "degree of reflective awareness of language" from the issue of control, "the question of how much ability the child has to sustain attention, resisting irrelevance while he considers implications" (p. 93). Ryan and Ledger, in their chapter in this book, likewise suggest a differentiation into two dimensions: analyzed knowledge and cognitive control. They point out that both Piagetian and information-processing views of cognitive development allow an explanation for metalinguistic performances.

This brief review of some theoretical positions indicates that the exact nature of the relationship between language awareness and other cognitive factors as well as the theoretical framework for the conceptualization of development still await clarification in further research.

The Development of the Concept of a Word

Word Unit Studies

Children's knowledge of the concept "word" as a unit of oral language has been measured with a variety of procedures:

(1) Counting words in sentences and tapping (Karpova, 1955); separating two-word sequences and reversing the two words (Huttenlocher, 1964)
(2) Interviews about the meaning of a word (Reid, 1966; Downing, 1970; Francis, 1973; Papandropoulou & Sinclair, 1974)
(3) Prompted segmentation of sentences (Fox & Routh, 1975), tapping a poker chip when uttering each word in a sentence (Holden & MacGinitie, 1972; Ehri, 1975)
(4) Distinguishing between orally presented real words and nonsense sounds without context (Downing & Oliver 1973-1974) or within a play situation (Sense and Silly Word Game, Ehri, 1979, p. 71)
(5) Recognizing and storing single words (Ehri, 1976), comparing the length of two orally presented words (Bosch, 1965; Katzenberger, 1967; Schmalohr, 1971; Lundberg & Tornéus, 1978)
(6) Giving examples of long and short words (Berthoud-Papandropoulou, 1978) and of ugly, difficult, or nice words (Januschek, Paprotté & Rohde, 1979a, b).

As could be predicted from Leontev's model of language ability, preschool children hardly show any knowledge of the linguistic unit "word." Karpova's study with Russian children aged three to seven showed a developmental sequence in the formation of the concept word:

(1) Inability to segment
(2) Identification of semantically salient units
(3) Distinction between noun phrases and verb phrases

(4) Differentation between component words on the one hand and prepositions and conjunctions on the other hand, although the latter were not always identified as words. The children of this age group were not able to divide a sentence into lexical units. They showed better results when the counting was connected with external support (removing a plate).

Berthoud-Papandropoulou (1978) reported similar stages in her study when children were asked to count words in spoken sentences (from an inability of the four to five year olds, who focused on the scene evoked by the sentence, to a segmentation into topic and comment, and then a differentiation into privileged semantic constituents). The separate counting of articles and other functors occurred only occasionally at the age of seven and systematically at the age of eleven. The interview study by Papandropoulou and Sinclair (1974) yielded indications of a similar sequence of response patterns.

In the study by Holden and MacGinitie (1972), children at the end of kindergarten had more difficulty isolating function words than words that had more lexical meaning. Most commonly, children compounded a function word with the following content word (as *The book/is in/the desk*). The authors note that some children spontaneously imposed a rhythmic pattern on the utterance and used rhythmic aspects as a basis for segmentation. But, even when children are confronted with sentences stripped of any suprasegmental rhythm and intonation, as in Ehri's (1975) study, they ignore function words more often than the semantically salient content words.

This difficulty in segmenting oral speech into lexical units is reflected in the spontaneous writings of preschool children who either show no concern for word boundaries or connect function words to one unit, as the example of the five-year-old son of Bissex (1980) demonstrates. He wrote: EFU WAUTH KLOZ I WEL GEVUA WAUTHENMATHEN (If you wash clothes I will give you a washing machine). In this connection, we may note that the segmentation of words in written text is a rather late historical achievement in alphabetic scripts. Early Greek texts are written without segmenting words (Coulmas, 1981, p. 108).

At first sight, the results obtained by Fox and Routh (1975) seem to contradict our hypothesis that children are unable to segment sentences into lexical units. The authors prompted children aged three to six with the statement: "I'm going to say something to you and I want you to say just a little bit of it. If you say just a little bit of it, you'll get a raisin." When the child responded with a multiple-phrase word, the experimenter repeated this phrase and prompted the child again to say only a little bit of it. It is doubtful, as also Ehri (1979) criticizes, whether the performance of the children can be regarded as a manifestation of lexical awareness, since the experimenter decided when the lexical analysis was complete. It is also highly probable that some children of this sample were already acquainted with print experience, since they were of far above average I.Q. and social status. Interestingly, the initial responses of the children reveal the same pattern as other studies have reported: Some of the children, especially when confronted with longer sentences, responded with the initial noun phrase ("A lady"; "The little boy"). Even with so much experimenter prompting, three year olds were not able to show perfect performance.

The Studies by Januschek, Paprotté, and Rohde

The German linguists Januschek, Paprotté, and Rohde (1979a,b) have explored what kind of meanings children connect with the term "word." Using Vygotsky's (1934) distinction between spontaneously acquired everyday concepts, which usually are unconscious and unsystematic, and the scientific concepts, which are acquired as a result of instruction, they emphasize that preschool children have had experiences only with the everyday concept of a word in connection with an action or a communication context. In German, the everyday term "word" (Wort) also relates to multiple-word utterances. Thus, one cannot expect that young children regard a word as a formal unit of semantic and syntactic features. The authors also criticize the methods of investigation used in many previous research studies because they did not take into consideration the fact that children find it difficult to dissociate their mental activity from the concrete situation. To overcome this problem, these German authors used a play situation where the task was embedded in a concrete situation. After this play activity, designed to produce a preliminary fix for the concept word, an interview was carried out.

The sample comprised 26 children at the end of kindergarten (ages five and six) and 32 children at the end of first grade. In the play activity the child had to tell the first word of either a three-word sequence or a whole sentence to a toy figure and was then allowed to "enter" a house and fetch chocolate. A short training period made sure that the children understood the task.

As expected, first graders showed significantly better performance in the play situation. Isolating the first word from a coherent text was for both groups significantly more difficult than from syntactically and semantically unrelated material. Also, in both age groups, significant differences were found between content words and functors. The interview revealed the following differences between the two groups:

(1) The number of children who explained "word" and gave an example for "word" increased with age, from about 30 to 50 percent.
(2) First graders tended to define "word" with regard to characteristics of the written language. Likewise, in their comments upon their examples for difficult, easy, long, or short words they referred to letters or spelling problems ("*Mehl* is a difficult word because of the silent letter *h*").
(3) First graders more often recognized plural nouns ("trees," "leaves") and mass nouns ("forest") as one word.
(4) There was no difference in judging the length of two words when the first word was part of the second ("Auto"-"Automobil," "die"-"diese"). But first graders recognized more often that "Knoten" (tie) was a longer word than "Band" (band).

The proportion of invective, taboo, and swear words was very high in the examples given for a "bad" or "ugly" word, thus reflecting the fact that children often encounter this meaning of "word" in their everyday life. At least in German, the following complaint is a quite common experience for children: "One does not say . . .; it is an ugly word."

In explaining why they regarded some words as long/short or difficult/easy, one third of the kindergarten children referred to attributes of the denoted object or to its social context of use ("Schrien" (crying) is a short word because "children are not allowed to cry for a long time"). Two thirds of the explanations referred to attributes of the language form, but none of the kindergarten children referred to the written form of the word as many first graders did (even when this strategy very often yielded wrong answers, e.g., "*Knoten* is shorter than *Band* because it contains fewer letters").

Januschek et al. (1979a) put forward the hypothesis that children at the same time have quite different concepts of "word," some features related to the everyday usage of "word," some features to a scholastic concept of "word." They observed that several children replied that "Blätter" (leaves) are many words and that the sentence "Die Blätter fallen" (Leaves are falling) contained three words. Even when confronted with the discrepancy of the answers, the children were not able to detect it. Discrepancies were also noted between the children's performances in the play situation and their answers in the interview. One kindergarten child, for instance, isolated the first words in the game almost without error and defined "word" as something that is "said," but gave as examples for long, difficult, or bad word objects with the corresponding attributes, and reported that "trees" are several words. Another child was merely unable to cope with the play task but gave correct examples for words. The authors conclude that the isolation of a word from word chains or sentences and defining and giving an example of words are different activities which do not necessarily refer to a uniform meaning of a word in the sense of a scientific or scholastic concept. Children seem to "switch between the everyday (intracommunicative-reflexive) and the scholastic (metacommunicative-descriptive) concepts of word, both of which depend on different functional attributes, criteria of decision, and on different strategies . . . among which children select quite ideosyncratically" (p. 252).

The authors postulate the following strategies which children use in dealing with the concept of a word:

(1) An articulatory-acoustic strategy where the number of syllables or phoneme clusters are used to distinguish between long or short words or to decide whether "Blätter" or "Gras" are one or several words

(2) A visual or visual motor strategy, using number of letters or spelling problems as criteria for decision

(3) A semantic strategy that relies on the meaning of the word for the child

(4) A pragmatic strategy. Some children gave, as examples for ugly words,

taboo words or invectives and explained them in terms of the intended results on the hearer (e.g., one boy reported "Arschgeige" as an ugly word, but then stated that the word was not ugly for him, but for the person to whom it is said).

The authors summarize their position as follows:

> The above results seem to indicate conclusively that the child's everyday concept of word draws on his experiences of language in use, of linguistic etiquette and verbal regimentation. Further, even younger children already reconstruct speakers' intentions, and reflect achieved results, conventions and norms of use: the functional core (cf. Nelson, 1974) of the child's everyday concept of word seems to rest firmly on pragmatic assumptions and generalizations of communicative experience. However, the above results also indicate that preschool children already master a multiple-functional-core concept of word even if "word" still contains all the functional attributes of "something spoken." Depending on the nature of the task and on its specific understanding, children select among alternative strategies which are bound to different attributes within the functional core of the concept (p. 251).

The authors further suggest that formal instruction in reading and writing in school leads to a restriction of the broad functional core in the child's concept of a word in the sense that literate persons are oriented mainly to visual strategies.

The study by Januschek et al. throws an interesting light on the question of how children build up semantic representations of concepts in general and points to a multiplicity of cognitive, linguistic, and pragmatic factors connected with this development. It also shows, that, through the influence of school, the child's concept of a word is restructured. Unfortunately, in this study many variables (age, cognitive development, and print experience) were confounded so that the exact influence of school and reading experience cannot be determined.

The Transformation of the Concept of "Word" Under the Influence of School

In a later study, Januschek and Rohde (1981) asked second and third graders to judge which of two words, each presented in a written form on a little card, was easier or more difficult. Again this task was embedded in a play situation. The percentage of the explanations referring to the written form of the word increased from about 15 among the second graders to more than 50 among the third graders. Half of the answers of the latter group referred to specific orthographic problems. Only 1 percent of the answers of the third graders, but about 15 percent of the explanations of the second graders, were related to the pronunciation of the word. This study seems to confirm the hypotheses of Januschek et al. (1979a,b) that the child's concept of a word is influenced by the written form of the word and changes under the influence of school instruction.

However, since this study required the subjects to read the two words, which were to be compared, the answers might be influenced by this mode of presentation. This study should be repeated with an oral presentation of words.

Results of other studies also indicate that the child's concept of a word gradually changes upon becoming acquainted with print. Ehri, in her chapter in this book, has outlined the impact of experiences with reading on the child's lexical awareness.

The results obtained in the interview study by Berthoud-Papandropoulou (1978) are also in agreement with the hypothesis of Januschek et al. She observed a "lack of homogeneity" in the responses of the children in the intermediate age group of four and five year olds, while the older subjects had "elaborated a clear concept of what words are" (p. 62), in the sense that it referred to a linguistic unit. Younger children had "not yet worked out how the different criteria can be combined, and see nothing contradictory in, for example, accepting *dormir* as a word . . . ('because you sleep in a bed'), and rejecting *la* . . . ('because there aren't enough letters') or accepting *it* . . . ('because you write it')." Berthoud-Papandropoulou stated that the graphic substance of words occupies a very special place in children's thinking after the age of six. These observations confirm the hypothesis of a multiple-core concept of a word in preschool children which is narrowed under the influence of the school. It should be noted, however, that Berthoud-Papandropoulou (1978) herself gives an alternative interpretation of her data in the sense that the "restrictive conception of a word" (p. 62) of young children becomes enlarged with increasing age and cognitive development.

That children restructure their everyday concept of an oral word under the influence of the school is also demonstrated by the results from Papandropoulou and Sinclair (1974). They asked children aged four to ten years for a definition of a word and stated with regard to the older age group: "Peculiarly enough, hearing is only rarely referred to. When questions are asked about the form of words, all subjects speak of words as being made up of letters" (p. 247). One seven-year-old child said "A word is made of letters, yes, even when one talks" (p. 247). A further observation by these authors confirms the hypothesis by Januschek et al. that the child's concept of a word is narrowed under the influence of personal experience: "Interestingly, the answers given by this age group (8 to 10 years) become remarkably uniform" (Papandropoulou & Sinclair, 1974, p. 247). As an explanation for these phenomena, they dismiss the influence of school as a superficial factor and regard features of the written language (which are learned in school, however!) as suited to make segmentations visible that do not exist in actual speech. A similar interpretation is provided by the authors in their chapter in this book.

Francis (1973), in order to trace children's comprehension of linguistic units, used a follow-up approach and interviewed infants school (kindergarten age) children at six-month intervals about their concepts of word, letter, and sentence. The outstanding characteristic was the almost universal reference to spelling, reading, and writing. As Francis reported, almost no replies indicated an awareness of the use of words or sentences in the spoken language. "It was as though the children had never thought to analyse speech, but in learning to read had been forced to recognize units and sub-divisions" (p. 22).

Trying to find further evidence for Francis' claim that young children's notions of units in speech are derived from the analysis of written language forms, Hall (1976) compared the performance of children between the ages of four years, eight months and six years in two segmentation tests, one requiring them to segment oral sentences into words, the other requiring the segmentation of written sentences into words. As expected, the children had significantly higher scores in the segmentation of print than in speech segmentation. Segmentation of print was positively correlated with two measures of print experience (the number of terms the child had been in school and reading experience as indicated by her or his progress in the basal reader series). Segmentation of speech was correlated significantly with the number of terms that the child had been at school. Since both segmentation tests were not significantly related, Hall concluded that children were applying different criteria of "word" to the two test situations. The major clue in recognizing written words was the spaces around the words. Hall observed: "In the testing situation . . . some of the very young children were not even concerned that the sentence they had to cut up was upside down and back to front. They were so cued up for spaces that, if the typographical representation of a word had two consecutive letters whose shape entailed a slightly larger gap than normal, the children cut into it" (p. 17). Some of the older children cut out words that were embedded in longer words. These observations seem to indicate that at least some young children's conception of word boundaries and spaces in written language was of a more figurative type (reaction to visual features without conceptual understanding).

Hall points out that the knowledge about spaces as cues for word boundaries may well be a result of learning to write rather than a result of learning to read. This is corroborated by the fact that young children who are beginning readers show in their spontaneous writings no concern for word boundaries (Bissex, 1980). On the other hand, they may well be able to include space when copying sentences (Clay, 1975, p. 54).

Other Factors Influencing the Development
of the Concept of "Word"

Besides scholastic influences and the impact of reading instruction, there are other factors which influence the development of word awareness. In some studies investigating the word consciousness of kindergarten children, bimodal distributions have been found (Ehri, 1979). Low scorers were those who knew little about printed language. Among high scorers were children with and without print experience. Some children obviously acquire some word knowledge before they learn to read. Watson, in his chapter in this book, has proposed general cognitive development as an explanatory factor and offered some evidence for this plausible suggestion. He refers to studies showing correlations between aspects of language awareness and operativity measures. In his longitudinal study he has demonstrated that seriation, as one aspect of conceptual reasoning, has direct effects on the child's understanding of units of print and that this effect is distinct from that of reading achievement and oral-language abilities.

Exposure to a second language has been considered to be another factor in promoting metalinguistic knowledge, because it may lead to an earlier separation of word sound from word meaning. Some research studies indeed seem to indicate that bilingual children show a slightly superior performance in tasks which require focusing on formal aspects of language (Ianco-Worrall, 1972; Ben-Zeev, 1977). Aronsson (1981) has provided a thorough analysis of previous studies on the effect of bilingualism. She concluded that the results are inconsistent with regard to the claim that bilinguals, due to their exposure to two different symbol systems, show a better understanding of the arbitrary relationships between word and referent. In two German studies the metalinguistic abilities of bilingual children were examined. Due to the smallness of the samples, the data must be viewed with caution, however. Karolije-Walz (1981) reported the superiority of five-year-old bilingual preschool children over monolingual children in their understanding of the arbitrariness of names. Bense (1981) observed better performance of bilingual children in the word-segmentation task used by Januschek et al. (1979a,b). From observations in her sample, Bense puts forward the interesting hypothesis that the attitude of the parents to the bilingualism of their child (active support or neglect) and the motivation of the child to learn the second language are important factors that influence the effect of bilingualism on the cognitive development of the child. Similarly, Rubinstein (1979) has shown that bilingualism acquired in a supportive, structured environment may be beneficial, but that bilingualism acquired in a stressful environment may result in serious deficits in linguistic and, presumably, cognitive functioning.

To summarize, the development of the concept "word" seems to be influenced by a multiplicity of factors. Scholastic influence (instruction in grammar and in reading and writing) seems to be the basic source for developing a scientific concept of a word. Other factors that might contribute to the development of an everyday concept of a word should not be overlooked, however (for instance, the child's experience with language in general). Parents emphasizing the appropriateness of utterances and their explicit statements about the adequacy of specific terms may focus children's attention on language forms and provide them with a metalinguistic vocabulary. But the degree to which children may profit from these stimulations may very much depend on their developmental level. Slobins' daughter Heida, for instance, who was exposed to foreign languages very early, already acquired a metalinguistic vocabulary after only two to three years. Interestingly, however, she was at the same time unable to distinguish English words from the concepts to which they referred (Slobin, 1978, p. 46). This example, together with the experimental evidence offered by Watson in his chapter, points to the importance of cognitive development. The child's construction of word knowledge may rest on a productive match between her or his conceptual level and the stimulations offered by home and school factors.

[1] I expressed the idea that his theory prevents us from understanding the process whereby the child learns grammatical forms. As is well-known, this process is at its most intense in early development. It would be completely impossible if orientation towards the sound form did not arise in the child while learning. One of my graduate students, M. I. Popova, carried out an experiment in which she showed that inflection for gender in the past tense of verbs could only be mastered if this orientation existed. Consequently this orientation takes place when grammatical forms are learnt, and, clearly, serves as a basis for learning language at an early age. This does not signify in the least that the "glass theory" does not correspond to reality. Even as adults, when we talk we do not notice the sound form in which we express our thoughts. The orientation that takes place in children when learning language is not at all conscious, but it is there (Elkonin, 1983).

CHAPTER 12

Awareness of Features and Functions of Language

Renate Valtin

This chapter's chief concern is with children's concepts of linguistic features and the child's understanding of the functions of language. The chapter begins with a review of research on phonemic awareness that has been conducted in several different languages by various authors. The extremely interesting array of research strategies, however, leaves us still uncertain about the precise nature of the roots of phonemic concepts. Turning to functional concepts, evidence will be presented that preschool children are implicitly aware of the communicative function of language. This may be inferred from their behavior that is adjusted to the needs of the listener. Moreover, as a study on metacommunicative behavior shows, preschool children are able to make judgments and verbal statements about inappropriate communication behavior. Finally, some considerations about methods of assessing aspects of language awareness will be presented and some pedagogical implications regarding different stages of the development of metalinguistic knowledge will be outlined.

Phonemic Awareness

In order to understand the alphabetic principle found in most written languages and to be able to profit from this principle in reading and writing, the child needs the apprehension that spoken language may be segmented into smaller units that are represented by letters. Synthesis of phonemes and syllables is one of the components of word-decoding subskills and a strategy to attack new words. Segmentation into syllables and phonemes is a component of spelling, at least of that spelling strategy that may be characterized as relying on letter-sound rules.

Once phonemic synthesis and analysis are mastered, they may become automatic and even be bypassed by fluent readers and spellers. Thus, they have relevance mainly for the acquisition phase of learning to read and spell. Singer, in his chapter in this book, as well as Doctor and Coltheart (1980), have provided convincing evidence that in the beginning-reading stage recognition of written words is, for the most part, mediated through phonological or phonetic recoding. Fluent readers, however, can bypass the phonetic recoding and directly extract meaning from print—at least at the word-recognition level (Barron, 1978). This type of phonetic recoding that serves as a means for word decoding or as a word attack subskill is not to be confounded with the phonetic recoding (or use of speech code) as it is referred to by Mattingly in his chapter. Phonetic recoding has different relevance for word decoding (identifying the semantic referents of words or word groups belonging together) and for language comprehension (analyzing the semantic/syntactic properties of language and combining concepts into meaningful units). While phonetic recoding may be bypassed in word recognition, some sort of speech code is indispensable for the processing of information in working, or short-term, memory (Barron, 1978; Massaro, 1978; Liberman, Liberman, Mattingly, & Shankweiler, 1980). In short-term memory the verbally encoded information is stored for several seconds while sentences are processed for meaning. Even readers of a nonalphabetic writing system (e.g., Chinese) have been shown to use phonetic recoding, either for temporary storage, or for comprehension processes, or both (Tzeng & Hung, 1981). This chapter deals only with phonetic recoding as a means to identify words by converting letters or letter patterns into sounds or sound patterns.

As already pointed out in the preceding chapter, it seems useful to differentiate two components of language awareness: (1) objectivation or the ability to reflect on language forms while abstracting from the meaning aspect and paying attention to sound properties of language and (2) the ability to identify phonemes and to manipulate phonemic segments. It was hypothesized that children need some minimum cognitive level in order to be able to abstract from the meaning aspect (in the sense of a decentration ability) and that some sort of training is required in order to acquire the concept of a phoneme. This training may come through direct phonemic segmentation tasks, as in the Elkonin experiment, or it may be the more or less direct effect of learning to read and to spell.

Brief Review of Anglo-American Studies

Some reviews of research studies on phonemic segmentation—referring mainly to Anglo-American literature—have been presented by Golinkoff (1978), Ehri (1979), and Ehri and Downing in their chapters in this book. The main findings may be summarized as follows:

(1) Phonemic segmentation is a difficult task due to the nature of the acoustic signal. In speech the phonemes are not discrete units but encoded at the acoustic level into larger units of approximately syllabic size (Liberman, Cooper, Shankweiler & Studdert-Kennedy, 1967, p. 203). Since phonemes are abstract units, phonemic segmentation and synthesis are thus not simple associative memory tasks but highly demanding conceptual tasks (Helfgott,

1976; Ehri, 1979). Syllable segmentation is easier than phonemic segmentation (Gleitman & Rozin, 1973; Goldstein, 1976; Fox & Routh, 1976).

(2) The difficulty of phonemic tasks varies with the complexity of the operations required, e.g., recognition, counting, partial or full segmentation, manipulation, and reversal of phonemic units (Golinkoff, 1978; Lewkowicz, 1980). The difficulty also depends on type and position of the phonemes. Continuants are easier to identify than stops (Marsh & Mineo, 1977), and phonemes in initial position easier than in terminal or middle position (Bruce, 1964; Zhurova, 1973). Initial consonants are easier to identify when followed by a vowel than by a consonant (Carver, 1967). While initial consonants seem to be more segmentable, final consonants seem to be easier to synthesize (Helfgott, 1976).

(3) Although phonemic segmentation may be trained in preschool children, not all children will learn it. Even after 80 trials and demonstrations, about one third of the kindergarten children in the study by Helfgott (1976) were unable to perform the segmentation of CVC words. In the study by McNeill and Stone (1965), kindergarten pupils trained to identify the presence or absence of two consonants in meaningful words did not perform above chance level in the posttest. This training was purely auditory, however. The effects of phonemic training are higher when visual aids are used to represent the sound sequence (Marsh & Mineo, 1977; Lewkowicz & Low, 1979). Using letters to visualize the phonemic task seems to be superior to using squares (Ehri, in her chapter).

(4) The greatest increase in phonemic segmentation abilities can be observed between kindergarten children and first graders (Liberman, Shankweiler, Fischer, & Carter, 1974; Rosner & Simon, 1971; Calfee, Lindamood & Lindamood, 1973). These findings led Ehri (1979) to the conclusion that "reading instruction may very well be the important factor enabling children to conduct this sort of analysis of words" (p. 92).

(5) The relationship between phonemic awareness and reading ability has been demonstrated by means of correlational studies, using concurrent or predictive or both kinds of correlations (Rosner & Simon, 1971; Calfee, Lindamood & Lindamood, 1973; Fox & Routh, 1975; Downing in his chapter in this book).

(6) The nature of this relationship remains unclear, however. There is no direct experimental evidence to specify the status of phonemic segmentation in the sense of a prerequisite, a facilitator, or a consequence of reading instruction. However, most of the researchers propose an interactive view in the sense that phonological sensitivity is both a contributor and a consequence of learning to read (Goldstein, 1976; Ehri, 1979).

A general problem of these studies is that they did not control for possible influence of children's preschool reading ability. Thus, their preschool phonemic segmentation abilities may already be a consequence of the familiarity with print and at least a glimpse of understanding of the alphabetic principle. Hiebert's (1981) study reveals how sophisticated some American three and four year olds can be about letter naming.

Ehri (1979) has made suggestions for experimental designs which could yield indications as to whether phonemic segmentation is to be regarded as a prerequisite or as a facilitator.

(7) Results presented by Treiman and Baron (1981) suggest that segmental analysis (ability to count phonemes) does not relate to reading ability in general but to a particular component of reading—the ability to use spelling-sound rules. Treiman and Baron differentiate two types of readers: "Phoenicians" who mainly use spelling-sound rules and "Chinese" who mainly depend on word-specific associations. "Phoenicians" seem to be better at phoneme analysis than "Chinese." The direction of the causal link remains unclear, however. Are children good at learning spelling-sound rules because they are good at segmental analysis? Or do children who know spelling-sound rules do well on phoneme analysis tests because they can imagine the spellings of words?

(8) There is some evidence that the relationship of phonemic segmentation to reading achievement is dependent on those components that are required in the specific reading instruction program used (Ehri, 1979, p. 100). In the study by Bruce (1964), children from a school with emphasis on phonics instruction were better in a task requiring deletion of a sound from various places in a word (S-NAIL, MON-K-EY, PART-Y) than children from a school that favored a sight word approach.

The literature referred to in the preceding passages also gives some indication of the need for a differentiation between objectivation of language (abstraction from meaning and paying attention to sound properties) and phonemic awareness tasks (ability to identify phonemic units and to manipulate them). In a similar way, Gleitman and Rozin (1973) have proposed that two critical cognitive problems should not be confounded in reading instruction: "The first is learning that English orthography directly maps sound rather than meaning, and the second is learning that orthographic units correspond to highly abstract and inaccessible phonological segments" (p. 479). From the strategies demonstrated in response to the task in the study of Bruce (1964), it may be inferred that kindergarten children are able to react to sound properties of speech, but are unable to identify and manipulate phonemes. Sinclair and Berthoud-Papandropoulou, in their chaper in this book, also provide evidence that preschool children are aware that language is audible and pronounceable and that they may react to sound properties but they are unable to perform phonemic analysis. Likewise, some five-year-old children may exhibit some awareness of the tactilokinesthetic sensations involved in speech production (Zei, 1979).

The literature also indicates that children do not use phonemic segments spontaneously, but require some training in order to perform such tasks. Zhurova's (1973) study with Russian subjects, ages three to seven, demonstrated that instruction is even needed in order to get children to isolate the first sound of a word. This segmentation task was embedded in a playful situation. (Children had to help various animals to cross a magic bridge by naming the first sound

of the animal's name to a wise crow). Children who "knew their letters pronounced the first letter as it is pronounced in the alphabet. Those children, however, who did not know the alphabet pronounced syllables" (Zhurova, 1973,
p. 148). Only when the experimenter repeated the first sound by intoning (iter-
ating the first sound) were the children able to respond with this intoned segment, but not with isolated sounds (Experimenter: "Here is belka. Who is
coming?" Child: "be." Experimenter: "What is the first sound in the word
's-s-sobaka'?" Subject: "S-s-s.") Without intoning, preschool children were
unable to single out a sound in a word.

The possible impact of letter knowledge on children's segmentation abilities
has been demonstrated in Zhurova's study. Ehri, in her chapter, has provided
more evidence for the claim that print experiences have an influence on spoken
language. She suggests that learning written language may provide a visual representational system for speech that may work various changes in children' competence with speech, both at the lexical and the phonemic level. In her experiments the active formative role of written language on the child's conceptualization of the sound-structure of language has been demonstrated. Sinclair and
Berthoud-Papandropoulou, in their chapter, also hypothesize that the objectivation of language in its written form may stimulate children's thinking about
language properties in general and that they may transpose the resulting concepts
to their reflections about spoken utterances.

Some European studies may be of interest in this context because they provide insight into the development of phonemic abilities in relation to reading and
spelling instruction across languages.

The French Study of Leroy-Boussion

In a longitudinal study, Leroy-Boussion and Martinez (1974) and Leroy-Boussion
(1975) investigated the development of syllable analysis and synthesis, as related
to I.Q. and number of years in school. A sample of 224 unselected pupils, representing a wide range of social backgrounds, was tested every three months during
three years, 5 and 6 (école maternelle or kindergarten), 6 to 7 (cours préparatoire
or first grade) and 7 to 8 (cours élémentaire or second grade). The measure of
syllable analysis consisted of 42 two-phoneme syllables that were spoken loudly
and clearly to the subjects who then had to answer with the two sounds ("pa"
is /p/ and /a/). In the synthesis task the experimenter spoke the two sounds and
the child had to respond with the syllable. The ∪rder of the tests was reversed
every time. Wrong answers of the children were corrected. Unfortunately, the
authors do not report in what way this perpetual correction that provided good
learning possibilities affected the performance of the children. Figure 12-1 represents the percentage of right answers of the children at different times of the
investigation. The data show that syllable synthesis is easier than syllable analysis.
Both skills show a gradual increase during the year in the kindergarten. No formal reading instruction is provided during this year, but the children learn to pay

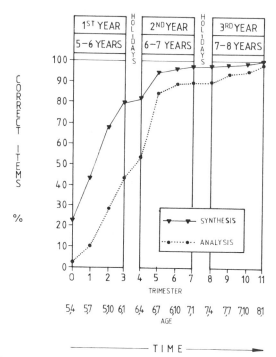

Figure 12-1. Percentage of right answers in the syllable analysis and synthesis tests at different times of the investigation (from Leroy-Boussion, 1975).

attention to the sound structure of speech and to read some sight words. With the commencement of reading instruction (after the fourth trimester) both abilities progress rapidly. Leroy-Boussion emphasizes that French first graders get simple dictations from this time on, parallel to the reading instruction. That means that children have to use some phonemic segmentation ability. Figure 12-2 shows the performance of the children as a function of different I.Q. levels. The higher the I.Q., the higher the analysis and synthesis abilities. If one regards syllable analysis and synthesis as important subskills of reading and writing, the deficiencies of the children with far below average I.Q.s in this sample become apparent. They do not seem to profit much from the ordinary classroom instruction. Unfortunately, no direct measures of the reading and writing abilities were used and the relationship with syllable synthesis and analysis remains unclear. Furthermore, no information is reported as to whether some children already possessed some degree of reading skill when entering the kindergarten.

It is important to note that the French authors have a purely maturational concept of the syllable analysis and synthesis. They regard these abilities as "mental abilities that develop 'naturally' and spontaneously under the effect of the linguistic stimulations provided in the everyday life of the child" and "follow the laws of the genetic development" (Leroy-Boussion, 1975, p. 185, translated by Valtin). According to their maturational view, the authors interpret their findings that not all children are "ready" for syllable analysis and synthesis

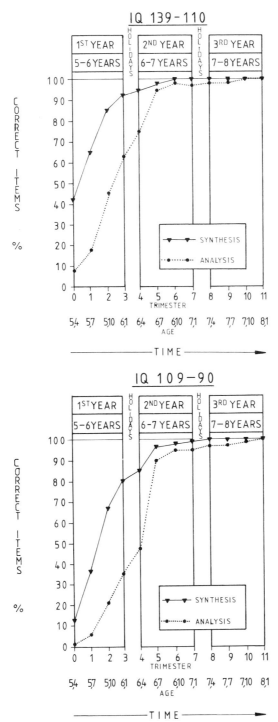

Figure 12-2. Performance of children at different I.Q. levels in the syllable analysis and synthesis tests at different times of the investigation (from Leroy-Boussion, 1975).

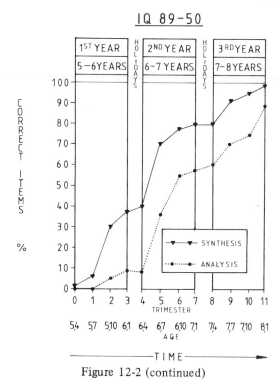

Figure 12-2 (continued)

when formal reading instruction begins as evidence for the need to postpone reading instruction. They suggest a differentiated beginning for children, according to their I.Q.s, and warn educators not to force or to boost these maturational abilities by too much formal instruction. However, since the authors did not consider preschool reading achievement of the children in their sample, no claim can be made that the development of syllable analysis and synthesis reflects purely maturational changes. Other data, provided by Leroy-Boussion (1975), show the influence of school reading instruction on these abilities. She investigated the syllable synthesis ability of five year olds who had received formal reading instruction during their kindergarten year. Figure 12-3 presents the data of 145 pairs of children with and without reading instruction in kindergarten matched in regard to I.Q., age, sex, and social status. Though the five-year-old early readers show better performance in the syllable synthesis test, they do not reach the same high level as the six-year-old control children after their first year of reading instruction. A differentiation according to various I.Q. levels shows that is was mainly the five year olds with a far above average I.Q. who profited from the early reading instruction.

The studies by these French authors point to the multiplicity of factors influencing the development of syllable analysis and synthesis abilities. A certain

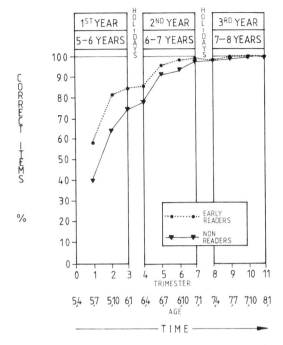

Figure 12-3. Performance of children with and without preschool reading ability in the syllable synthesis test at different times of the investigation (from Leroy-Boussion, 1975).

minimal cognitive level seems to be a prerequisite for learning these subskills and for being able to profit from ordinary reading instruction. The influence of specific reading instruction procedures on the development of these abilities has not been determined in this French work.

The Dutch Study "Preventie van leesmoeilijkheden"

During the past five years a comprehensive project, "prevention of reading disabilities," has been designed and carried out in the Netherlands (Van Dongen, 1979; Van Dongen, Bosch & Mommers, 1981; Van Dongen & Van Leent, 1981; Van Dongen & Wolfhage, 1982; Mommers, 1982; Van Leent, 1982, 1983). This longitudinal study, based on a sample of more than 600 subjects, has three aims: (1) the study of the development of various subskills in oral and silent reading and in spelling over three years, (2) the prediction of reading and spelling abilities from a comprehensive battery of tests carried out in kindergarten (with measures of reading readiness, personality, and social factors) and (3) to observe the development of children defined as children "at risk" on the basis of their scores in the kindergarten tests compared with children "non-at-risk." This latter group comprised 96 children, 48 at risk and 48 non-at-risk. The development of auditory analysis and synthesis of this group has been studied in detail. During the first year of school, beginning with the first week and then every month, two tests of auditory synthesis and auditory analysis have been carried out. These tests included words with two, three, and four phonemes, but the material was not identical in the two tests as in the study of Leroy-Boussion. An examination of the distribution tables of the scores of auditory analysis and synthesis at various points of time during the first year of school (Mommers, 1982, Fig. XII) shows, first, that the synthesis scores were consistently higher than the analysis scores (as was found also in the study of Leroy-Boussion) and, second, that all children (except those with a percentile less than 10 in the auditory analysis test at the beginning of the school year) reached the maximum scores at the end of the year. A differentiation of the group according to different performance levels at the beginning of the school year demonstrated that the children with the lowest performance at the beginning needed the longest time to reach the maximum score. The development of the auditory synthesis and analysis subskills was not a gradual process. In the analysis test, for all groups with various performance levels at the beginning, at a certain point in time, a dramatic increase in the test scores occurred: during one month the scores increased more than 30 points in all groups. This increase can be interpreted as indicating that the children suddenly gained an important insight about how to perform the tasks. This insight took place very early for children with a high performance level (during the first two months) and later for children with a low performance level at the beginning (the third to sixth month). The development of the auditory synthesis subskill revealed a similar picture, with a rapid increase of 25 points in two months.

A comparison of children at risk and non-at-risk (Van Leent, 1982, Fig. 3) shows that children of the latter group started with some knowledge in this test (median score of 15), progressed rapidly, and achieved the maximum test score at the end of the fourth month. Children at risk started with a low score (median of 4), needed more time to progress (median of 15 at the end of the third month), had a rapid increase of scores between the third and fifth months (up to a median score of 37), and reached the maximum score at the seventh month. The children in the non-at-risk group had a mean I.Q. of 110, whereas the group of children at risk had a mean I.Q. of only 94 (Van Dongen, personal communication). That children with a lower I.Q. need more time to develop the subskills of auditory analysis and synthesis has been demonstrated also in the study by Leroy-Boussion.

Unfortunately, in the Dutch project no direct measures of preschool reading ability have been used. Presumably some of the children in the non-at-risk group already possessed some reading skill and thus showed a quicker progression in the analysis and synthesis tests. But this is only a hypothesis. Up to now, the authors of this project have not reported the scores of the various reading and spelling subskills and the relation of these scores to the auditory analysis and synthesis abilities.

Van Leent (1982) discussed the question whether auditory synthesis can be regarded as a prerequisite for learning to read, as had been supposed by some

Dutch researchers. He concluded: "Our results indicate that many children do not master auditory synthesis at the moment when reading instruction starts. They acquire this ability mainly during the first months of school. Auditory synthesis is perhaps a prerequisite for mastering the reading process, but not a prerequisite for the beginning of reading instruction, because in our exploratory investigation many of those children (with no auditory synthesis skill at the beginning) have learned to read normally" (Van Leent, 1982, p. 14, translated by Valtin). Although this conclusion is plausible and reflects the ideas of the present author, it is not based on empirical results, since the relationship of the auditory analysis and synthesis subskills to reading achievement was not reported in the available literature about this project.

In his review of research on auditory analysis and learning to read, Van Leent (1983) specifies his theoretical position. He states that, during the first phases of the learning-to-read process, the child has to gain insight into the specific character of our alphabetic script. The child has to realize that a spoken word can be segmented into phonemes and that the temporal sequence of sounds corresponds to the spatial sequence of letters in a written word. The ability of phonemic segmentation may thus be viewed as a good indicator of a child's insight into the structuring of the speech flow upon which alphabetic script is based. If children do not have this phonemic awareness or the understanding that words may be segmented into smaller parts at the beginning of their reading instruction, they may not be able to learn related subskills such as phonemic analysis or synthesis which are highly emphasized in Dutch learning to read programs. Only children with this insight are able to profit from ordinary reading instruction. A lack of phonemic awareness should not be taken as a reason for postponing reading instruction, however, but as a challenge for the teacher to develop this insight in children. Since all subjects in this project reached a high level of skill in auditory synthesis and analysis during the first year of school, Dutch teachers seem to be able to provide this aspect of cognitive clarity.

In the prediction study of the Dutch project, three tests of auditory or phonemic subskills were used. These were paper and pencil tests based on pictures and administered as group tests. In the auditory synthesis test the experimenter spoke isolated sounds, and the child had to blend the sounds silently and circle a picture representing this word. Five pictures were presented for each word, thus allowing a 20 percent chance of guessing correctly. The presentation of the target word also facilitates the task. In the phonemic analysis test the child had to identify the name of a picture and then find, out of a row of three other pictures, the word with the same beginning sound. Again, there is a 33 percent chance to guess correctly. In the auditory discrimination test the experimenter presented a pair of similar words and the child had to circle a picture. The chance of guessing correctly is 50 percent in this task. Another problem of these tests was the high number of unscorable items because many subjects had circled more alternatives or given no answer at all. This high number suggests that some subjects either did not understand the task requirements or were unable to deal with phonemes. An inspection of the distribution tables of the analysis test (Van Dongen, Bosch

& Mommers, 1981, p. 34) reveals that nearly a quarter of the sample had answers on the chance level. Despite the fact that these tests are relatively easy with regard to the required operation (recognition) and the number of phonemes involved (mainly one, and up to three), many preschool children were unable to cope with the task demands.

Since no preschool reading measures were used, it is unclear whether the ability to deal with phonemes in these tests reflect some acquaintance with the reading process.

In the whole sample of about 550 subjects, the correlation of the auditory synthesis test with reading achievement in first grade was 0.34 (oral reading) and 0.32 (reading comprehension). Phonetic analysis had a slightly higher correlation, 0.35 with oral reading and 0.43 with reading comprehension. The best single predictor was a test of visual letter perception (a series of three and four letters had to be identified out of a row of four alternatives). This test correlated 0.44 with oral reading and 0.49 with reading comprehension (Mommers 1982, p. 12). The whole battery of kindergarten measures, including social, personality, intelligence, and reading-readiness measures, accounted for 30 to 34 percent of the explained variance of oral reading, for 40 percent of reading comprehension, and for 31 percent of spelling at the end of first grade. At the end of second grade the variance accounted for by the battery dropped to 22 percent with regard to oral reading, 30 percent with regard to reading comprehension, and 24 percent with regard to spelling (Van Dongen & Wolfhage, 1982, Table 1). In an attempt to reduce the number of variables in the predictive battery, Van Dongen and Wolfhage (1982, Table 2) found that, for the prediction of reading and spelling ability at the end of first grade, kindergarten teacher ratings of the child's intelligence, school readiness, and school progress accounted for the highest amount of explained variance (18 to 25 percent). The visual letter-perception test accounted for an additional 6 to 9 percent of the variance, and the phonemic analysis test accounted for only an additional 1 to 4 percent.

Van Leent (1983) has provided some more data concerning the predictive validity of the phonemic analysis test. After three months of reading instruction, 38 percent of the children with low scores in phonemic analysis (below the lowest quartile) also belonged to the lowest quartile on their reading measure, and 72 percent belonged to the last half of the sample. To put it the other way round, despite poor phonemic abilities at the beginning of reading instruction, 28 percent showed above average reading performance after three months of instruction.

One word of caution should be uttered with regard to predictive validity. The underlying rationale of these predictive studies is the assumption that children's scholastic achievement is a direct function of the abilities that they exhibit when entering school and that instruction does not provide a systematic but rather a random effect. The role assigned to the teacher and her teaching methods seems to be merely ephemeral. Basically this implies a pessimistic picture of education. Whoever believes education to have some beneficial effect and considers compensatory effects as desirable must be suspicious of predictive correlations between preschool abilities and scholastic achievement.

Apparently, the Dutch researchers started their project with the idea that linguistic awareness is a predictor in the sense of a prerequisite for learning to read and spell. In their latest publication, as has already become obvious in the above citation of Van Leent (1982), they prefer to speak of linguistic awareness not as a "conditio sine qua non," but as a facilitative factor. A German study also casts doubt on the prerequisite hypothesis.

The German Study by Röhr

Röhr (1978) investigated more than 200 kindergarten children with a battery of predictive tests. He used psychometric tests (intelligence, language, and perception), social and biographic data (social status, number of siblings, etc.), and a questionnaire for the kindergarten teachers (ratings concerning the child's intelligence, personality, and social behavior). Table 12-1 gives an overview of the results of the prediction of reading and spelling ability at the end of second grade. The last row contains the results of a reading test administered in grade one. Grade-one reading-test scores were the best predictors for the achievement in reading and spelling in grade two. As in the Dutch study, the kindergarten teacher ratings had a relatively good predictive value compared with psychometric tests. Of special interest is the relationship between low auditory synthesis scores at the beginning of reading instruction and the later success in reading and spelling. Of the 82 children of this sample with low scores in the subtest "sound blending" of the Illinois Test of Psycholinguistic Abilities (ITPA), 73 percent showed average or above average performance in reading at the end of second grade and 83 percent showed above average and average performance in spelling. Röhr (1978, p. 320) concludes that auditory synthesis abilities, at least those as measured with this subtest of the ITPA, are not a necessary prerequisite for learning to read and to spell.

The Swedish Studies of Lundberg

Lundberg, Wall, and Olofsson (1980) and Lundberg (1982a) reported a study designed for the prediction of reading and spelling abilities from a battery of tests used in kindergarten, including metalinguistic tests and tests of nonlinguistic decentration. The follow-up study started with about 200 children from kindergarten (6 to 7 years old) who were tested again at the end of first grade

Table 12-1. Determinants (in percentage) of Reading and Spelling Achievement at the End of Second Grade (according to Röhr, 1978).

Predictors	Determination of Spelling			Determination of Reading		
Psychometric Tests	28			54		
Social-biographic data	28	54		24	63	75
Teacher Rating	ca. 30		73	ca. 40		
Reading (first grade)	64			63		

and in the middle of second grade. At this time, 133 subjects were available. In Sweden, children begin school relatively late (about seven years old). Reading instruction during the first school years is rather uniform, with an emphasis on phonics. The orthography can be characterized as regular.

Figure 12-4 presents an outline of the causal model underlying this investigation. The authors describe their model as follows:

> The main determinant of reading and spelling ability in school is assumed to be a set of metalinguistic skills measured in kindergarten which included analysis as well as synthesis of phonemes and syllables. Necessary prerequisites for these skills to develop are assumed to be general intelligence and more specific ability to analyze and decenter in nonverbal tasks. The latter factors may also directly influence the reading and spelling ability (Lundberg et al., 1980, p. 161).

In the kindergarten study a battery of tests was used. The measures for word synthesis and analysis included:

(1) Synthesis of syllables and of phonemes, concretely represented by pegs on a board (SYNSYLC and SYNPHONC)
(2) Direct auditory synthesis of syllable and of phonemes (SYNSYLD and SYNPHOND). Syllables and phonemes were presented in isolation, using a play situation
(3) Segmentation into syllables and into phonemes, concretely represented (ANSYLC and ANPHONC)
(4) Analysis of phoneme position (ANPHONPOS). The child had to judge whether a spoken word started, ended with, or included a given sound (/s/, /m/, /r/ and /i/)

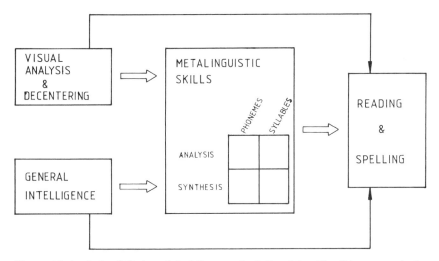

Figure 12-4. A simplified model of the causal relationships. The thin arrows designate weaker causal links (from Lundberg et al., 1980).

(5) Reversals of phonemes (ANPHONREV). Two- and three-phoneme words had to be reversed and pronounced
(6) Rhyme production.

Lundberg regards the attention shift from content to form as the central aspect of linguistic awareness and suggests a relationship to perceptual decentration abilities. In order to assess this relationship, two nonlinguistic decentration tests were used. In the Target Identification test (TIT) the child has to identify a specific visual figure that is embedded in a drawing of a lively scene. In the Picture Integration test (PIT) the child has to pay attention to two independent meaningful aspects of an object (e.g., an array of fruits that constitutes a man).

A preschool reading test (PREAD) was also used, but it was a "quick and rather crude screening device" (Lundberg et al., 1980, p. 164). Since preschool reading ability (at least in the view of the present author) plays a rather important role in this study, details of the tests should be noted. The test included four parts: (1) five simple CVC words, (2) three simple three-word sentences, (3) a long sentence, and (4) a long sentence with a higher difficulty level. Children were scored pass/fail on these parts. A score of 0 designated failure on the first level and a score of 4 designated success on all items. During first and second grade a silent word reading test (SWREAD), a spelling test, and an intelligence test (RAVEN) were administered.

In order to assess the relative importance of the predictor variables the authors used path analysis. Figure 12-5 presents the results with regard to the silent word reading test in grade one as the last dependent variable. Lundberg et al. (1980) conclude, "The most powerful determinant of reading achievement in grade one is the ability in kindergarten to analyze phonemes and reverse their order" (p. 166). This conclusion seems to be at variance with our hypothesis that phonemic-segmentation abilities are a result of learning to read. This hypothesis can be maintained, however, when it can be demonstrated that the successful segmenters in the Lundberg study were already readers when entering school. Preschool reading ability and its correlations with the other measures must therefore be considered further. Table 12-2 shows the correlations between preschool reading ability (PREAD), metalinguistic tasks, and reading and spelling achievement in grade one. As can be seen from Table 12-2, the correlations between preschool reading ability and the metalinguistic tasks are higher than the correlations between the metalinguistic tasks and the reading and spelling abilities in grade one and, furthermore, preschool reading ability correlates higher with later reading achievement than the ability to reverse phonemes. Thus, metalinguistic abilities may be seen, in large part, as reflecting preschool reading abilities.

Since seldom in the literature have correlations between different measures of metalinguistic abilities been reported, these data may be of some interest. While the correlations between the different phonemic tasks are moderate to high (between 0.68 and 0.86), the correlations between the syllabic tasks are not very impressive, possibly because of a ceiling effect (maximum score in these tasks is 4). The low correlation between the two perceptual decentration tests (0.21)

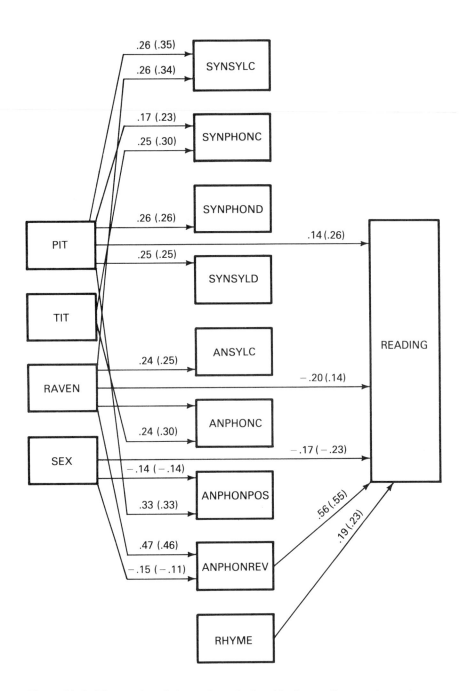

Figure 12-5. The results of the path analysis with the reading test in grade one as the last dependent variable. Above each arrow, the path coefficient and, within parentheses, the product moment correlation are given (from Lunderg et al., 1980).

Table 12-2. Intercorrelations Among Metalinguistic Tasks and Measures of Reading and Spelling (from Lundberg et al., 1980, p. 165).

	2	3	4	5	6	7	8	9	10	11	12	13	14
1 SYNSYLC	0.58	0.42	0.49	0.18	0.54	0.63	0.49	0.18	0.17	0.33	0.46	0.33	0.40
2 SYNPHONC		0.48	0.79	0.02	0.77	0.78	0.75	0.21	0.30	0.23	0.72	0.50	0.46
3 SYNSYLD			0.50	0.04	0.40	0.42	0.45	0.11	0.11	0.25	0.44	0.24	0.24
4 SYNPHOND				0.04	0.68	0.69	0.69	0.13	0.23	0.26	0.64	0.45	0.39
5 ANSYLC					0.03	0.09	0.16	0.24	0.07	0.13	0.18	0.13	0.06
6 ANPHONC						0.86	0.64	0.18	0.27	0.20	0.58	0.47	0.48
7 ANPHONPOS							0.69	0.15	0.26	0.33	0.61	0.46	0.51
8 ANPHONREV								0.16	0.26	0.26	0.78	0.55	0.37
9 RHYME									0.10	0.19	0.13	0.24	0.29
10 TIT										0.21	0.23	0.19	0.21
11 PIT											0.29	0.27	0.19
12 PREAD												0.65	0.41
13 READING													0.46
14 SPELLING	1												

Note: $r \geqslant 0.16$; $p < 0.05$; $r \geqslant 0.21$; $p < 0.01$.

and the low correlations of the metalinguistic tasks with these tests are rather unexpected.

For a better interpretation of the correlation matrix presented by Lundberg et al. (1980, p. 165), a factor analysis was performed by the present author, including all the tests of the kindergarten battery and the tests of silent word reading (SWREAD) and spelling (SPELL), each administered at the end of grades one and two. Table 12-3 presents the data.

The first four components with an eigenvalue ⩾ 1 account for 66 percent of the total variance. The largest factor, 1, shows high weights in all those metalinguistic tasks requiring the analysis and synthesis of phonemes and syllables (except ANSYLC) and in the preschool reading test. It has low loadings in the two spelling tests. Factor 2 has high loadings in the three reading tests and moderate loadings in the spelling tests. Factor 3 has a high loading in the Raven intelligence test and moderate loadings in the two nonlinguistic decentration tasks, in two syllable tasks, and in the rhyming task. With caution one might suggest that the three last tasks may represent the ability to objectify language which would also require a certain level of decentration. Referring to a differentiation proposed by Valtin (1979) between the ability to reflect on form properties of language ("Vergegenständlichung von Sprache" or objectivation) and explicit knowledge about linguistic units of language, Van Leent (1983) suggests that tasks like rhyming, analyzing words into syllables, and comparing the length of spoken words measure this ability of objectivation. He also points to a relationship with

Table 12-3. Factor Analysis of the Variables in the Lundberg study.*

Variables	Factor Loadings			
	I	II	III	IV
SYNSYLC	0.57	0.09	0.34	0.21
SYNPHONC	0.85	0.30	0.10	0.14
SYNSYLD	0.54	0.06	0.15	-0.03
SYNPHOND	0.78	0.26	0.10	0.05
ANSYLC	-0.03	0.11	0.43	-0.04
ANPHONC	0.76	0.24	0.09	0.37
ANPHONPOS	0.79	0.23	0.18	0.36
ANPHONREV	0.73	0.38	0.31	-0.05
RHYME	0.04	0.14	0.35	0.21
TIT	0.22	0.07	0.29	0.09
PIT	0.22	0.14	0.37	0.07
PREAD	0.64	0.57	0.23	-0.18
SWREAD 1	0.28	0.80	0.17	0.12
SPELL 1	0.32	0.37	0.24	0.41
RAVEN	0.24	0.04	0.67	0.02
SWREAD 2	0.18	0.88	0.18	0.14
SPELL 2	0.39	0.53	0.34	0.09
% variance	75	12	8	5

*Principal factor solution, varimax rotation. N = 133.

decentration ability in the sense of Piaget. Factor 3 might be viewed as a substantiation for this claim and as a certain validation for the hypothesis that it is indeed meaningful to differentiate at least between these two aspects of language awareness. The last factor, which accounts for only 5 percent of the explained variance, has moderate weights in the grade one spelling test and in two tasks that require phonemic segmentation, a subskill that certainly is a component of learning to spell.

The finding that the prereading test has loadings on both the metalinguistic and the reading and spelling factors is in agreement with the hypothesis that the abilities of analyzing and synthesizing phonemes may be regarded as a reflection of experience with print.

It is profitable to examine more closely the metalinguistic abilities of the children who at the end of kindergarten showed no apparent reading abilities and those who demonstrated some reading skill, keeping in mind that the prereading test was only a very crude measure. Table 12-4 allows a comparison of early readers and nonreaders.

The tests requiring the synthesis, segmentation, and reversal of phonemes show the highest amount of difference between the two groups. An inspection of the distribution tables of the scores of the nonreaders (data from Lundberg, personal communication) reveals that, in the phoneme tasks, a considerable proportion of nonreaders completely failed. Only three of the 48 nonreaders had a score higher than 1 in the task that required the ability to analyze phonemes and to reverse their order. The result is very plausible since it is hard to conceive how such a task might be accomplished in the absence of a mental representation, for instance, an orthographic image of the sounds. The nonreaders achieved rather high scores in the task analysis of phoneme position when they had to indicate whether a specific sound occurred at the beginning, at the end, or in the middle

Table 12-4. Comparison of Readers and Nonreaders in the Lundberg Study (data from Lundberg, Personal Communication).

Tests	Nonreaders ($N = 48$)		Readers ($N = 85$)		Difference
	M	S	M	S	
SYNSYLC	2.6	1.3	3.7	0.7	+
SYNSYLD	2.6	0.9	3.4	0.8	+
SYNPHONC	1.6	2.3	6.6	2.2	+
SYNPHOND	1.0	1.8	4.5	2.4	+
ANSYLC	2.0	1.7	2.1	1.6	n.s.
ANPHONC	2.8	3.3	7.1	1.5	+
ANPHONPOS	5.7	3.6	10.7	2.1	+
ANPHONREV	0.3	1.1	3.6	2.4	+
RHYME	6.7	1.8	6.9	1.8	n.s.
PIT	10.5	2.5	11.3	2.3	n.s.
TIT	4.3	2.4	5.0	2.5	n.s.

Note: + = significant difference, using the t test (Kreiszig, 1968, p. 219).

of a word. It is not clear whether this rather high score is due to the fact that there is a 33 percent chance of guessing or due to the fact that "easy" consonants (continuants and a vowel) had to be recognized. In any case, some of the non-readers were able to analyze two- and three-phoneme words and to synthesize phonemes. As can be seen from Table 12-5, these segmentation abilities of the children in the nonreading group show a moderate correlation with reading and spelling achievement in grades one and two and with spelling achievement in grade three.

In the training study described by Olofsson and Lundberg (1983), the non-readers also outnumbered the children with low pretest scores in the phonemic tests. At least two explanations are possible. First, the preschool reading test was too crude a measure and the segmenters already had a glimpse of understanding of the alphabetic principle. Second, the segmenters had reached a high cognitive level that allowed them to perform these tasks (decentration). Since Swedish children start school relatively late, this view that the segmentation abilities of the nonreaders might be a manifestation of cognitive growth seems plausible. The mean age of this sample at the time of administering the metalinguistic tasks was six years and eleven months, with a standard deviation of three months (Lundberg, personal communication). The scores of the nonreaders in the cognitive decentration tasks were relatively high, indicating that these children had already reached a high level of decentration. From the data available to the present author this hypothesis cannot be tested, however.

Open Questions: The Roots of Phonological Knowledge

When learning to read and write in the languages discussed here, the child has to gain some knowledge about the alphabetic principle or, in other words, an awareness of phonemic segments, at least in the acquisition phase of this learning process. The present literature is far from clear with regard to the nature and origins of these phonemic segments. There are at least three possible interpretations concerning the nature of these segments. Each of these will be discussed as an alternative hypothesis.

Table 12-5. Correlations Among Variables in the Group of Children With No Preschool Reading Ability (Lundberg, Personal Communication).

Predictors	SWREAD1	SPELL1	SWREAD2	SPELL2	SWREAD3	SPELL3
ANPHONC	0.33[+]	0.30[+]	0.37[++]	0.30[+]	0.19	0.60[++]
ANPHONPOS	0.39[++]	0.38[++]	0.42[++]	0.40[++]	0.28	0.38[++]
ANPHONREV	0.13	0.21	0.27	0.22	0.21	0.33[+]
RAVEN 1	0.07	0.27	-0.08	0.26	-0.03	0.23
PIT	0.19	0.12	0.19	0.25	0.13	0.09
TIT	0.15	0.20	-0.05	0.06	-0.22	0.32[+]

+ = $p < 0.05$
++ = $p < 0.01$

The first hypothesis in that phonemic awareness is implicit knowledge that has become explicit, either spontaneously or as a result of reading experiences which bring these units relatively easily into awareness. Read (1978), for instance, suggests that "awareness" should be taken as "focusing one's attention on something that one knows" (p. 70) and that the segmentation training studies with preschool children changed the awareness of these segments for the children. A variant of this first hypothesis is that children, when entering school, have not yet fully achieved a mature command of the phonological structure of their language, that they do not yet possess "full knowledge of the sound structure system that corresponds to the orthography" (Chomsky, 1970, p. 301), and that orthography will facilitate in many cases the child's awareness of the internal organization of the sound system of the language (p. 300).

The second hypothesis is that phonology must be taught (Gleitman & Rozin, 1973) by making children consciously aware of it because, as Levelt, Sinclair, and Jarvella (1978, p. 6) suggest, linguistic awareness has no "special 'hot-line' to implicit knowledge." For children, phonemes are not natural units; rather, they tend to rely on phonetic cues. Andresen, referring to studies on invented spellings by Read (1975) and Eichler (1976), suggests that children use other classifications of speech sounds than adults and demonstrate in their spelling strategies that they are nearer to the phonetic than to the phonemic level. She proposes that children, when learning to spell, must be guided from a phonetic transcription to phonemic writing, taking into account morphophonological, semantic, syntactic, and historical knowledge. Similarly, Chomsky (1970) has argued for the need of a "shift in emphasis from a phonetic to a lexical interpretation of the spelling system" (p. 297).

More recently, Read (1978) has presented additional evidence that shows "how close children are to the level of sounds when they first encounter reading and writing. Their judgments are perfectly reasonable and phonetically correct" (Read, 1978, p. 76). Furthermore, he points out that these judgments from a phonetic point of view deteriorate in adulthood and concludes, "Familiarity with the written form eventually drives the phonetic relationships out of our awareness, it seems" (p. 78). From this statement, one may wonder if phonemic knowledge is derived from written language experience, but earlier in his chapter Read argues in favor of an implicit knowledge. Günther (1979), in a response to Andresen (1979), argues that the conceptual bases of the phonetic classification of speech sounds used by preschool children are articulatory cues. Some support for this claim comes from the observations of Beers (1980), who investigated developmental strategies of children's spellings and demonstrated that young children rely on their knowledge of articulation features.

Common to both of the above outlined hypotheses is the view that the child has the "phonological machinery already in his head" (Gleitman & Rozin, 1973, p. 456), be it more or less accessible or more or less well developed. This presupposition has been questioned by Treiman and Baron (1981). As an explanation for the fact that younger children appear to lack the knowledge that a spoken word consists of smaller units, they offer two propositions. The first is that even young children represent speech in terms of segments, but that they do not know that they have such representations. This implicit knowledge comes to consciousness through developmental change. The second explanation, preferred by the authors, is that young children "do not represent speech in terms of segments, either consciously or unconsciously. Rather, they represent the sound of a word or syllable 'integrally,' as an indivisible whole" (p. 192). According to this view, the development consists in the way one represents spoken words. "The wholistic representation of the child becomes the piece-wise representation of the adult" (p. 193). Consistent with their second explanation were their findings that kindergarten children tended to classify syllables on the basis of overall similarity and that the tendency to classify syllables on the basis of shared phonemes increased with age. As regards the roots of this piece-wise representation, Baron and Treiman (1980, p. 186) suggest an effect of learning spelling-sound rules.

Francis (1979), in her reaction to Mattingly's (1979a) paper, finds it debatable whether the tacit knowledge of a child "can be modelled at any stage on an explicit theory of linguistic structure" (p. 5). She argues that the child after five years of language experience has "a highly developed discrimination of features which firm up functional and semantic distinctions in language use." The linguist can "characterize this skill in terms of a theory of phonology, but the theory is not a statement of the mechanisms behind the skill. It is a metaphor" (p. 3). These statements lead over to the third hypothesis.

This third hypothesis is that phonemes are not "natural" units, either given or deeply buried in the tacit knowledge of the child, but artificial concepts of sound structuring. These concepts may be mediated through instruction in reading and spelling. Phonemic hearing could thus be regarded as an illusion of the fluent reader/speller who imposes his orthographic knowledge on speech. Günther (1979) discusses various conceptions regarding the theoretical status of the unit phoneme and suggests that one may conceptualize the phoneme as a symbol ("Repräsentant") of a concrete speech sound. The phonemic notation, then, is a means by which the continuous speech flow may be represented as a sequence of discrete segments. Using this concept, it is not necessary to presuppose that children already know phonemes implicitly. He argues that children are able to distinguish speech sounds, but, in order to learn to read and write, they must be able to objectify the continuous speech flow into segments. Thus, children have to learn a new internal system for the representation of speech sounds that does not correspond to their phonetic classifications in a one-to-one fashion. Alphabetic script offers the child a new external representational system that is phonemically (or morphophonemically) structured. Günther suggests two possible alternatives: "The child either has to acquire a new internal representational system for his phonetic knowledge or to develop strategies for relating the new external (phonemically structured) representational system to his already existing, phonetically oriented classification system" (Günther, 1979, p. 59, translated by Valtin).

In this context a study by Marcel (1980) seems of interest. Marcel observed a specific spelling problem in some adult illiterates, schoolchildren, and three neurological patients. In initial consonant clusters containing a liquid, the liquid was omitted (*thoat* for *throat*) or misplaced inconsistently (*fiurt* for *fruit*) and the voicing of the stop was sometimes changed. In terminal clusters, nasal and lateral consonants were omitted (*wet* for *went*). Marcel stresses the similarity of these misspellings to deviations from adult phonology which appear at certain stages of children's speech and to spontaneous spellings of preschool children studied by Read (1975). Ehri in her chapter in this book reports similar observations with first graders: "Novices do not detect and represent as many phonetic segments as mature spellers." As an explanation for these misspellings, Ehri suggests the insufficient experience with conventional print that helps to devide the sound stream properly.

Marcel considers two slightly different hypotheses to explain the findings of his study. His subjects had no trouble in sound perception, sound production, and auditory discrimination of two acoustically similar words, but demonstrated difficulties in phonemic judgment and segmentation (subjects tended to underestimate the number of phonemes in clusters, compared with single consonant items). His first hypothesis concerns the role of linguistic awareness since "a conscious representation of speech which can be intentionally manipulated" is required for the learning of reading and spelling (p. 390). Marcel draws a distinction between the automatic processing of sensory input and the conscious percepts, and states that the conscious representation of sensory information involves an act of "recovery." Tests requiring phonemic judgment and segmentation "reflect recovery of unconscious information, as opposed to the normal processing of that information, and it is in this recovery, or the availability of the relevant information for recovery, that these people are 'impaired' " (pp. 392-293). Marcel suggests that some poor spellers are impaired in becoming aware of particular phonemes in certain contexts; in other words, these people are impaired in their locutionary awareness. The second hypothesis considered by Marcel is that abnormal spellers, like young children, are using a different set of features from normal adults to analyze and produce speech due to slight deviancies in speech production and perception. The hypothesis that normal adults classify speech in terms of phonemes and the deviant spellers in terms of distinctive features or acoustic cues is plausible in explaining certain misspellings, e.g., omissions of liquids and nasals in certain contexts, changes in articulatory voicing ($/k/{\rightarrow}/g/$), and placing values ($/tr/{\rightarrow}/g/$) of letters. Further validation of this hypothesis by analyzing the speech of such adults in a more acoustic manner (e.g., spectrographically) is needed, however.

While Marcel favors his second hypothesis, he still stresses the importance of reflexive awareness in learning to read and spell. Marcel thus emphasizes two important factors that might cause problems for the learning of spelling: "that the speller may have problems in conscious recovery of what is to be spelled," and "that the speech code used by the speller may not be the same code as that used by the teacher" (p. 401).

Marcel cites evidence from other studies that children acquiring adult phonology are using acoustic features. The question again arises as to how children achieve adult phonological distinctions. Marcel suggests lexical, semantic, morphemic, and syntactic knowledge as contributing factors. Marcel's findings also point to the importance of print experience as another factor because his subjects had different error patterns in words and nonwords: errors of omissions were more frequent in nonwords and errors of misplacing of liquids and nasals were more frequent in words, thus suggesting an effect of printed words. This would corroborate Ehri's hypothesis that orthographic images help the child to become aware of phonological segments in words. Marcel arrives at the interesting suggestion that the representation of speech on which the alphabet relies— the phoneme—is rather unnatural and concludes: "In whatever way the alphabet was first invented, it is possible that, for each learner today, *the concept of the phoneme* (tacit if not explicit) *comes from rather than leads to the particular alphabetic system with which he or she is confronted*" (pp. 401-402).

Whether knowledge of how to segment speech and how to classify speech sounds is implicitly given and only awaits "awakening," or whether a new system of representing speech sounds must be learned, in any case, the teacher cannot expect beginning readers to already possess a complete system of phonemic categories nor that those categories are easily detected or acquired. Marsh and Mineo (1977), in their training study, found that performance was phoneme specific, and without explicit instruction did not generalize either to new phonemes of the same class or to the phonemes of another class. As Scheerer-Neumann (1982) puts it, "A comprehensive phonemic classification system is the end and not the beginning of the learning process" (p. 3, translated by Valtin). There is enough evidence now to show that orthography plays an active part in the conceptualization of the sound structure of words by children and adults (see Ehri in her chapter in this book). But the exact nature of the roots of phonemic concepts awaits further investigation.

The Child's Awareness of the Communicative Function of Language

Piaget's (1926) findings on egocentric speech and communication difficulties of preoperational children are often interpreted as an indication that young children have little awareness of the communication function of language. The global concept of egocentrism, however, has been criticized lately. In reviewing research on young children's communication abilities, Donaldson (1978) comes to the conclusion that there is "little indication of the existence of egocentrism as a serious barrier to communication" (p. 30). Rather, it seems as if Piaget's communication tasks requiring the retelling of a complex story and the explanation of a mechanical object have been cognitively too demanding. An analysis of recent research on role-taking abilities (Shantz, 1975; Valtin, 1982b) reveals that this ability is clearly related to task-specific requirements. Given simpler experimental conditions—embeddedness in a play situation that makes sense for the young child, familiarity with test materials and patterns of events of the task— even young children are able to take the role of the other.

Käsermann (1979) has shown that children as young as two years old are sensitive to failures of their speech intentions and provide not only a repetition but also a modified version of their previous utterances. These modifications or repair strategies may yield some information about the child's implicit knowledge about language. Käsermann also points to the fact that children's failures and their subsequent attempts to reorganize their utterances may be instigators of developmental progress.

As with language awareness, it seems important to make a clear conceptual distinction between implicit awareness of communication requirements—what may be inferred from the overt behavior of the child—and conscious and explicit knowledge about communication, its conditions, requirements, or even rules. Only very few studies have addressed this metacommunicative knowledge of children. Flavell (1976) has outlined a framework for the development of this ability. Robinson and Robinson (1976, 1977, 1978) have investigated in various experiments how children explain communication failure. They had children observe or participate in a referential communication game and asked the subjects who was to blame for a communication failure (choosing a wrong card). Young children tended to blame the listener even when the speaker's message was inadequate. Similarly, Singer and Flavell (1981) and Flavell, Speer, Green, and August (1981) had children evaluate the communication clarity of brief oral instructions carried out by a puppet or by the children themselves. These studies also demonstrated the young child's difficulty in accurately assessing the communicative adequacy of messages.

In a study by Valtin (for details see Valtin, Dovifat & Thomalla, 1982), the communication failure was embedded in a more familiar everyday situation. This study employed a clinical interview method to examine children's recognition and explanation of inappropriate communication behavior that might be interpreted as a violation of a communication rule. The rules were mainly derived from Grice's (1975) "be-cooperative" principle, and referred to the maxims "be perspicuous," "be as informative as necessary," and "be relevant" and to turn-taking behavior in conversation (one part at a time, summons-answer routine, interrupting before the topic of speech is known). Thirteen film scenes (super 8 sound) were developed, each containing a face-to-face, two-party communication situation where one participant shows inappropriate behavior because of rule violation. Each film was shown twice and the children were interviewed individually after each film scene. The sample consisted of children five, six, seven, eight, and ten years old. The results of the interview showed a clear age dependency of scores. In this sample, the subjects—even those at the preoperational level, as the five year olds supposedly were—had no difficulties evaluating inappropriate communication behavior if the message referred directly to an ongoing concrete and simple action (e.g., a child states that she is not hungry and at the same time grabs a sandwich and eats it). All children detected such inconsistencies between statement and behavior or between the stated intention to do something and the actual behavior. The following example may illustrate this.

Film scene 10: Two girls are sitting at the table. **Katja:** "Do you want to see my new Mickey Mouse Book?" **Michi:** "Yes." **Katja** (goes and fetches the book): "Here it is." **Michi:** "What shall I do with it?" **Katja:** "But you wanted to see it!" **Michi:** "But not now!"

All children recognized the inappropriate behavior, were able to give an adequate explanation, and blamed the speaker. Robinson and Robinson observed in their earlier experiments (1976) that younger children tended to blame the listener for a communication failure which was caused by the inadequate message of the speaker. The authors concluded that the listener-blamers had no understanding that messages could be inadequate. In a later study (1977), however, they revised their assumption, after having shown that the listener-blamers were able to detect the message as inadequate. The authors now suggest that the difficulties of the younger children lie in judging messages. The findings of the study reported here are in agreement with the later hypothesis. If there is a clear relation of the message to an ongoing concrete and very simple action, young children are able to notice discrepancies. They have more problems, however, if these discrepancies are more subtle and if the behavior of the speaker deviates from her or his previously stated intentions embedded in a speech act, for instance, if the speaker's utterance implies a contract to behave in a certain way: the demand for an explication obliges one to listen; after her or his summons has been answered, the speaker is obligated to give the topic. If the message clearly related to the verbal plane only, young children are unable to deal with it, as the following example shows.

In one film a girl (Katja) is asked by her friend Michi to retell a TV story she had seen. Katja gives a report with many pronouns and demonstrative adjectives which do not clearly indicate to what they are referring. The report cannot be understood. The five-year-old children in this sample did not understand the situation and that Michi was unable to grasp the incoherent report of Katja. Interestingly, some of them said that they themselves had understood the report of Katja. These results are in agreement with Piaget's (1926) findings on verbal communication. In Piaget's experiment, a child had to retell a story and explain a mechanical object to another child. The younger speakers (six and seven years old) used verbalizations which were like Katja's report (undefined pronouns and demonstrative adjectives and omissions of relevant parts). Even under these circumstances, the listeners in Piaget's study nearly always felt quite confident that they had understood what the speaker said. Piaget explains this fact using the concept of distorted assimilation; the listener is unable because of her or his egocentrism to make a critical evaluation of the message itself and assimilates the statements to her or his own schemata, elaborates on them, and feels confident that she or he has understood. Some of our five-year-old children showed this behavior.

Markman (1977), in her study of children's awareness of their comprehension failure, also observed that first graders are insensitive to their own comprehension failure, and attributed this to a lack of constructive processing. Children in her study realized their lack of understanding better if they enacted the information or observed a demonstration of the action embedded in the information.

The majority of the six-year-old children in our study said that they did not understand what Katja said, but mostly they could not say why. Most of them, however, thought that Michi understood the report. Si, a six-year-old girl, said: "Katja made a fine report. But I could not understand. I am so forgetful." Jo, an eight-year-old boy, after being asked if Katja had told the story well or badly, answered: "Well, I cannot say that she told the story badly, that it was badly told." He himself, however, had not understood the film: "Perhaps I am too young for that, perhaps it is only for ten or twelve year olds." It was mainly the older children who recognized that the story of Katja was not well told, but many of them had difficulties in explaining why. Ke, a six-year-old girl, said: "Katja spoke so unintelligibly (undeutlich), she should speak as we do, we don't speak unintelligibly." Asked to give an example of how "we speak" she told about a film she had seen one day: "I saw a film where a man was and he was on a ship. And they (!) went away, and the woman (!) saw the man, and she helped him, the man, to lay this, the planks." Ke's report itself is a nice example of context-bound speech. While she recognized—herself in the role of the listener —that Katja's report was not understandable, she was not able to take the role of *her* listener and to judge that her report also was unintelligible. Another boy, seven years old, said: "Katja didn't speak clearly, so Michi didn't understand it." The experimenter asked the boy if he himself had understood the film and he said: "Yes, but I knew the film." In this study, it was mainly the ten year olds who could clearly designate the communicative fault and give some valid explanation for it. When trying to explain Katja's communicative fault, nearly all eight year olds used a physical concept: Katja spoke "too fast." This result resembles findings in person-perception studies. When children are asked to describe other persons they tend to use physical or overt characteristics. With increasing age—and the age of about eight seems to be a critical period—the range and frequency of psychological (inferential, abstract, and covert) categories used have been found to increase (for a review see Shantz, 1975). In her interview study on "thinking about language," Sinclair (1981) found that kindergarten children proposed the following strategies for dealing with incomprehension: "shout," "talk louder," repeat the message in order to improve the phonological quality of the message or correct the speech errors. Older children (above the age of eight) referred mainly to a lack of comprehension as a cause of communication failure, due to ambiguity, missing words, or unclear reference.

Our study reported above yields some evidence that five year olds are aware of the communication function of language. They are able to detect communication failures in simple, everyday situations, and some are also able to make explicit verbal judgments. Some findings of this study also demonstrated the difficulties that young children have in focusing on the form of language and in differentiating what is said from what is meant. In two films the children's understanding of the indirect speech act was investigated. In one film scene, the girl, after having drunk all her milk, holds out her empty glass to the second girl and says: "Kommst du an die Flasche heran?" (Can you reach the bottle?). The second girl stretches out her arm, reaches the bottle and says "Yes." Some children, five years of age, in retelling the scene, spontaneously reported that the first girl had said that she wanted something to drink. They only reacted to the illocutionary aspect of the utterance, and thus were also not able to explain why the second girl stretched out her arm. Similar results were obtained in a film scene where the neighbor rings and asks the child who has opened the door, "Is your mother at home?," and the child closes the door after having stated, "Yes, my mother is at home." As in the dialogue presented above with the little boy who spontaneously interpreted "carpet" as "rug," most of the younger children in this study reacted spontaneously to the meaning aspect (in this case the function of the utterance in this specific context) and had difficulties in concentrating on and/or rehearsing the particular phonetic or syntactic form. This observation is in agreement with the interpretation of Hirsh-Pasek et al. (1978) that the early stages of linguistic processing, where phonological and syntactical stimuli are perceived, decay rather rapidly and thus are relatively inaccessible to reflection (p. 104).

It was mostly the older children of this sample (about age eight) who were able to make explicit distinctions between the linguistic form of the utterance and the intention of the speaker. Olson (1982) has proposed that written language may be the basic source for the distinction between what is said and what is meant, because "writing preserves the surface structure, what was said, independently of the intention it expresses, what was meant. In oral language, what was said is ephemeral; what is preserved is the meanings and intentions of the speaker" (p. 12). This new sensitivity begins "with the attention to language fostered by literacy, whether through children's encounters with books and learning to read, or through the speaking practices of literate adults" (p. 14). Forrest-Pressley and Gillies (1983) also observed in their study that the tendency to recognize the difference between what a word "says" and what it "means" and the ability to give an adequate explanation increased dramatically with grade and reading ability.

Some Comments on Methods of Assessing Language Awareness

The testing procedures used in assessing language-awareness abilities may be categorized as belonging to two broadly defined approaches: a "deficit" and a "constructive" approach. Whereas the deficit approach might be characterized as focusing on children's "errors committed because of a lack of knowledge" (Ferreiro, 1978, p. 25), the constructive approach may be characterized as emphasizing "original productions constructed by the child" (p. 25).

As regards the *deficit approach,* the children are asked questions about scientific (linguistic) concepts and the answers are scored pass/fail. The general outcome is that children are deficient in their understanding of linguistic concepts. The studies by Januschek et al. (1979a,b) and Bruce (1964), among others, have demonstrated the usefulness of interpreting the reactions of the children in terms of underlying strategies and of connecting the tasks with a verbal interview. It surely is problematic to rely only on verbal reponses and judgments, especially when the subjects are children, but, together with other measures, interviews may provide and substantiate new insights into the thought processes of young children.

Additional interviews with children will also yield evidence as to whether they react on the same premises as those expected by the experimenter. There is ample evidence in the literature that this is not always the case. The following example given by Chomsky (1979) is a good illustration:

> I presented a series of grammatical and ungrammatical sentences to a four year old, trying to get him to tell me which ones were no good, or strange, or odd in some way. The problem is a familiar one. Even after I thought he understood the task, his responses still appeared to have no relation to the degree of well-formedness of the sentences. I finally gave him the unequivocal sentence: *John read the book,* just to check the criteria he was using. "That's no good," he replied. "What's the matter with it?" I asked. "You can't say it," came the reply, "it's not a good sentence." "Why?" I persisted. "Because maybe John doesn't know how to read," he answered simply (p. 3).

Some authors (Januschek et al., 1979a,b; Paprotté, 1981) emphasize that children's performances should be regarded from the perspective of what they actively construct in relation to their everyday experiences in metalinguistic tasks. From the perspective of a child (not so from the perspective of a linguist, though), his or her behavior in linguistic tasks may be "meaningful" and not deficient. Aronsson (1981) points out that in the everyday life of a child, reflections on meaning are of great importance whereas reflections on form are not yet of great pragmatic value. Hence, the child's preference pattern for meaning is "indeed most rational if taken from the perspective of the child" (p. 106). Aronsson further postulates that the young child has a concrete conception of names such that "to each word there is one object and to each object one word" (p. 103). This is a rational theory of concrete reference from the child's perspective and, on her or his premises, a sensible one. Following from this perspective, it seems plausible that the young child is unwilling to substitute lexicon words with other words and to see the arbitrary nature of names. The child's apparent difficulties with the differentiation of word and referent need not be explained in terms of a reified view of language on the part of the child.

This view sees children's performances in metalinguistic tasks as functional for their everyday experiences and for the conditions of the typical communication situation in which speech is embedded. The fact that children do not differentiate the sound and the meaning aspects of speech spontaneously contains nothing unusual if we consider that we normally interpret sounds immediately without paying much attention to the internal structure of the sound event itself (music is one of the exceptions). It is functional for our everyday requirements that we identify sounds by interpreting them immediately. Given appropriate test conditions as Sinclair and Berthoud-Papandrcpoulou (in their chapter in this book) have provided them, children are able to judge formal properties of language using length of the utterance and repetitiveness of words as criteria. Instead of stating deficits in childrens's knowledge, studies belonging to the *constructive approach* try to trace the development of knowledge which children are in the process of constructing about print and the relation between written and oral language (Ferreiro, 1978; Ferreiro & Teberosky, 1979) or their knowledge about concepts of writing (Bissex, 1980; Clay, 1975). These approaches seem to be very promising because the child's conceptions and the instructional procedures should yield a productive match. Downing's revisions to his cognitive clarity theory seem to place it on an orbit moving from the deficit approach to the constructive approach (see Downing's chapter in this book).

Pedagogical Consequences

The pedagogical consequences outlined in this book refer to three aspects of learning to read and write:

(1) Understanding the nature of reading and writing, e.g., that written language represents—or at least corresponds to—speech
(2) Grasping the alphabetic principle of our script
(3) Conscious and deliberate mastery of language.

This differentiation corresponds to the three metalinguistic stages proposed by Ryan and Ledger in their chapter in this book: (1) analyzed knowledge of the functions of print (i.e., to convey meaning), (2) analyzed knowledge of the features of print and the control to relate them to units of oral language, and (3) the control to coordinate forms and meanings to achieve the goal of extracting meaning from text.

Usually children entering school are aware of the communicative function of oral language. Some, however, who are deprived of experience of literacy in the home, must be helped in recognizing the special functions and the usefulness of written language. Francis (1982), in her longitudinal observation study of children from an English school, found that the slow learners (children who received no support at home for learning to read) did not know what reading and writing were all about and could make no sense of the set of prereading activities in school. She concluded, "The major problems of understanding the nature of reading and learning to read ... were based in the combination of finding the task in school somewhat incomprehensible and of having no particularly relevant prior or background experience to draw on" (p. 136). She points out that, for those children, a sense for the nature of the acts of reading and writing and of their communicative function must be established within school. Children without this understanding are inclined toward learning by rote, but, if the number of sight words is too large, passive memorizing will fail. Other authors (Downing, Pidgeon and Johns in their respective chapters in this book, and Donaldson, 1978) also point to the importance of making the child aware that the squiggles on the page are written versions of speech and that they serve a communicative function.

As regards the second aspect, grasping the alphabetic principle, it seems useful to rehearse the different positions suggested by the authors of this book regarding the relationship of phonemic awareness and learning to read. Three explanations have been provided: Phonemic awareness is seen as a prerequisite by Mattingly and—at best—as a consequence of learning to read by Singer. Downing and Ehri each present an interactive view. Interestingly, this reciprocal relation was already pointed out by Vygotsky (1934) about 50 years ago. He emphasized the conscious control of language that is required in reading and writing and stated "that the development of the psychological foundations for instruction in basic subjects does not precede instruction but unfolds in a continuous interaction with the contributions of instruction" (p. 101). Vygotsky's statement implies that the notion of reading readiness is a relative one, dependent upon the components and requirements of the specific instructional procedures.

The findings outlined in this book do not seem to support the prerequisite hypothesis. Though no direct experimental evidence has been presented up to now, the results fit the theoretical position that phonemic awareness is mainly the result of learning to read. If some children already possess phonemic segmentation abilities (mainly because of some preschool reading ability), this may facilitate the first stages of learning to decode words. Children with no or low phonemic segmentation abilities when entering school may generally acquire these by learning during the first months of school. Although it is possible to train segmentation abilities prior to reading instruction, as Russian studies show, there seem to be some arguments against this procedure. First, phonemes may not be identified simply through auditory perception or articulation. Second, by means of acoustic and articulatory analysis of speech sounds, the child will not become aware of the relevant speech units which are represented by graphemes. At best the child will arrive at allophones. According to German linguists (Eichler & Bünting, 1976), a German adult speaker produces in his speech about 120 to 150 perceptually discriminable speech sounds whereas only 38 to 40 phonemes have the function of indicating a difference in meaning. The child must learn what class of speech sound corresponds to a specific phoneme. Third, the child may learn a wrong spelling strategy while relying too much on phonetic cues. In Germany the spelling teaching method "spell as you speak" is still widely in use and is undoubtedly responsible for many spelling errors. The children need to shift from phonetic transcriptions to a transcription of deeper structures of language. Fourth, this technique seems to be uneconomic because segmentation training with visual markers, especially letters, are more effective. Ehri, in her chapter, has provided evidence that letters were helpful in enabling learners to distinguish the correct size and identity of the sound units.

The question as to how explicitly the child should be introduced to phonology is answered differently in the literature. Donaldson (1978) argues that children should be helped in understanding the nature of the system but that they must detect by themselves all the individual patterns of the relationship between oral and written language. This procedure may not be effective for all children, however. Francis (1982), in her longitudinal study, observed that some children were able to read without phonic knowledge, while relying on contextual cues and generalized knowledge about orthographic patterns. She proposed that children without a literacy background and thus with little experience of written language, as well as those with poor visual memory, may profit from some phonics teaching. The method should not be based on individual letter-sound conversions but on letter patterns and sound correspondences so that the child may learn orthographic patterns. Singer, in his chapter in this book, argues convincingly against phonetic instruction in reading that requires children to learn and state the rules for sounding out printed words. He has nothing against phonics instruction, however, "which is merely learning to relate sounds to print, that is, ability to segment and relate phonemes, combinations of phonemes, syllables, and words to corresponding segments of print. "

The answer to the question– How much phonemic segmentation training is required?– may well depend on the answer to another question– Does the orthography of the alphabetic script in a given language represent deep or shallow phonology? In countries like Finland, Sweden, the Netherlands, and Germany, where the orthography is not as remote from the phonetic representation as it is in English, phonemic segmentation training (analysis and synthesis) seems to play an important part in reading instruction (Kyöstiö, 1973; Lundberg, 1982b; Van Leent, 1983; Biglmaier, 1973; Dathe, 1981). After a period of synthetic methods of learning to read and a subsequent period of analytic methods, in present-day Germany so-called analytic-synthetic methods are widely used. In the German Democratic Republic (East Germany) this method is the only one officially permitted and there exists only one basal reader for the whole country. This primer, like many of the new primers in the Federal Republic of Germany (West Germany), begins with simple and regular words which the child reads in various contexts. Then the words are analyzed into their visual, phonemic, and articulatory components so that the child may become aware of the function of graphemes and phonemes in a word. Much emphasis is placed on articulatory cues to support the segmentation of words into smaller units. Stretched pronunciation is a very common activity used in the early stages of reading in the German Democratic Republic (Dathe, 1981). Some authors of basal readers state that phonemic analysis should always start with the written form of the word which is subsequently segmented into smaller units and that orthographic patterns (*Sch-n-ee, Uh-r*) should be respected (Dathe, 1979, 1981; Will-Beuermann, Hinnrichs & Valtin, 1980). Pidgeon has argued in his chapter for the opposite approach, to start with spoken words and then to proceed to the learning of the corresponding graphemes. The arguments raised against segmentation training prior to learning to read may be applied against this approach, however.

Phonological recoding is only one aspect of the word-decoding subskill, mainly related to beginning-reading acquisition. Neither the component of phonological recoding nor the word-decoding subskill should be overemphasized in reading instruction at the expense of other reading subskills and strategies. If teachers mainly stress word decoding, it is not surprising to find the kind of results reported by Johns in his chapter on students' perceptions of reading. When explaining reading, these students emphasized learning words and sounding techniques and seldom referred to meaning or comprehension. It would be of interest to compare these students' perceptions of reading with their teachers' perceptions and their teaching methods. Possibly the students' perceptions are simply a reflection of the teaching strategies.

As Singer has pointed out in his chapter, the mature reader is able to make flexible use of various types of subskills depending on the purpose of reading. Ryan and Ledger, in their chapter, emphasize the metacognitive aspect: the control to coordinate form and meaning to achieve the goal of extracting meaning from text. This executive control function has recently gained attention in theoretical frameworks of the reading process that relate reading to metacognitive processes as selecting, monitoring, and modifying cognitive processes and strategies (Paris & Meyers, 1981; Paris & Lipson, 1982). Forrest-Pressley and Waller (1982) and Forrest-Pressley and Gillies (1983) have demonstrated that mature readers are more flexible in their use of different reading strategies (rereading, skimming, paraphrasing, summarizing, and identifying important ideas) depending on the purpose of reading and that they are more able to monitor comprehension. In their chapter, Ryan and Ledger cited some evidence for the hypothesis that poor readers are less likely to use active organizing strategies necessary for good comprehension. Paris and Lindauer (1982) have outlined strategic differences between good and poor readers in three domains: decoding, comprehension monitoring, and planning for study and memory. They hypothesize that the basic source of the "lack of strategic intervention during reading is the lack of awareness and understanding about reading" (p. 22).

Consequently, Paris and his associates (Paris, 1982; Cross, 1982; Jacobs, 1982; Lipson, 1982) have designed a project to test how children's understanding of the nature of reading activities influences their actual reading behavior. The program was designed to increase students' understanding of the goals of reading, strategies which are useful in promoting comprehension, and strategies that may be used to remedy comprehension failures. The goal of this intervention program was not only to instruct children in these strategies but also to persuade them that these strategies are useful and worth employing. Third- and fifth-grade students received two lessons each week for four months. The experimental groups scored significantly higher on the posttest measures, including reading tasks and awareness of metacognitive subskills of evaluating, planning, and regulating reading comprehension. Cross (1982) points to the fact that there is still some ambiguity in explaining the causal mechanisms. It is possible that the improvement in reading is due to an indirect effect (mediated through the learning-about-reading strategies) or to a direct effect of this training. In any case, this type of training seems to be suited for promoting children's reading comprehension. (For more information about reading strategy training and procedures for fostering comprehension, see Forrest-Pressley & Gillies, 1983; Lunzer & Gardner, 1979). Brown (1979) has provided a useful overview about reading strategies that teachers should consider:

(1) Clarifying the purposes of reading, that is, understanding the task demands, both explicit and implicit
(2) Identifying the aspects of a message that are important
(3) Allocating attention so that concentration can be focused on the major content area rather than on trivia
(4) Monitoring ongoing activities to determine whether comprehension is occurring
(5) Engaging in review and self-interrogation to determine whether goals are being achieved
(6) Taking corrective action when failures in comprehension are detected
(7) Recovering from disruptions and distractions—and many more deliberate, planful activities that render reading an efficient information-gathering activity (p. 456).

The third aspect, the deliberate and conscious mastery of language, has only been considered so far in its usefulness for reading and writing, and undoubtedly it is also useful for effective communication and for learning foreign languages. But it is not only a means for those communicative ends. It also has effects on the thinking and self-awareness of the individual and on the culture of a society. Donaldson (1978) stresses the importance of such effects of written language: those "features of the written words which encourage awareness of language may also encourage awareness of one's own thinking and be relevant to the development of intellectual self-control, with incalculable consequences for the development of the kinds of thinking which are characteristic of logic, mathematics and the sciences" (p. 95).

Olson (1977) has outlined this impact of characteristics of written language (especially the alphabetic writing system) on the development of the individual and the culture in greater detail: "The bias of written language toward providing definitions, making all assumptions and premises explicit, and observing the formal rules of logic produces an instrument of considerable power for building an abstract and coherent theory of reality. The development of this explicit, formal system accounts, I have argued, for the predominant features of Western culture and for our distinctive ways of using language and our distinctive modes of thought" (p. 278). Another statement by Olson (1977) in this context is suited to constitute a dignified conclusion both for this chapter and for this book:

> The faculty of language stands at the center of our conceptions of mankind; speech makes us human and literacy makes us civilized (p. 278).

Acknowledgments

The study on metacommunicative abilities was carried out with financial support from the Central Institute of Education Science and Curriculum Development of the Free University of Berlin, Germany. Ingvar Lundberg kindly provided me with original data from his study.

References

Adams, M. J., & Collins, A. A schema-theoretic view of reading (Tech. Rep. No. 32). Champaign, Ill.: Center for Study of Reading, 1977.

Agnew, A. T. Using children's dictated stories to assess code consciousness. *The Reading Teacher*, 1982, *35*, 450-454.

Allen, R. V. *Language experiences in communication.* Boston: Houghton Mifflin, 1976.

Anderson, R. C., Spiro, R. J., & Montague, W. (Eds.). *Schooling and the acquisition of knowledge.* Hillsdale, NJ: Erlbaum, 1977.

Andresen, H. Die Bedeutung auditiver Wahrnehmungen und latenter Artikulation für das Anfangsstadium des Schrifterwerbs—Ein linguistischer Beitrag zur Diskussion über die Beziehungen zwischen lautsprachlicher und schriftsprachlicher Tätigkeit. *Osnabrücker Beiträge zur Sprachtheorie* (Universität Osnabrück, Germany), 1979, *13*, 28-58.

Andresen, H. Aspekte der Vergegenständlichung von Sprache: Spontane Sprachreflexion bei Kindern und die Bedeutung des Schriftspracherwerbs für die Entstehung von Sprachbewußtsein. Paper presented at the Institut für Phonetik und sprachliche Kommunikation der Universität München, Munich, Germany, 1982.

Andresen, H. Was Menschen hören können, was sie lernen können, zu "hören", und was sie glauben, zu hören. Reflexionen über die Bedeutung der Lautstruktur des Deutschen für den Schriftspracherwerb. *Osnabrücker Beiträge zur Sprachtheorie* (Universität Osnabrück, Germany), 1983, *24*.

Andresen, H., & Januschek, F. Sprachreflexion und Rechtschreibunterricht. *Diskussion Deutsch*, 1983, *74*.

Andrews, R. J. *St. Lucia Graded Word Reading Test.* Brisbane, Queensland: Teaching & Testing Resources, 1969.

Arlin, P. K. Piagetian tasks as predictions of reading and math readiness in grades K-1. *Journal of Educational Psychology*, 1981, *73*, 712-721.

Aronsson, K. Nominal realism and bilingualism. A critical review of studies on word: Referent differentiation. *Osnabrücker Beiträge zur Sprachtheorie* (Universität Osnabrück, Germany), 1981, *20*, 100-113.

Athey, I. Research in the affective domain. In H. Singer & R. B. Ruddell (Eds.), *Theoretical models and processes of reading*. Newark, Del.: International Reading Association, 1976.

Au, K. H. Using the experience-text-relationship method with minority children. *The Reading Teacher*, 1979, *32*, 677-679.

Ayers, D., & Downing, J. Testing children's concepts of reading. *Educational Research*, 1982, *24*, 277-283.

Bailey, E. J. *Academic activities for adolescents with learning disabilities*. Evergreen, Col.: Learning Pathways, 1975.

Baird, B. *The art of the puppet*. New York, NY: Macmillan, 1965.

Baltes, P. B., & Nesselroade, J. R. History and rationale of longitudinal research. In J. R. Nesselroade & P. B. Baltes (Eds.), *Longitudinal research in the study of behavior and development*. New York, NY: Academic Press, 1979.

Bamberger, R. Literature and development in reading. In J. E. Merritt (Ed.), *New horizons in reading*. Newark, Del.: International Reading Association, 1976.

Banks, W. P., Oka, E., & Shugarman, S. Recoding of printed words to speech: Does recoding come before lexical access? In O. J. L. Tzeng & H. Singer (Eds.), *Perception of print: Reading research in experimental psychology*. Hillsdale, NJ: Erlbaum, 1981.

Baron, J., & Treiman, R. Use of orthography in reading and learning to read. In J. F. Kavanagh & R. L. Venezky (Eds.), *Orthography, reading and dyslexia*. Baltimore, Md.: University Park Press, 1980.

Barr, R. Instructional pace differences and their effect on reading acquisition. *Reading Research Quarterly*, 1973-74, *9*, 528-554.

Barron, R. W. Access to the meanings of printed words: Some implications for reading and for learning to read. In F. B. Murray (Ed.), *The recognition of words*. Newark, Del.: International Reading Association, 1978.

Barton, A. H., & Wilder, D. E. Research and practice in the teaching of reading: A progress report. In M. B. Miles (Ed.), *Innovation in education*. New York, NY: Teachers College Press, 1964.

Barton, D., & Hamilton, M. E. Awareness of the segmental structure of English in adults of various literacy levels. Unpublished manuscript, Stanford University Dept. of Linguistics, Calif., 1980.

Barton, D., Miller, R., & Macken, M. A. Do children treat clusters as one unit or two? *Papers and reports on child language development* (Stanford University Dept. of Linguistics, Calif.), 1980, *18*, 105-137.

Bayley, N. Development of mental abilities. In P. H. Mussen (Ed.), *Carmichael's manual of child psychology* (Vol. 1). New York, NY: Wiley, 1970.

Beers, C. S. The relationship of cognitive development to spelling and reading abilities. In E. H. Henderson & J. W. Beers (Eds.), *Developmental and cognitive aspects of learning to spell*. Newark, Del.: International Reading Association, 1980.

Beers, J. W., & Henderson, E. H. A study of developing orthographic concepts among first graders. *Research in the Teaching of English*, 1977, *11*, 133-148.

Beilin, H. *Studies in the cognitive basis of language development*. New York, NY: Academic Press, 1975.

Bell, A. M. *Visible speech: The science of universal alphabetics*. London, England: Simkin Marshall, 1867.

Belmont, J. M., & Butterfield, E. C. The instructional approach to developmental cognitive research. In R. V. Kail, Jr. & J. W. Hagen (Eds.), *Perspectives on the development of memory and cognition.* Hillsdale, NJ: Erlbaum, 1977.

Bense, E. Der Einfluß von Zweisprachigkeit auf die Entwicklung der metasprachlichen Fähigkeiten von Kindern. *Osnabrücker Beiträge zur Sprachtheorie* (Universität Osnabrück, Germany), 1981, *20*, 114-138.

Ben-Zeev, S. Mechanisms by which childhood bilingualism affects understanding of language and cognition structures. In P. A. Hornky (Ed.), *Bilingualism: Psychological, social and educational implications.* New York, NY: Academic Press, 1977.

Berman, L. M. Helping students understand their thought processes. *Elementary English,* 1963, *40*, 21-24.

Berthoud-Papandropoulou, I. An experimental study of children's ideas about language. In A. Sinclair, R. J. Jarvella, & W. J. M. Levelt (Eds.), *The child's conception of language.* New York, NY: Springer-Verlag, 1978.

Berthoud-Papandropoulou, I. La réflexion métalinquistique chez l'enfant. Doctoral thesis, University of Geneva, 1980.

Bever, T. G. The cognitive basis for linguistic structures. In J. R. Hayes (Ed.), *Cognition and the development of language.* New York, NY: Wiley, 1970.

Bialystok, E. Explicit and implicit judgments of L2 grammaticality. *Language Learning,* 1979, *29*, 81-104.

Bialystok, E., & Ryan, E. B. A metacognitive framework for the development of first and second language skills. In D. L. Forrest-Pressley, T. G. Waller, & G. E. MacKinnon (Eds.), *Metacognition, cognition, and human performance.* New York, NY: Academic Press, in press.

Biglmaier, F. Germany. In J. Downing (Ed.), *Comparative reading.* New York, NY: Macmillan, 1973.

Bissex, G. *Gnys at wrk: A child learns to read and write.* Cambridge, Mass.: Harvard University Press, 1980.

Block, J. H. (Ed.) *Mastery learning: Theory and practice.* New York, NY: Holt, Rinehart, & Winston, 1971.

Bloom, B. S. Learning for mastery. U.C.L.A.-C.S.E.I.P. *Evaluation Comment,* 1968, *1*, 2.

Bloom, B. S. Mastery learning and its implications for curriculum development. In E. W. Eisner (Ed.), *Confronting curriculum reform.* New York, NY: Little, Brown, 1971. (a)

Bloom, B. S. Affective consequences of school achievement. In J. H. Block (Ed.), *Mastery learning: Theory and practice.* New York, NY: Holt, Rinehart, & Winston, 1971. (b)

Bloom, B. S. *Human characteristics and school learning.* New York, NY: McGraw-Hill, 1976.

Bloomfield, L. *Language.* New York, NY: Holt, Rinehart, & Winston, 1933.

Bloomfield, L. Linguistics and reading. *Elementary English Review,* 1942, *19*, 125-130 and 183-186.

Bohannon, J. N. III. The relationship between syntax discrimination and sentence imitation in children. *Child Development,* 1975, *46*, 444-451.

Bohannon, J. N. III. Normal and scrambled grammar in discrimination, imitation and comprehension. *Child Development,* 1976, *47*, 669-681.

Bohannon, J. N. III. Word order discrimination and reading. Paper presented at Biennial Meeting of the Society for Research in Child Development, San Francisco, Calif., 1979.

Borger, R., & Seaborne, A. E. M. *The psychology of learning.* Harmondsworth, England: Penguin, 1966.

Borkowski, J. G., Kurtz, B., & Reid, M. The reliability and validity of metamemory. Unpublished manuscript, University of Notre Dame, Ind., 1980.

Bosch, B. *Grundlagen des Erstleseunterrichts.* Ratingen, Germany: Henn, 1965.

Bradley, L., & Bryant, P. E. Difficulties in auditory organization as a possible cause of reading backwardness. *Nature,* 1978, *271,* 746-767.

Braine, M. D. S., & Wells, R. S. Caselike categories in children: The actor and some related categories. *Cognitive Psychology,* 1978, *10,* 100-122.

Bransford, J. D., & Franks, J. J. The abstraction of linguistic ideas. *Cognitive Psychology,* 1971, *2,* 331-350.

Bright, W. Linguistic change in some Indian caste dialects. In C. A. Ferguson & J. J. Gumperz (Eds.), *Linguistic diversity in South Asia. International Journal of American Linguistics,* 1960, *26,* 19-26.

Bright, W., & Ramanujan, A. K. Sociolinguistic variation and language change. In H. G. Lunt (Ed.), *Proceedings of the Ninth International Congress of Linguists.* Cambridge, Mass.: Mouton, 1962.

Brodzinsky, D. M. Children's comprehension and appreciation of verbal jokes in relation to conceptual tempo. *Child Development,* 1977, *48,* 960-967.

Brooks, L. Visual patterns in fluent word identification. In A. S. Reber & D. L. Scarborough (Eds.), *Toward a psychology of reading.* Hillsdale, NJ: Erlbaum, 1977.

Brown, A. L. Metacognitive development and reading. In R. J. Spiro, B. Bruce, & W. F. Brewer (Eds.), *Theoretical issues in reading comprehension.* Hillsdale, NJ: Erlbaum, 1979.

Brown, A. L., & French, L. A. The zone of potential development: Implications for intelligence testing in the year 2000. In R. J. Sternberg (Ed.), *Human intelligence.* Norwood, NJ: Ablex Publishers, 1979.

Bruce, D. J. The analysis of word sounds by young children. *British Journal of Educational Psychology,* 1964, *34,* 158-170.

Bruner, J. S. *The relevance of education.* London, England: Allen and Unwin, 1971.

Bruner, J. S. The process of concept attainment. In J. M. Anglin (Ed.), *Beyond the information given.* New York, NY: Norton, 1973.

Bruner, J. S. The role of dialogue in language acquisition. In A. Sinclair, R. J. Jarvella, & W. J. M. Levelt (Eds.), *The child's conception of language.* New York, NY: Springer-Verlag, 1978.

Burling, R. *English in black and white.* New York, NY: Holt, Rinehart, & Winston, 1973.

Burton, W. H. *Reading in child development.* New York, NY: Bobbs-Merrill, 1956.

Buswell, G. T. *Fundamental reading habits: A study of their development.* (*Supplementary Educational Monographs,* No. 21.) Chicago, Ill.: University of Chicago Press, 1922.

Buswell, G. T. *Diagnostic studies in arithmetic.* (*Supplementary Educational Monographs*, No. 30.) Chicago, Ill.: University of Chicago Press, 1926.

Buswell, G. T. *Non-oral reading: A study of its use in the Chicago public schools.* (*Supplementary Educational Monographs*, No. 66.) Chicago, Ill.: University of Chicago Press, 1945.

Byrne, B., & Shea, P. Semantic and phonetic memory codes in beginning readers. *Memory and Cognition,* 1979, *7*, 333-338.

Calfee, R., Chapman, R., & Venezky, R. How a child needs to think to learn to read. In J. L. Gregg (Ed.), *Cognition in learning and memory.* New York, NY: Wiley, 1972.

Calfee, R. C., Lindamood, P., & Lindamood, C. Acoustic-phonetic skills and reading—Kindergarten through twelfth grade. *Journal of Educational Psychology,* 1973, *64*, 293-298.

Campbell, D. T., & Fiske, D. W. Convergent and discriminant validation by the multitrait-multimethod matrix. *Psychological Bulletin,* 1959, *56*, 81-105. Reprinted and abridged in D. A. Payne & R. F. McMorris. *Educational and psychological measurement.* Morristown, NJ: General Learning Press, 1975.

Canney, G., & Winograd, P. *Schemata for reading and reading comprehension performance.* Champaign, Ill.: University of Illinois at Urbana-Champaign, 1979. (ERIC Document Reproduction Service No. ED 109 520).

Carroll, J. B. A model of school learning. *Teachers College Record,* 1963, *64*, 723-733.

Carroll, J. B., & Sapon, S. M. *Modern language aptitude test.* New York, NY: The Psychological Corporation, 1959.

Carver, C. The aural analysis of sounds. *British Journal of Educational Psychology,* 1967, *37*, 379-380.

Cazden, C. B. *Child language and education.* New York, NY: Holt, Rinehart, & Winston, 1972.

Cazden, C. B. Play with language and metalinguistic awareness: One dimension of language experience. *The Urban Review,* 1974, *7*, 28-39.

Chall, J. *Learning to read: The great debate.* New York, NY: McGraw-Hill, 1967.

Chappell, P. R. Early development of awareness of lexical units. Paper presented at the annual meeting of the American Educational Research Association, Chicago, Ill., 1968.

Chomsky, C. Reading, writing and phonology. *Harvard Educational Review,* 1970, *40*, 287-310.

Chomsky, C. Stages in language development and reading exposure. *Harvard Educational Review,* 1972, *42*, 1-33.

Chomsky, C. Approaching reading through invented spelling. In L. B. Resnick & P. A. Weaver (Eds.), *Theory and practice of early reading* (Vol. 2). Hillsdale, NJ: Erlbaum, 1977.

Chomsky, C. Consciousness is relevant to linguistic awareness. Paper presented at the University of Victoria/IRA Research Seminar on Linguistic Awareness and Learning to Read, Victoria, B. C., 1979.

Chomsky, N. *Syntactic structures.* The Hague: Mouton, 1957.

Chomsky, N. Review of Skinner's "Verbal Behavior." *Language,* 1959, *35*, 26-58.

Chomsky, N. *Aspects of the theory of syntax.* The Hague: Mouton, and New York, NY: Harper & Row, 1965.

Chomsky, N. *Language and mind*. New York, NY: Harcourt, Brace & World, 1968.

Chomsky, N. The case against B. F. Skinner: Review of "Beyond freedom and dignity." *New York Times, 1971, 17,* 18-24.

Chomsky, N. *Reflections on language*. Glasgow, Scotland: Fontana & Collins, 1975.

Chomsky, N. *Essays on form and interpretation*. New York, NY: North Holland, 1977.

Chomsky, N., & Halle, M. *The sound pattern of English*. New York, NY: Harper & Row, 1968.

Christensen, F. *Notes toward a new rhetoric*. New York, NY: Harper & Row, 1967.

Clark, E. V. Awareness of language: Some evidence from what children say and do. In A. Sinclair, R. J. Jarvella, & W. J. M. Levelt (Eds.), *The child's conception of language*. New York, NY: Springer-Verlag, 1978.

Clay, M. M. *Reading: The patterning of complex behaviour*. Auckland, New Zealand: Heinemann, 1972.

Clay, M. M. *What did I write?* Auckland, New Zealand: Heinemann, 1975.

Clymer, T. What is "reading"? Some current concepts. In H. M. Robinson (Ed.), *The sixty-seventh yearbook of the National Society for the Study of Education* (Pt. II). Chicago, Ill.: University of Chicago Press, 1968.

Cohen, A. S. Oral reading errors of first grade children taught by a code emphasis approach. *Reading Research Quarterly,* 1974-75, *10,* 616-650.

Cohen, D. H. The effect of literature on vocabulary and reading achievement. *Elementary English,* 1968, *45,* 209-213, 217.

Cohen, M. *La grande invention de l'écriture et son évolution*. Paris, France: Imprimerie Nationale, 1958.

Cole, R. A., & Jakimik, J. Understanding speech: How words are heard. In G. Underwood (Ed.), *Strategies of information processing*. London, England: Academic Press, 1978.

Combs, W. E. Sentence-combining practice aids reading comprehension. *Journal of Reading,* 1977, *21,* 18-24.

Conrad, R. Short-term memory processes in the deaf. *British Journal of Psychology,* 1970, *61,* 179-195.

Conrad, R. Speech and reading. In J. F. Kavanagh & I. G. Mattingly (Eds.), *Language by ear and by eye*. Cambridge, Mass.: MIT Press, 1972.

Conrad, R., & Rush, M. On the nature of short-term memory encoding by the deaf. *Journal of Speech and Hearing Disorders,* 1965, *30,* 336-343.

Cooper, C. R., & Petrosky, A. R. A psycholinguistic view of the fluent reading process. *Journal of Reading,* 1976, *20,* 184-207.

Coulmas, F. *Über Schrift*. Frankfurt, Germany: Suhrkamp, 1981.

Cronbach, L. J. *Educational psychology* (3rd ed.). New York, NY: Harcourt Brace Jovanovich, 1977.

Cross, D. Individual differences in the acquisition of reading skill. Paper presented at the annual convention of the International Reading Association, Chicago, Ill., 1982.

Cushman, R. C. The Kurzweil reading machine. *The Wilson Library Bulletin,* 1980, *54,* 311-315.

Daniels, J. C., & Diack, H. *Progress in reading*. Nottingham, England: University of Nottingham Institute of Education, 1956.

Danks, J. H. Comprehension in listening and reading: Same or different? In F. B. Murray (Ed.), *Reading and understanding*. Newark, Del.: International Reading Association, 1980.

Dathe, G. (Ed.) *Unterrichtshilfen Deutsch. Klasse 1*. Berlin, Germany: Volk und Wissen, 1979.

Dathe, G. *Erstleseunterricht*. Berlin, Germany: Volk und Wissen, 1981.

Denner, B. Representational and syntactic competence of problem readers. *Child Development*, 1970, *41*, 881-887.

Denny, T., & Weintraub, S. First-graders' responses to three questions about reading. *The Elementary School Journal*, 1966, *66*, 441-448.

Department of Education and Science. *A language for life* (The Bullock Report). London, England: Her Majesty's Stationary Office, 1975.

Desberg, P., Elliott, D., & Marsh, G. American Black English and spelling. In U. Frith (Ed.), *Cognitive processes in spelling*. London, England: Academic Press, 1980.

DeStefano, J. S. *Some parameters of register in adult and child speech*. Louvain, Belgium: Institute of Applied Linguistics, 1972.

deVilliers, J. G., & deVilliers, P. A. Competence and performance in child language: Are children really competent to judge? *Journal of Child Language*, 1974, *1*, 11-22.

deVilliers, P. A., & deVilliers, J. G. Early judgments of semantic and syntactic acceptability by children. *Journal of Psycholinguistic Research*, 1972, *1*, 299-310.

Diack, H. *In spite of the alphabet*. London, England: Chatto & Windus, 1965.

Dickenson, D. K., & Weaver, P. A. Remembering and forgetting: Story recallabilities of dyslexic children. Paper presented at the annual meeting of the American Educational Research Association, San Francisco, Calif., 1979.

Doctor, E. A., & Coltheart, M. Children's use of phonological encoding when reading for meaning. *Memory and Cognition*, 1980, *8*, 195-209.

Donaldson, M. *Children's minds*. London, England: Fontana, 1978.

Downing, J. Children's concepts of language in learning to read. *Educational Research*, 1970, *12*, 106-112.

Downing, J. Children's developing concepts of spoken and written language. *Journal of Reading Behavior*, 1971-72, 7, 1-19.

Downing, J. (Ed.) *Comparative reading*. New York, NY: Macmillan, 1973.

Downing, J. Linguistics for infants. *Reading*, 1977, *11*, 36-45.

Downing, J. *Reading and reasoning*. New York, NY: Springer-Verlag, and Edinburgh: Chambers, 1979.

Downing, J. A source of cognitive confusion for beginning readers: Learning in a second language. *The Reading Teacher*, in press.

Downing, J., Ayers, D., & Schaefer, B. Conceptional and perceptual factors in learning to read. *Educational Research*, 1978, *21*, 11-17.

Downing, J., Ayers, D., & Schaefer, B. *Linguistic awareness in reading readiness (LARR) test*. Windsor, England: NFER-Nelson, 1983.

Downing, J., & Downing, M. Metacognitive readiness for literacy learning. *Papua New Guinea Journal of Education*, in press.

Downing, J., & Leong, C. K. *Psychology of reading.* New York, NY: Macmillan, 1982.

Downing, J., & Oliver, D. The child's conception of a word. *Reading Research Quarterly,* 1973-74, *9,* 568-582.

Downing, J., Ollila, L., & Oliver, P. Cultural differences in children's concepts of reading and writing. *British Journal of Educational Psychology,* 1975, *45,* 312-316.

Downing, J., Ollila, L., & Oliver, P. Concepts of language in children from differing socioeconomic backgrounds. *Journal of Educational Research,* 1977, *70,* 277-281.

Dunn, L. M. *Peabody picture vocabulary test.* Minneapolis, Minn.: American Guidance Service, 1959.

Educational Testing Service. *Co-operative primary reading test.* Princeton, NJ: Educational Testing Service, 1967.

Edwards, D. L. Reading from the child's point of view. *Elementary English,* 1958, *35,* 239-241.

Edwards, D. L. The relation of concept of reading to intelligence and reading achievement scores of fifth grade children. Unpublished doctoral dissertation, University of Buffalo, NY, 1961.

Edwards, D. L. Teaching beginners the purpose of reading. *Elementary English,* 1962, *39,* 194-195, 215.

Eeds-Kniep, M. The frenetic fanatic phonic backlash. *Language Arts,* 1979, *56,* 909-917.

Egorov, T. G. *The psychology of mastering the skill of reading* (in Russian). Moscow, U.S.S.R.: Academy of Pedagogical Sciences, R.S.F.S.R., 1953.

Ehri, L. C. Word consciousness in readers and prereaders. *Journal of Educational Psychology,* 1975, *67,* 204-212.

Ehri, L. C. Word learning in beginning readers and prereaders: Effects of form class and defining contexts. *Journal of Educational Psychology,* 1976, *68,* 832-842.

Ehri, L. C. Beginning reading from a psycholinguistic perspective: Amalgamation of word identities. In F. B. Murray (Ed.), *The development of the reading process.* (International Reading Association Monograph No. 3.) Newark, Del.: International Reading Association, 1978.

Ehri, L. C. Linguistic insight: Threshold of reading acquisition. In T. G. Waller & G. E. MacKinnon (Eds.), *Reading research: Advances in theory and practice* (Vol. 1). New York, NY: Academic Press, 1979.

Ehri, L. C. The development of orthographic images. In U. Frith (Ed.), *Cognitive processes in spelling.* London, England: Academic Press, 1980. (a)

Ehri, L. C. The role of orthographic images in learning printed words. In J. F. Kavanagh & R. Venezky (Eds.), *Orthography, reading and dyslexia.* Baltimore, Md.: University Park Press, 1980. (b)

Ehri, L. C., & Wilce, L. S. The mnemonic value of orthography among beginning readers. *Journal of Educational Psychology,* 1979, *71,* 26-40.

Ehri, L. C., & Wilce, L. S. Do beginners learn to read function words better in sentences or in lists? *Reading Research Quarterly,* 1980, *15,* 451-476. (a)

Ehri, L. C., & Wilce, L. S. The influence of orthography on readers' conceptualization of the phonemic structure of words. *Applied Psycholinguistics,* 1980, *1,* 371-385. (b)

Ehri, L. C., & Wilce, L. S. The salience of silent letters in children's memory for word spellings. *Memory and Cognition,* 1982, *10,* 155-166.

Ehri, L. C., & Wilce, L. S. Development of word identification speed in skilled and less skilled beginning readers. *Journal of Educational Psychology,* in press.

Eichler, W. Zur linguistischen Fehleranalyse von Spontanschreibungen bei Vor- und Grundschulkindern. In A. Hofer (Ed.), *Lesenlernen: Theorie und Unterricht.* Düsseldorf, Germany: Schwann, 1976.

Eichler, W., & Bünting, K.-D. *Deutsche Grammatik.* Kronberg/Ts., Germany: Scriptor, 1976.

Elkind, D. Discrimination, seriation, and numeration of size and dimensional differences in young children: Piaget replication study VI. *Journal of Genetic Psychology,* 1964, *104,* 275-296.

Elkind, D. Cognitive development and reading. *Claremont Reading Conference 38th Yearbook.* Claremont, Calif.: Claremont Graduate School, 1974.

Elkind, D. Perceptual development in children. *American Scientist,* 1975, *63,* 533-541.

Elkind, D. *Child development and education: A Piagetian perspective.* New York, NY: Oxford University Press, 1976.

Elkind, D., Larson, M., & Van Doorminck, W. Perceptual decentration learning and performance in slow and average readers. *Journal of Educational Psychology,* 1965, *56,* 50-56.

Elkind, D., & Scott, L. Studies in perceptual development, I: The decentration of perception. *Child Development,* 1962, *33,* 619-630.

Elkonin, D. B. Development of speech. In A. V. Zaporoshetz & D. B. Elkonin (Eds.), *The psychology of pre-school children.* Cambridge, Mass.: MIT Press, 1971.

Elkonin, D. B. USSR. In J. Downing (Ed.), *Comparative reading* New York, NY: Macmillan, 1973. (a)

Elkonin, D. B. Further remarks on the psychological bases of the initial teaching of reading (in Russian). *Sovetskaia Pedagogika,* 1973, 14-23. (b)

Elkonin, D. B. Personal communication to J. Downing, 30 October, 1983.

Emerson, H. F. Children's comprehension of "because" in reversible and non-reversible sentences. *Journal of Child Language,* 1979, *6,* 279-300.

Emerson, H. F. Children's judgements of correct and reversed sentences with "if." *Journal of Child Language,* 1980, *7,* 135-155.

Engin, A. W. The relative importance of the subtests of the Metropolitan Readiness Test in the prediction of first grade reading and arithmetic achievement criteria. *Journal of Psychology,* 1974, *88,* 289-298.

Erickson, D., Mattingly, I. G., & Turvey, M. Phonetic coding of Kanji. *Journal of the Acoustical Society of America,* 1972, *52,* 132.

Erickson, D., Mattingly, I. G., & Turvey, M. Phonetic activity in reading: An experiment with Kanji. *Language and Speech,* 1977, *20,* 384-403.

Evanechko, P., Ollila, L., Downing, J., & Braun, C. An investigation of the reading readiness domain. *Research in the Teaching of English,* 1973, *7,* 61-78.

Evans, M., Taylor, N., & Blum, I. Children's written language awareness and its relation to reading acquisition. *Journal of Reading Behavior,* 1979, *11,* 7-19.

Farnham-Diggory, S. D. Symbol and synthesis in experimental reading. *Child Development,* 1967, *38,* 223-231.

Feldman, J. M. Wh* N**ds V*w*ls? In F. B. Murray (Ed.), *The recognition of words*. Newark, Del.: International Reading Association, 1978.

Ferguson, N. Pictographs and prereading skills. *Child Development*, 1975, *46*, 786-789.

Ferreiro, E. Vers une théorie génétique de l'apprentissage de la lecture. *Revue Suisse de Psychologie Pure et Appliquée*, 1977, *36*, 109-130.

Ferreiro, E. What is written in a written sentence? A developmental answer. *Journal of Education*, 1978, *160*, 25-39.

Ferreiro, E. Qu'est-ce qui est écrit dans une phrase écrite? Une réponse psycho-génétique. *Institut Romand de Documentation Pédagogique*, Recherche 1979 (No. 5), 1979.

Ferreiro, E., & Teberosky, A. *Los sistemas de escritura en el desarrollo del niño*. Mexico: Siglo Veintiuno Editores, SA. 1979. English translation, *Literacy before schooling*. Exeter, NH: Heinemann, 1982.

Fitts, P. M. Factors in complex skill training. In R. Glaser (Ed.), *Training research and education*. Pittsburgh, Penn.: University of Pittsburgh Press, 1962.

Fitts, P. M., & Posner, M. J. *Human performance*. Belmont, Calif.: Brooks-Cole, 1967.

Flahive, D. E., & Carrell, P. Metalinguistic awareness and cognitive development. In D. Lance & D. Gulstad (Eds.), *Papers from the 1977 Mid-America Linguistics Conference*. Columbia, Mo.: University of Missouri, 1978.

Flavell, J. H. The development of metacommunication. Paper presented at the Twenty-first International Congress of Psychology, Paris, France, 1976.

Flavell, J. H. *Cognitive development*. Englewood Cliffs, NJ: Prentice-Hall, 1977.

Flavell, J. H., Speer, J. R., Green, F. L., & August, D. L. The development of comprehension monitoring and knowledge about communication. *Monographs of the Society for Research in Child Development*, 1981, *46* (5, Serial No. 192).

Flesch, R. *Why Johnny can't read and what you can do about it*. New York, NY: Harper & Brothers, 1955.

Flexman, R. E., Matheny, W. G., & Brown, E. L. Evaluation of the school link and special methods of instruction. *University of Illinois Bulletin*, 1950, *47* (No. 80).

Fodor, J. A., Bever, T. G., & Garrett, M. F. *The psychology of language*. New York, NY: McGraw-Hill, 1974.

Forrest-Pressley, D. L. *Cognitive and meta-cognitive aspects of reading*. New York, NY: Springer-Verlag, 1983.

Forrest-Pressley, D. L., & Gillies, L. A. Children's flexible use of strategies during reading. In M. Pressley & J. Levin (Eds.), *Cognitive strategy training and research*. New York, NY: Springer-Verlag, 1983.

Forrest-Pressley, D. L., & Waller, T. G. Knowledge of cognitive processes related to reading. Paper presented at the annual convention of the International Reading Association, Chicago, Ill., 1982.

Fox, B., & Routh, D. K. Analyzing spoken language into words, syllables, and phonemes: A developmental study. *Journal of Psycholinguistic Research*, 1975, *4*, 331-342.

Fox, B., & Routh, D. K. Phonemic analysis and synthesis as word-attack skills. *Journal of Educational Psychology*, 1976, *68*, 70-74.

Fox, B., & Routh, D. K. Phonemic analysis and severe reading disability in children. *Journal of Psycholinguistic Research,* 1980, *9,* 115-119.

Francis, H. Children's experience of reading and notions of units in language. *British Journal of Educational Psychology,* 1973, *43,* 17-23.

Francis, H. *Language in childhood.* London, England: Paul Elek, 1975.

Francis, H. Children's strategies in learning to read. *British Journal of Educational Psychology,* 1977, *47,* 117-125.

Francis, H. Linguistic awareness, analytic ability and learning to read—a reply to Mattingly. Paper presented at the University of Victoria/IRA Research Seminar on Linguistic Awareness and Learning to Read, Victoria, B.C., 1979.

Francis, H. *Learning to read.* London, England: Allen & Unwin, 1982.

French, L. A., & Brown, A. L. Comprehension of before and after in logical and arbitrary sequences. *Journal of Child Language,* 1977, *4,* 247-256.

Fries, C. C. *Linguistics and reading.* New York, NY: Holt, Rinehart, & Winston, 1963.

Frith, U. From print to meaning and from print to sound, or how to read without knowing how to spell. *Visible Language,* 1978, *12,* 43-54.

Frostig, M., Lefever, W., & Whittlesey, J. R. B. *The Marianne Frostig development test of visual perception* (3rd ed.). Palo Alto, Calif.: Consulting Psychologists Press, 1964.

Gagné, R. M. *The conditions of learning.* New York, NY: Holt, Rinehart, & Winston, 1965.

Gambrell, L. B., & Heathington, B. S. Adult disabled readers' metacognitive awareness about reading tasks and strategies. *Journal of Reading Behavior,* 1981, *13,* 215-222.

Garson, C. C. Use of contextual knowledge by skilled and unskilled readers in first and second grade. Unpublished M.A. thesis, University of Notre Dame, Ind., 1980.

Garton, S., Schoenfelder, P., & Skriba, P. Activities for young word bankers. *The Reading Teacher,* 1979, *32,* 453-457.

Gelb, I. J. *A study of writing.* Chicago, Ill.: University of Chicago Press, 1952.

Gelman, R. Conservation acquisition: A problem of learning to attend to relevant attributes. *Journal of Experimental Child Psychology,* 1969, *7,* 167-187.

Gibson, E. J. Learning to read. *Science,* 1965, *148,* 1066-1972.

Gibson, E. J., Shurcliff, A., & Yonas, A. Utilization of spelling patterns by deaf and hearing subjects. In H. Levin & J. Williams (Eds.), *Basic studies on reading.* New York, NY: Basic Books, 1970.

Glass, G. G. Students' misconceptions concerning their reading. *The Reading Teacher,* 1968, *21,* 765-768.

Gleason, H. A. *An introduction to descriptive linguistics.* New York, NY: Holt, Rinehart, & Winston, 1961.

Gleitman, H., & Gleitman, L. R. Language use and language judgments. In C. Fillmore, D. Kempler, & S-Y. Wang (Eds.), *Individual differences in language ability and language behavior.* New York, NY: Academic Press, 1979.

Gleitman, L. R. Metalinguistics is not kid-stuff (comments on Mattingly). Paper presented at the University of Victoria/IRA Research Seminar on Linguistic Awareness and Learning to Read, Victoria, B. C., 1979.

Gleitman, L. R., & Gleitman, H. *Phrase and paraphrase.* New York, NY: Norton, 1970.

Gleitman, L. R., Gleitman, H., & Shipley, E. F. The emergence of the child as grammarian. *Cognition,* 1972, *2,* 137-164.

Gleitman, L. R., & Rozin, P. Teaching reading by use of a syllabary. *Reading Research Quarterly,* 1973, *8,* 447-483.

Gleitman, L. R., & Rozin, P. The structure and acquisition of reading, I: Relations between orthographies and the structure of language. In A. S. Reber & D. L. Scarborough (Eds.), *Toward a psychology of reading.* Hillsdale, NJ: Erlbaum, 1977.

Goldman, S. R. Reading skill and the minimum distance principle: A comparison of listening and reading comprehension. *Journal of Experimental Child Psychology,* 1976, *22,* 123-142.

Goldstein, D. M. Cognitive-linguistic functioning and learning to read in preschoolers. *Journal of Educational Psychology,* 1976, *68,* 680-688.

Goldstein, H. Some models for analysing longitudinal data on educational attainments. *Journal of the Royal Statistical Society,* 1979, *142,* 407-442.

Golinkoff, R. M. Critique: Phonemic awareness skills and reading achievement. In F. B. Murray & J. J. Pikulski (Eds.), *The acquisition of reading: Cognitive, linguistic and perceptual prerequisites.* Baltimore, Md.: University Park Press, 1978.

Golinkoff, R. M., & Rosinski, R. Decoding, semantic processing, and reading comprehension skill. *Child Development,* 1976, *47,* 252-258.

Goodacre, E. J. *Children and learning to read.* London, England: Routledge & Kegan Paul, 1971.

Goodman, K. S. Dialect barriers to reading comprehension. *Elementary English,* 1965, *42,* 853-860.

Goodman, K. S. A psycholinguistic view of reading comprehension. In G. B. Schick & M. M. May (Eds.), *New Frontiers in College-Adult Reading.* Milwaukee, Wis.: National Reading Conference, 1966.

Goodman, K. S. Reading: A psycholinguistic guessing game. In H. Singer & R. B. Ruddell (Eds.), *Theoretical models and processes of reading.* Newark, Del.: International Reading Association, 1970.

Goodman, K. S., & Buck, C. Dialect barriers to reading comprehension revisited. *The Reading Teacher,* 1973, *27,* 6-12.

Goodman, Y. M. The roots of literacy. In M. P. Douglass (Ed.), *Reading: A humanizing experience, Claremont Reading Conference 44th Yearbook.* Claremont, Calif.: Claremont Graduate School, 1980.

Gough, P. One second of reading. In H. Singer & R. B. Ruddell (Eds.), *Theoretical models and processes of reading.* Newark, Del.: International Reading Association, 1976.

Gough, P. Personal communication to H. Singer. Chicago, Ill., May 1982.

Gray, C. T. *Types of reading ability as exhibited through tests and laboratory experiments. (Supplementary Educational Monographs,* No. 5). Chicago, Ill.: University of Chicago Press, 1917.

Grice, H. P. Logic and conversation. In P. Cole & J. L. Morgan (Eds.), *Syntax and semantics.* New York, NY: Academic Press, 1975.

Groff, P. The topsy-turvy world of "sight" words. *The Reading Teacher,* 1974, *27,* 572-578.

Groff, P. Children's spelling of features of Black English. *Research in the Teaching of English,* 1978, *12,* 21-28.

Groff, P. A critique of teaching reading as a whole task venture. *The Reading Teacher,* 1979, *32,* 647-652.

Günther, H. Vom Verschriften zum Schreiben. Zur Diskussion des Beitrages von Helga Andresen. *Osnabrücker Beiträge zur Sprachtheorie* (Universität Osnabrück, Germany), 1979, *13,* 57-60.

Hakes, D. T. *The development of metalinguistic abilities in children.* New York, NY: Springer-Verlag, 1980.

Hakes, D. T. The development of metalinguistic abilities: What develops? In S. A. Kuczaj (Ed.), *Language development: Language, thought and culture.* Hillsdale, NJ: Erlbaum, 1982.

Hall, N. A. Children's awareness of segmentation in speech and print. *Reading,* 1976, *10,* 11-19.

Hall, P., & Pacey, C. *Paragraph understanding test.* Sydney, N.S.W.: Department of Education, 1978.

Halliday, M. A. K. *Explorations in the functions of language.* London, England: Edward Arnold, 1973.

Harris, A. J., & Sipay, E. R. *How to increase reading ability* (7th ed.). New York, NY: Longman, 1980.

Harris, J. W. *Spanish phonology.* Cambridge, Mass.: MIT Press, 1969.

Heeschen, V. The metalinguistic vocabulary of a speech community in the highlands of Irian Jaya (West New Guinea). In A. Sinclair, J. Jarvella, & W. J. M. Levelt (Eds.), *The child's conception of language.* New York, NY: Springer-Verlag, 1978.

Heilman, A. W. *Principles and practices of teaching reading* (4th ed.). Columbus, Oh.: Merrill, 1977.

Helfgott, J. Phonemic segmentation and blending skills of kindergarten children: Implications for beginning reading acquisition. *Contemporary Educational Psychology,* 1976, *1,* 157-169.

Henderson, E. H. Developmental concepts of word. In E. H. Henderson & J. W. Beers (Eds.), *Developmental and cognitive aspects of learning to spell.* Newark, Del.: International Reading Association, 1980.

Henderson, E. H., & Long, B. H. Predictors of success in beginning reading among negroes and whites. In J. A. Figurel (Ed.), *Reading goals for the disadvantaged,* Newark, Del.: International Reading Association, 1970.

Henderson, L. *Orthography and word recognition in reading.* New York, NY: Academic Press, 1982.

Hiebert, E. H. The relationship of logical reasoning ability, oral language comprehension and home experiences to preschool children's print awareness. *Journal of Reading Behavior,* 1980, *12,* 313-324.

Hiebert, E. H. Developmental patterns and interrelationships of preschool children's print awareness. *Reading Research Quarterly,* 1981, *16,* 236-260.

Hirsh-Pasek, K., Gleitman, L., & Gleitman, H. What did the brain say to the mind? A study of the detection and report of ambiguity by young children. In A. Sinclair, R. J. Jarvella, & W. J. M. Levelt (Eds.), *The child's conception of language.* New York, NY: Springer-Verlag, 1978.

Hjelmquist, E. A note on adults' conception of language and communication. *Osnabrücker Beiträge zur Sprachtheorie* (Universität Osnabrück, Germany), 1981, *20*, 62-73.

Hohn, W. E., & Ehri, L. C. *Do alphabet letters help prereaders acquire phonemic segmentation skill?* Manuscript submitted for publication, 1983.

Holden, M. H. Word awareness, reading and development. *Perceptual and Motor Skills*, 1977, *44*, 203-206.

Holden, M. H., & MacGinitie, W. H. Children's conceptions of word boundaries in speech and print. *Journal of Educational Psychology*, 1972, *63*, 551-557.

Holden, M. H., & MacGinitie, W. H. Metalinguistic ability and cognitive performance in children from five to seven. Paper presented at the annual meeting of the American Educational Research Association, New Orleans, February, 1973. (ERIC Document Reproduction Service No. ED 078 436.)

Holmes, J. A. The substrata-factor theory of reading: Some experimental evidence. In H. Singer & R. B. Ruddell (Eds.), *Theoretical models and processes of reading*. Newark, Del.: International Reading Association, 1970.

Holmes, J. A., & Singer, H. Theoretical models and trends towards basic research in reading. *Review of Educational Research*, 1964, *34*, 127-155.

Hook, P. E., & Johnson, D. J. Metalinguistic awareness and reading strategies. *Bulletin of the Orton Society*, 1978, *28*, 62-78.

Hoskisson, K. Successive approximation and beginning reading. *Elementary School Journal*, 1975, *75*, 443-451. (a)

Hoskisson, K. The many facets of assisted reading. *Elementary English*, 1975, *52*, 312-315. (b)

Hoskisson, K., & Krohm, B. Reading by immersion: Assisted reading. *Elementary English*, 1974, *51*, 832-836.

Householder, F. *Linguistic speculations*. Cambridge, England: Cambridge University Press, 1971.

Howe, H. E., & Hillman, D. The acquisition of semantic restrictions in children. *Journal of Verbal Learning and Verbal Behavior*, 1973, *12* 132-139.

Huey, E. B. *The psychology and pedagogy of reading*. New York, NY: Macmillan, 1908. (Republished, Cambridge, Mass.: MIT Press, 1968.)

Hull, C. H., & Nie, N. H. *SPSS update 7-9*. New York, NY: McGraw-Hill, 1981.

Hunt, D. E. Person-environment interaction: A challenge found wanting before it was tried. *Review of Educational Research*, 1975, *45*, 209-330.

Hurwitz, A. V., & Goddard, A. *Games to improve your child's English*. New York, NY: Simon & Schuster, 1969.

Hutson, B. A. Macro-theory and micro-theory in the study of concepts about language and reading. Paper presented at the University of Victoria/IRA Research Seminar on Linguistic Awareness and Learning to Read, Victoria, B.C., 1979.

Huttenlocher, J. Children's language: Word phrase relationship. *Science*, 1964, *143*, 264-265.

Ianco-Worrall, A. D. Bilingualism and cognitive development. *Child Development*, 1972, *43*, 1390-1400.

Inhelder, B., & Piaget, J. *The early growth of logic in the child*. New York, NY: Harper & Row, 1964.

Jacobs, J. Children's reports about cognitive aspects of reading comprehension. Paper presented at the annual convention of the International Reading Association, Chicago, Ill., 1982.

Jaeger, J. J. Vowel shift rule vs. spelling rules: Which is psychologically real? Paper presented at the annual meeting of the Linguistic Society of America, Los Angeles, Calif., 1979.

Jakimik, J., Cole, R. A., & Rudnicky, A. I. The influence of spelling on speech perception. Paper presented at the annual meeting of the Psychonomic Society, St. Louis, Mo., 1980.

James, S., & Miller, J. Children's awareness of semantic constraints in sentences. *Child Development*, 1973, *44*, 69-76.

Januschek, F., Paprotté, W., & Rohde, W. The growth of metalinguistic knowledge in children. In M. Van de Velde & W. Vandeweghe (Eds.), *Sprachstruktur, Individuum und Gesellschaft. Akten des 13. Linguistischen Kolloquiums, Gent, 1978*, Vol. 1. Tübingen, Germany: Niemeyer, 1979, 243-254. (a)

Januschek, F., Paprotté, W., & Rohde, W. Zur Ontogenese sprachlicher Handlungen. *Osnabrücker Beiträge zur Sprachtheorie*, 1979, *10*, 37-69. (b)

Januschek, F., & Rohde, W. Untersuchung von Sprechhandlungsstrategien zur Aneignung von Wortbedeutungen bei Primarstufenschülern. *Osnabrücker Beiträge zur Sprachtheorie* (Universität Osnabrück, Germany), 1981, *20*, 75-99.

Jenkinson, M. D. Sources of knowledge for theories of reading. *Journal of Reading Behavior*, 1969, *1*, 11-29.

Jensen, H. *Sign, symbol and script: An account of man's efforts to write.* London, England: Allen & Unwin, 1970.

Johns, J. L. Reading: A view from the child. *The Reading Teacher*, 1970, *23*, 647-648.

Johns, J. L. Concepts of reading among good and poor readers. *Education*, 1974, *95*, 58-60.

Johns, J. L., & Ellis, D. W. Reading: Children tell it like it is. *Reading World*, 1976, *16*, 115-128.

Judd, C. H., & Buswell, G. *Silent reading: A study of various types.* (*Supplementary Educational Monographs*, No. 23). Chicago, Ill., University of Chicago Press, 1922.

Jung, U. O. H. Zur auditiven Diskrimination legasthener und normaler Schüler. *Linguistik und Didaktik*, 1977, *31*, 210-218.

Karolije-Walz, P. Metasprachliche Fähigkeiten bilingualer Kinder. *Osnabrücker Beiträge zur Sprachtheorie* (Universität Osnabrück, Germany), 1981, *20*, 139-157.

Karpova, S. N. Realization of the verbal composition of speech by a preschool child (in Russian). *Voprosy Psikhologii*, 1955, *4*, 43-55.

Karpova, S. N. The preschooler's realisation of the lexical structure of speech. In F. Smith & G. A. Miller (Eds.), *The genesis of language: A psycholinguistic approach.* Cambridge, Mass.: MIT Press, 1966.

Käsermann, M.-L. Language acquisition and interaction: Some determinants of modifications in non-understanding situations. Paper presented at the University of Victoria/IRA Research Seminar on Linguistic Awareness and Learning to Read, Victoria, B.C., 1979.

Katz, I. C. The effects of instructional methods on reading acquisition systems. Unpublished doctoral dissertation, University of California, Riverside, Calif., 1980.

Katz, I. C., & Singer, H. The substrata-factor theory of reading: Differential development of subsystems underlying reading comprehension in the first years of instruction. In J. Niles & L. Miller (Eds.), *New inquiries in reading research and instruction*. Rochester, NY: National Reading Conference, 1982.

Katz, I. C., & Singer, H. Substrata-factor theory of reading: Subsystem patterns underlying achievement in beginning reading. Submitted for publication, 1983.

Katzenberger, L. Schulanfänger und Lesenlernen. *Schule und Psychologie*, 1967, *14*, 345-359.

Kavanagh, J. F. (Ed.), *Communicating by language: The reading process*. Bethesda, Md.: NICHD, U.S. Dept. of Health, Education and Welfare, 1968.

Kavanagh, J. F., & Mattingly, I. G. (Eds.), *Language by ear and by eye*. Cambridge, Mass.: MIT Press, 1972.

Kavanagh, J. F., & Strange, W. (Eds.), *Speech and language in the laboratory, school and clinic*. Cambridge, Mass.: MIT Press, 1978.

Kavanagh, J. F., & Venezky, R. L. (Eds.), *Orthography, reading and dyslexia*. Baltimore, Md.: University Park Press, 1980.

Keeton, A. Children's cognitive integration and memory processes for comprehending written sentences. *Journal of Experimental Child Psychology*, 1977, *23*, 459-471.

Keil, F. Development of the ability to perceive ambiguities: Evidence for the task specificity of a linguistic skill. *Journal of Psycholinguistic Research*, 1980, *9*, 219-229.

Kerek, A. The phonological relevance of spelling pronunciation. *Visible Language*, 1976, *10*, 323-338.

Kerlinger, F. N., & Pedhazur, E. J. *Multiple regression in behavioral research*. New York, NY: Holt, Rinehart, & Winston, 1973.

Kessel, F. S. The role of syntax in children's comprehension from ages six to twelve. *Monographs of the Society for Research in Child Development*, 1970, *35*, 1-95.

Killey, J. C., & Willows, D. M. Good-poor readers' differences in detecting, pinpointing and correcting errors in orally presented sentences. Paper presented to the American Educational Research Association, Boston, Mass., 1980.

Kingston, A. J., Weaver, W. W., & Figa, L. E. Experiments in children's perception of words and word boundaries. In F. P. Greene (Ed.), *Investigations relating to mature reading*. Milwaukee, Wis.: National Reading Conference, 1972.

Kintsch, W., & Buschke, H. Homophones and synonyms in short-term memory. *Journal of Experimental Psychology*, 1969, *80*, 403-407.

Kintsch, W., & Van Dijk, T. A. Toward a model of text comprehension and production. *Psychological Review*, 1978, *85*, 363-394.

Kirk, S. A., McCarthy, J. J., & Kirk, W. E. *Illinois test of psycholinguistic abilities*. Urbana, Ill.: University of Illinois Press, 1968.

Kleiman, G. N. Speech recoding in reading. *Journal of Verbal Learning and Verbal Behavior*, 1974, *14*, 323-339.

Kligman, D. S., Cronnell, B. A., & Verna, G. B. Black English pronunciation and spelling performance. *Elementary English*, 1972, *49*, 1247-1253.

Klima, E. S. How alphabets might reflect language. In J. F. Kavanagh & I. G. Mattingly (Eds.), *Language by ear and by eye*. Cambridge, Mass.: MIT Press, 1972.

Kreiszig, E. *Statistische Methoden und ihre Anwendungen*. Göttingen, Germany: Vandenhoek & Ruprecht, 1968.

Kuczaj, S. A. Children's judgments of grammatical and ungrammatical irregular past tense verbs. *Child Development*, 1978, *49*, 319-326.

Kuntz, M. H. The relationship between written syntactic attainment and reading ability in seventh grade. Unpublished doctoral dissertation, University of Pittsburgh, Penn., 1975.

Kyöstiö, O. K. Finland. In J. Downing (Ed.), *Comparative reading*. New York, NY: Macmillan, 1973.

LaBerge, D., & Samuels, S. J. Toward a theory of automatic information processing in reading. *Cognitive Psychology*, 1974, *6*, 293-323.

Lamb, P. *Guiding children's language learning* (2nd ed.). Dubuque, Ia.: Wm. C. Brown, 1971.

Lansdown, R. *Reading: Teaching and learning*. London, England: Pitman, 1974.

Laubach, F. C., & Laubach, R. *Toward world literacy: The each one teach one way*. Syracuse, NY: University of Syracuse Press, 1960.

Lavine, L. Differentiation of letterlike forms in pre-reading children. *Developmental Psychology*, 1977, *13*, 89-94.

Ledger, G. W., & Ryan, E. B. Semantic integration: Effects of imagery, enaction and sentence repetition training on prereaders' recall for pictograph sentences. Paper presented at the annual meeting of the Psychonomic Society, Philadelphia, Penn., 1981.

Ledger, G. W., & Ryan, E. B. The effects of semantic integration training on recall for pictograph sentences. *Journal of Experimental Child Psychology*, 1982, *33*, 39-54.

Lee, A. Thematic structure and the comprehension of text. Unpublished doctoral dissertation, University of California, Riverside, Calif., 1979.

Lee, D. M., & Allen R. V. *Learning to read through experience* (2nd ed.). New York, NY: Appleton-Century-Crofts, 1963.

Leonard, L. B., Bolders, J. G., & Curtis, R. A. On the nature of children's judgements of linguistic features: Semantic relations and grammatical morphemes. *Journal of Psycholinguistic Research*, 1977, *6* 233-245.

Leontev, A. A. Die psychophysischen Mechanismen der Rede. In B. A. Serébrennikov (Ed.), *Allgemeine Sprachwissenschaft*. München, Germany: Fink, 1973.

Leontev, A. A. *Psycholinguistische Einheiten und die Erzeungung sprachlicher Äußerungen*. Berlin, Germany: Akademie-Verlag, 1975.

Leontev, A. A. *Psychology and the language learning process*. Oxford, NY: Pergamon Press, 1981.

Leroy-Boussion, A. Une habileté auditivo-phonétique nécessaire pour apprendre à lire: La fusion syllabique. Nouvelle étude génétique entre 5 et 8 ans. *Enfance*, 1975, *2*, 164-190.

Leroy-Boussion, A., & Martinez, F. Un pré-requis auditivo-phonétique pour l'apprentissage du langage écrit: L'analyse syllabique. Etude génétique longitudinale entre 5 et 8 ans. *Enfance*, 1974, *1-2*, 111-130.

Lesgold, A. M., McCormick, C., & Golinkoff, R. M. Imagery training and children's prose learning. *Journal of Educational Psychology*, 1975, *67*, 663-667.

Lessler, K., Schoeninger, D. W., & Bridges, J. S. Prediction of first grade performance. *Perceptual and Motor Skills*, 1970, *31*, 751-756.

Levelt, W. J. M., Sinclair, A., & Jarvella, R. J. Causes and functions of linguistic
awareness in language acquisition: Some introductory remarks. In A. Sinclair,
J. Jarvella, & W. J. M. Levelt (Eds.), *The child's conception of language*. New
York, NY: Springer-Verlag, 1978.

Levin, J. R. Inducing comprehension in poor readers: A test of a recent model.
Journal of Educational Psychology, 1973, *65*, 19-24.

Lewkowicz, N. K. Phonemic awareness training: What to teach and how to teach
it. *Journal of Educational Psychology*, 1980, *72*, 686-700.

Lewkowicz, N. K., & Low, L. Y. Effects of visual aids and word structure on
phonemic segmentation. *Contemporary Educational Psychology*, 1979, *4*,
238-252.

Liberman, A. M., Cooper, F. S., Shankweiler, D. P., & Studdert-Kennedy, M.
Perception of the speech code. *Psychological Review*, 1967, *74*, 431-461.

Liberman, I. Y. Segmentation of the spoken word and reading acquisition.
Bulletin of the Orton Society, 1973, *23*, 65-77.

Liberman, I. Y., Liberman, A. M., Mattingly, I., & Shankweiler, D. Orthography
and the beginning reader. In J. F. Kavanagh & R. L. Venezky (Eds.), *Orthography, reading, and dyslexia*. Baltimore, Md.: University Park Press, 1980.

Liberman, I. Y., & Shankweiler, D. Speech, the alphabet and teaching to read. In
L. B. Resnick & P. A. Weaver (Eds.), *Theory and practice of early reading*.
Hillsdale, NJ: Erlbaum, 1979.

Liberman, I. Y., Shankweiler, D., Fischer, F. W., & Carter, B. Explicit syllable
and phoneme segmentation in the young child. *Journal of Experimental Child
Psychology*, 1974, *18*, 201-212.

Liberman, I. Y., Shankweiler, D., Liberman, A. M., Fowler, C., & Fischer, F. W.
Phonetic segmentation and recoding in the beginning reader. In A. S. Reber &
D. L. Scarborough (Eds.), *Toward a psychology of reading*. Hillsdale, NJ: Erl-
baum, 1977.

Lieberman, P. On the development of vowel production in young children. In
G. H. Yeni-Komshian, J. F. Kavanagh, & C. A. Ferguson (Eds.), *Child phonology, Vol. 1: Production*. New York, NY: Academic Press, 1980.

Liles, B. Z., Shulman, M. D., & Bartlett, S. Judgments of grammaticality by normal and language-disordered children. *Journal of Speech and Hearing Disorders*, 1977, *42*, 199-209.

Lipson, M.Y. Promoting children's metacognition about reading through direct
instruction. Paper presented at the annual convention of the International
Reading Association, Chicago, Ill., 1982.

Loban, W. *Language development: Kindergarten through grade twelve*. (Research
Rept. No. 18). Urbana, Ill.: National Council of Teachers of English, 1976.

Locke, J. L. Short-term memory encoding strategies of the deaf. *Psychonomic
Science*, 1970, *18*, 233-234.

Lundberg, I. Aspects of linguistic awareness related to reading. In A. Sinclair,
R. J. Jarvella, & W. J. M. Levelt (Eds.), *The child's conception of language*.
New York, NY: Springer-Verlag, 1978.

Lundberg, I. Linguistic awareness as related to dyslexia. In Y. Zotterman (Ed.),
Dyslexia, neuronal, cognitive & linguistic aspects. New York, NY: Pergamon
Press, 1982. (a)

Lundberg, I. Longitudinal studies of reading and its difficulties in Sweden. Unpublished manuscript. University of Umeå, Sweden, 1982. (b)

Lundberg, I., & Tornéus, M. Nonreaders' awareness of the basic relationship between spoken and written words. *Journal of Experimental Child Psychology*, 1978, *25*, 404-412.

Lundberg, I., Wall, S., & Olofsson, A. Reading and spelling skills in the first school years predicted from phonemic awareness skills in kindergarten. *Scandinavian Journal of Psychology*, 1980, *21*, 159-173.

Lunzer, E. A. Construction of a standardized battery of Piagetian tests to assess the development of effective intelligence. *Research in Education*, 1970, *3*, 53-72.

Lunzer, E. A., Dolan, T., & Wilkinson, J. E. The effectiveness of measures of operativity, language and short term memory in the prediction of reading and mathematical understanding. *British Journal of Educational Psychology*, 1976, *46*, 295-305.

Lunzer, E., & Gardner, K. *The effective use of reading*. London, England: Heinemann, 1979.

Luria, A. R. On the pathology of grammatical operations (in Russian). *Izvestija APN RSFSR* (No. 17), 1946.

Luria, A. R. *Cognitive development: Its cultural and social foundations*. Cambridge, Mass.: Harvard University Press, 1976.

MacGinitie, W. H. Using the spoken language analyzer to understand written language. Paper presented at the University of Victoria/IRA Research Seminar on Linguistic Awareness and Learning to Read, Victoria, B.C., 1979.

Malmquist, E. Problems of reading and readers: An international challenge. In D. K. Bracken & E. Malmquist (Eds.), *Improving reading ability around the world*. Newark, Del.: International Reading Association, 1971.

Malmquist, E. Sweden. In J. Downing (Ed.), *Comparative reading*. New York, NY: Macmillan, 1973.

Malt, L. G. A skills analyst takes a look at reading schemes. *Reading*, 1977, *11* (2), 12-28.

Mandler, J. M., & Johnson, N. S. Remembrance of things passed, story structure and recall. *Cognitive Psychology*, 1977, *9*, 111-151.

Marcel, T. Phonological awareness and phonological representation: Investigation of a specific spelling problem. In U. Frith (Ed.), *Cognitive processes in spelling*. London, England: Academic Press, 1980.

Markman, E. M. Realizing that you don't understand: A preliminary investigation. *Child Development*, 1977, *48*, 986-992.

Marsh, G., & Mineo, R. J. Training preschool children to recognize phonemes in words. *Journal of Educational Psychology*, 1977, *69*, 748-753.

Martin, S. Nonalphabetic writing systems: Some observations. In J. F. Kavanagh & I. G. Mattingly (Eds.), *Language by ear and by eye*. Cambridge, Mass.: MIT Press, 1972.

Mason, G. E. Preschoolers' concepts of reading. *The Reading Teacher*, 1967, *21*, 130-132.

Massaro, D. W. A stage model of reading and listening. *Visible Language*, 1978, *12*, 3-26.

Mattingly, I. G. Reading, the linguistic process, and linguistic awareness. In J. F. Kavanagh & I. G. Mattingly (Eds.), *Language by ear and by eye*. Cambridge, Mass.: MIT Press, 1972.

Mattingly, I. G. Reading, linguistic awareness and language acquisition. Paper presented at the University of Victoria/IRA Research Seminar on Linguistic Awareness and Learning to Read, Victoria, B.C., 1979. (a)

Mattingly, I. G. The psycholinguistic basis of linguistic awareness. In M. L. Kamil & A. J. Moe (Eds.), *Reading research: Studies and applications*. Clemson, SC: National Reading Conference, 1979. (b)

Mavrogenes, N. A. The language development of the disabled secondary reader. Paper presented at the annual convention of the International Reading Association, Miami Beach, Fl., 1977. (ERIC Document Reproduction Service No. 141 763).

Mavrogenes, N. A. The language development of disabled secondary readers: II. Paper presented at the Great Lakes Regional Conference of the International Reading Association, Cincinnati, Oh., 1978. (ERIC Document Reproduction Service No. 172 135).

Mayer, M. *A boy, a dog, and a frog*. New York, NY: Dial, 1967.

McCall, R. B. Challenges to a science of developmental psychology. *Child Development*, 1977, *48*, 333-344.

McCarr, J. E. *Lessons in syntax*. Lake Oswego, Ore.: Dormac, 1973.

McDonald, F. J. *Educational psychology*. Belmont, Calif.: Wadsworth, 1965.

McLeish, J., & Martin, J. Verbal behaviour: A review and experimental analysis. *Journal of Genetic Psychology*, 1975, *93*, 3-66.

McNeill, J. D., & Stone, J. Note on teaching children to hear separate sounds in spoken words. *Journal of Educational Psychology*, 1965, *56*, 13-15.

Meltzer, N. S., & Herse, R. The boundaries of written words as seen by first graders. *Journal of Reading Behavior*, 1969, *1*, 3-14.

Menagh, H. B. *Creative dramatics in guiding children's language learning*. Dubuque, Ia.: Wm. C. Brown, 1967.

Menyuk, P. *Sentences children use*. Cambridge, Mass.: MIT Press, 1969.

Menyuk, P. Language development and reading. In J. Flood (Ed.), *Understanding reading comprehension*. Newark, Del.: International Reading Association, 1981.

Messer, S. Implicit phonology in children. *Journal of Verbal Learning and Verbal Behavior*, 1967, *6*, 609-613.

Mommers, M. I. C. Linguistisch bewustzijn en leren lezen. Unpublished manuscript, Nijmegen, The Netherlands: Instituut voor Onderwijskunde, 1982.

Moore, T. E. Linguistic intuitions of 12-year-olds. *Language and Speech*, 1975, *18*, 213-218.

Morais, J., Cary, L., Alegria, J., & Bertelson, P. Does awareness of speech as a sequence of phones arise spontaneously? *Cognition*, 1979, *7*, 323-331.

Moskowitz, B. A. On the status of vowel shift in English. In T. E. Moore (Ed.), *Cognitive development and the acquisition of language*. New York, NY: Academic Press, 1973.

Muskopf, A. F. The beginning reader's concept of reading as related to intelligence, reading achievement, and the method of introducing reading. Unpublished Master's thesis, University of Chicago, Ill., 1962.

Nelson, K. Concept, word and sentence: Interrelations in acquisition and development. *Psychological Review*, 1974, *4*, 267-285.

Nie, N. H., Hull, C. H., Jenkins, J. G., Steinbrenner, K., & Bert, D. H. *Statistical package for the social sciences* (2nd ed.). New York, NY: McGraw-Hill, 1975.

Nixon, M. *Children's classification skills.* Hawthorn, Victoria: Australian Council for Educational Research, 1971.

O'Hare, F. *Sentencecraft.* Boston, Mass.: Ginn, 1975.

Ollila, L., Johnson, T., & Downing, J. Adapting Russian methods of auditory discrimination training for English. *Elementary English*, 1974, *51*, 1138-1141, 1145.

Olofsson, A., & Lundberg, I. Can phonemic awareness be trained in kindergarten? *Scandinavian Journal of Psychology*, 1983, *24*, 35-44.

Olson, D. R. From utterance to text: The bias of language in speech and writing. *Harvard Educational Review*, 1977, *47*, 257-281.

Olson, D. R. *The nature and consequences of literacy.* Book in preparation, 1981.

Olson, D. R. What is *said* and what is *meant* in speech and writing. Unpublished manuscript, Ontario Institute for Studies in Education, Toronto, Ontario, 1982.

Papandropoulou, I., & Sinclair, H. What is a word? Experimental study of children's ideas on grammar. *Human Development*, 1974, *17*, 241-258.

Paprotté, W. Zum Wortbegriff des Vorschul- und Schulkindes. In M. Hartig & B. Leuschner (Eds.), *Kongreßbericht der 9. Jahrestagung der GAL e.V., Mainz, 1978* (Bd. II). Heidelberg, Germany: Julius Groos, 1979.

Paprotté, W. Überlegungen zu einer Theorie des Erwerbs metasprachlicher Handlungen. *Osnabrücker Beiträge zur Sprachtheorie* (Universität Osnabrück, Germany), 1981, *20*, 9-43.

Paris, S. G. Combining research and instruction on reading comprehension in the classroom. Paper presented at the annual convention of the International Reading Association, Chicago, Ill., 1982.

Paris, S. G., & Lindauer, B. K. The development of cognitive skills during childhood. In B. Wolman (Ed.), *Handbook of developmental psychology*. Englewood Cliffs, NJ: Prentice-Hall, 1982.

Paris, S. G., & Lipson, M. Y. Metacognition and reading comprehension. Paper presented at the annual convention of the International Reading Association, Chicago, Ill., 1982.

Paris, S. G., & Meyers, M. Comprehension monitoring, memory and study strategies of good and poor readers. *Journal of Reading Behavior*, 1981, *13*, 5-22.

Peck, V., & Borkowski, J. G. Intelligence and giftedness. Unpublished manuscript, University of Notre Dame, Ind., 1980.

Piaget, J. *Le langage et la pensée chez l'enfant.* Neuchatel, Switzerland: Delachaux et Niestlé, 1923.

Piaget, J. *The language and thought of the child.* London, England: Routledge & Kegan Paul, 1926. (Second edition, 1959).

Piaget, J. The mental development of the child. In D. Elkind (Ed.), *Six psychological studies.* New York, NY: Random House, 1967.

Piaget, J. *La prise de conscience.* Paris, France: Presses Universitaires de France, 1974. (a)

Piaget, J. *Réussir et comprendre.* Paris, France: Presses Universitaires de France, 1974. (b)

Piaget, J. Article in B. Inhelder & H. H. Chipman (Eds.), *Piaget and his school.* New York, NY: Springer-Verlag, 1976.

Pidgeon, D. A. Logical steps in the process of learning to read. *Educational Research,* 1976, *18*, 174-181.

Pidgeon, D. A. Why put the cart before the horse? In D. Thackray (Ed.), *Growth in reading.* London, England: Ward Lock, 1979.

Pidgeon, D. A. Study of the effectiveness of a new programme for teaching children to read (Report No. HR 4177). London, England: The Social Science Research Council, 1981.

Pienaar, P. T. Using the language experience approach in special classes in South Africa. *The Reading Teacher,* 1977, *31*, 60-66.

Postman, L., & Senders, V. Incidental learning and generality of set. *Journal of Experimental Psychology,* 1946, *36*, 153-165.

Potter, R. K., Kopp, G. A., & Green, H. C. *Visible speech.* New York, NY: Van Nostrand, 1947.

Powers, J. E., & Gowie, C. J. Children's strategies in processing active- and passive-voice sentences: Use of semantic and syntactic information. *Genetic Psychology Monographs,* 1977, *96*, 337-355.

Pressley, G. M. Mental imagery helps eight-year-olds remember what they read. *Journal of Educational Psychology,* 1976, *68*, 355-359.

Randel, M. A., Fry, M. A., & Ralls, E. M. Two readiness measures as predictors of first- and third-grade reading achievement. *Psychology in the Schools,* 1977, *14*, 37-40.

Rasmussen, C. *Let's say poetry together.* Minneapolis, Minn.: Burgess, 1962.

Read, C. Pre-school children's knowledge of English phonology. *Harvard Educational Review,* 1971, *41*, 1-34.

Read, C. *Children's categorization of speech sounds in English.* (Research Rept. No. 17). Urbana, Ill.: National Council of Teachers of English, 1975.

Read, C. Children's awareness of language, with emphasis on sound systems. In A. Sinclair, J. Jarvella, & W. J. M. Levelt (Eds.), *The child's conception of language.* New York, NY: Springer-Verlag, 1978.

Reber, A. S. Implicit learning of artificial grammars. *Journal of Verbal Learning and Verbal Behavior,* 1967, *6*, 855-863.

Reddy, R. Speech recognition by machine: A review. *Proceedings of the IEEE,* 1976, *64*, 501-531.

Reder, S. The written and the spoken word: Influence of Vai literacy on Vai speech. In S. Scribner & M. Cole (Eds.), *The psychology of literacy.* Cambridge, Mass.: Harvard University Press, 1981.

Reed, D. W. A theory of language, speech and writing. *Elementary English,* 1965, *42*, 845-851.

Reid, J. F. Learning to think about reading. *Educational Research,* 1966, *9*, 56-62.

Rickards, J. P., & August, G. J. Generative underlining strategies in prose recall. *Journal of Educational Psychology,* 1975, *67*, 860-865.

Robinson, E. J., & Robinson, W. P. The young child's understanding of communication. *Developmental Psychology,* 1976, *12*, 328-333.

Robinson, E. J., & Robinson, W. P. Development in the understanding of causes of success and failure in verbal communication. *Cognition,* 1977, *5,* 363-378.

Robinson, E. J., & Robinson, W. P. The role of egocentrism and of weakness in comparing in children's explanations of communication failure. *Journal of Experimental Child Psychology,* 1978, *26,* 147-160.

Rogosa, D. Causal models in longitudinal research. In J. K. Nesselroade & P. B. Baltes (Eds.), *Longitudinal research in the study of behavior and development.* New York, NY: Academic Press, 1979.

Röhr, H. *Voraussetzungen zum Erlernen des Lesens und Rechtschreibens.* Dissertation, Universität Münster, Germany, 1978.

Rosner, J. *Phonetic analysis training and beginning reading skills.* Pittsburgh, Penn.: University of Pittsburgh, Learning Research and Development Center, 1972.

Rosner, J. Auditory analysis training with pre-readers. *The Reading Teacher,* 1974, *27,* 379-384.

Rosner, J., & Simon, D. P. *The auditory analysis test: An initial report.* Pittsburgh, Penn.: University of Pittsburgh, Learning Research and Development Center, 1971.

Rozin, P., & Gleitman, L. The structure and acquisition of reading, II: The reading process and the acquisition of the alphabetic principle. In A. S. Reber & D. L. Scarborough (Eds.), *Toward a psychology of reading.* Hillsdale, NJ: Erlbaum, 1977.

Rozin, P., Poritsky, S., & Sotsky, R. American children with reading problems can easily learn to read English represented by Chinese characters. *Science,* 1971, *171,* 1264-1267.

Rubinstein, R. A. The cognitive consequences of bilingual education in northern Belize. *American Ethnologist,* 1979, *6,* 583-601.

Ruddell, R. B. A longitudinal study of four programs of reading instruction varying in emphasis on regularity of grapheme-phoneme correspondences and language structure on reading achievement in grades two and three. (Final Report, Project Nos. 3099 and 78085). Berkeley, Calif.: University of California, 1968.

Ruddell, R. B. *Reading-language instruction: Innovative practices.* Englewood Cliffs, NJ: Prentice-Hall, 1974.

Ruddell, R. B. Psycholinguistic implications for a systems of communication model. In H. Singer & R. B. Ruddell (Eds.), *Theoretical models and processes of teaching reading.* Newark, Del.: International Reading Association, 1976.

Ruddell, R. B. (Senior Ed.). *Pathfinder: Allyn and Bacon Reading Program.* Boston, Mass.: Allyn & Bacon, 1977.

Rumelhart, D. E. Towards an interactive model of reading. In S. Dornič (Ed.), *Attention and performance* (Vol. 6). Hillsdale, NJ: Erlbaum, 1977. (a)

Rumelhart, D. E. Understanding and summarizing brief stories. In D. LaBerge & S. J. Samuels (Eds.), *Basic processes in reading: Perception and comprehension.* Hillsdale, NJ: Erlbaum, 1977. (b)

Russell, D. H. *The dynamics of reading.* Waltham, Mass.: Ginn-Blaisdell, 1970.

Ryan, E. B. Metalinguistic development and reading. In L. H. Waterhouse, K. M. Fischer, & E. B. Ryan (Eds.), *Language awareness and reading.* Newark, Del.: International Reading Association, 1980.

Ryan, E. B., & Bialystok, E. The development of grammatical sensitivity in monolingual and bilingual children. Unpublished manuscript, McMaster University, 1983.

Ryan, B., & Ledger, G. W. Differences in syntactic skills between good and poor readers in the first grade. Paper presented at the meeting of the Midwest Psychological Association, Chicago, Ill., 1979. (a)

Ryan, E. G., & Ledger, G. W. Grammaticality judgments, sentence repetitions, and sentence corrections of children learning to read. *International Journal of Psycholinguistics*, 1979, *6*, 23-40. (b)

Ryan, E. B., & Ledger, G. W. Assessing sentence processing skills in prereaders. In B. Hutson (Ed.), *Advances in language reading research* (Vol. 1). Greenwich, Conn.: J. A. I. Press, 1982.

Ryan, E. B., Ledger, G. W., & Robine, D. M. Effects of semantic integration training on the recall for pictograph sentences of kindergarten and first grade children. *Journal of Educational Psychology*, in press.

Ryan, E. B., Ledger, G. W., Short, E. J., & Weed, K. A. Promoting the use of active comprehension strategies by poor readers. In B. Y. L. Wong (Ed.), *Special issue on metacognition in learning and learning disabilities. Topics in Learning and Learning Disabilities*, 1982, *2* 53-60.

Ryan, E. B., Ledger, G. W., & Weed, K. A. Acquisition and transfer of an integrative imagery strategy by 5 year olds. Paper presented at the annual meeting of the American Educational Research Association, Montreal, Quebec, 1983.

Ryan, E. B., McNamara, S. R., & Kenney, M. Linguistic awareness and reading performance in beginning readers. *Journal of Reading Behavior*, 1977, *9*, 399-400.

Sales, B. D., Haber, R. N., & Cole, R. A. Mechanisms of aural encoding, IV: Hear-see, say-write interactions for vowels. *Perception and Psychophysics*, 1969, *6*, 385-390.

Samuels, S. J. Hierarchical subskills in the reading acquisition process. In J. T. Guthrie (Ed.), *Aspects of reading acquisition*. Baltimore, Md.: Johns Hopkins University Press, 1976. (a)

Samuels, S. J. Modes of word recognition. In H. Singer & R. B. Ruddell (Eds.), *Theoretical models and processes of reading*. Newark, Del.: International Reading Association, 1976. (b)

Savin, H. B., & Bever, T. G. The non-perceptual reality of the phoneme. *Journal of Verbal Learning and Verbal Behavior*, 1970, *9*, 295-302.

Schank, R. C., & Abelson, R. P. *Scripts, plans, goals and understanding*. Hillsdale, N J: Erlbaum, 1977.

Scheerer-Neumann, G.: Kognitionspsychologische Überlegungen zum Rechtschreiben nach Diktat. *IRA/D-Beiträge*, 1982, *2*, 2-7.

Schmalohr, E. *Psychologie des Erstlese- und Schreibunterrichts*. München, Germany: Ernst Reinhardt, 1971.

Schneckner, P. J. The concepts of reading of selected first and third grade children and the relationship of these concepts to the children's intelligence and reading achievement. Unpublished doctoral dissertation, University of Northern Colorado, 1976.

Schneeberg, H. Listening while reading: A four year study. *The Reading Teacher*, 1977, *30*, 629-635.

Scholl, D. M., & Ryan, E. B. Child judgments of sentences varying in grammatical complexity. *Journal of Experimental Child Psychology*, 1975, *20*, 274-285.

Scholl, D. M., & Ryan, E. B. Development of metalinguistic performances in the early school years. *Language and Speech*, 1980, *23*, 199-211.

Segui, J., Frauenfelder, U., & Mehler, J. Phoneme monitoring, syllable monitoring and lexical access. *British Journal of Psychology*, 1981, *72*, 471-477.

Seidenberg, M. S., & Tanenhaus, M. K. Orthographic effects on rhyme monitoring. *Journal of Experimental Psychology: Human Learning and Memory*, 1979, *5*, 546-554.

Shantz, C. U. The development of social cognition. In M. Hetherington (Ed.), *Review of child development research* (Vol. 5). Chicago, Ill.: University of Chicago Press, 1975.

Shipley, J. *Word play*. New York, NY: Hawthorn Books, 1972.

Short, E. J., & Ryan, E. B. Metacognitive differences between skilled and less skilled readers: Remediating deficits through story grammar and attribution training. *Journal of Educational Psychology*, in press.

Shultz, T. R., & Horibe, F. Development of the appreciation of verbal jokes. *Developmental Psychology*, 1974, *10*, 13-20.

Shultz, T. R., & Pilon, R. Development of the ability to detect linguistic ambiguity. *Child Development*, 1973, *44*, 728-733.

Siegler, R. S., & Liebert, R. M. Effects of presenting relevant rules and complete feedback on the conservation of liquid quantity task. *Developmental Psychology*, 1972, *7*, 133-138.

Sinclair, A. Thinking about language: An interview study of children aged three to eight. *International Journal of Psycholinguistics*, 1980, *4-7*, 19-40.

Sinclair, A. Thinking about language: An interview study of children aged eight to eleven. *Osnabrücker Beiträge zur Sprachtheorie* (Universität Osnabrück, Germany), 1981, *20*, 44-61.

Sinclair, A. Some recent trends in the study of language development. *Journal of the International Society for Research in Behavioral Development*, 1982, *5*, 413-431.

Sinclair, A., Jarvella, R. J., & Levelt, W. J. M. (Eds.), *The child's conception of language* New York, NY: Springer-Verlag, 1978.

Sinclair, H. Conceptualization and awareness in Piaget's theory and its relevance to the child's conception of language. In A. Sinclair, R. J. Jarvella, & W. J. M. Levelt (Eds.), *The child's conception of language*. New York, NY: Springer-Verlag, 1978.

Sinclair de Zwart, H. Developmental psycholinguistics. In D. Elkind & J. H. Flavell (Eds.), *Studies in cognitive development*. New York, NY: Oxford University Press, 1969.

Singer, H. Conceptualization in learning to read. In G. B. Schick & M. M. May (Eds.), *New frontiers in college-adult reading*. Milwaukee, Wis.: National Reading Conference, 1966.

Singer, H. Teaching word recognition. In S. Rausch (Ed.), *The volunteer tutor*. Newark, Del.: International Reading Association, 1968.

Singer, H. Conceptualization in learning to read. In H. Singer & R. B. Ruddell (Eds.), *Theoretical models and processes of reading*. Newark, Del.: International Reading Association, 1976. (a)

Singer, H. Substrata-factor theory of reading: Theoretical design for teaching reading. In H. Singer & R. B. Ruddell (Eds.), *Theoretical models and processes of reading*. Newark, Del.: International Reading Association, 1976. (b)

Singer, H. IQ is and is not related to reading. In S. Wanat (Ed.), *Issues in evaluating reading*. Arlington, Va.: Center for Applied Linguistics, 1977. (a)

Singer, H. Resolving curricular conflicts in the 1970's: Modifying the hypothesis, "It's the teacher who makes the difference in reading achievement." *Language Arts*, 1977, *54*, 158-163. (b)

Singer, H. Developmental changes in reading instruction: From learning to read to learning from text. *The Florida Reading Quarterly*, 1978, *14*, 10-13.

Singer, H. Review of "Reading research: Advances in theory and practice" (Vol. I by T. G. Waller & G. E. MacKinnon). *The Reading Teacher*, 1981, *35*, 114-119.

Singer, H. An integration of instructional approaches for teaching reading and learning from text. In L. Reed & S. Ward (Eds.), *2. Basic skills: Issues and choices*. St. Louis, Mo.: CEMREL, Inc., 1982.

Singer, H. The substrata-factor theory of reading: Its history and conceptual relationship to interaction theory. In L. Gentile, M. Kamil, & J. Blanchard (Eds.), *Reading research revisited*. Columbus, Oh.: Merrill, 1983.

Singer, H., & Beasley, S. Motivating a disabled reader. In M. P. Douglass (Ed.), *Claremont Reading Conference 34th Yearbook*. Claremont, Calif.: Claremont Graduate School, 1970.

Singer, H., & Donlan, D. *Reading and learning from text*. Boston, Mass.: Little, Brown, 1980.

Singer, H., & Lucas, M. Dialect in relation to oral reading achievement: Recoding, encoding, or merely a code? *Journal of Reading Behavior*, 1975, *7*, 137-148.

Singer, J. B., & Flavell, J. H. Development of knowledge about communication. Children's evaluations of explicitly ambiguous messages. *Child Development*, 1981, *52*, 1211-1215.

Skinner, B. F. *Verbal behavior*. New York, NY: Appleton-Century-Crofts, 1957.

Skousen, R. English spelling and phonemic representation. *Visible Language*, 1982, *16*, 28-38.

Slobin, D. J. A case study of early language awareness. In A. Sinclair, J. Jarvella, & W. J. M. Levelt (Eds.), *The child's conception of language*. New York, NY: Springer-Verlag, 1978.

Smith, F. *Understanding reading*. New York, NY: Holt, Rinehart, & Winston, 1971.

Smith, F. Phonology and orthography: Reading and writing. *Elementary English*, 1972, *49*, 1075-1088.

Smith, F. *Psycholinguistics and reading*. New York, NY: Holt, Rinehart, & Winston, 1973.

Smith, N. B. *American reading instruction*. Newark, Del.: International Reading Association, 1965.

Snowling, M. J. The development of grapheme-phoneme correspondence in normal and dyslexic readers. *Journal of Experimental Child Psychology*, 1980, *29*, 294-305.

Southgate, V. *Beginning reading*. London, England: University of London Press, 1972.

Sperling, G. The information available in brief visual presentations. *Psychological Monographs*, 1960, *74* (No. 498).

Standish, E. J. Readiness to read. *Educational Research*, 1959, *2*, 29-38.

Stanovich, K. E. Toward an interactive-compensatory model of individual differences in the development of reading fluency. *Reading Research Quarterly*, 1980, *16*, 32-71.

Stauffer, R. G. *The language-experience approach to the teaching of reading* (2nd ed.). New York, NY: Harper & Row, 1980.

Stotsky, S. L. Sentence-combining as a curricular activity: Its effect on written language development and reading comprehension. *Research in the Teaching of English*, 1975, *9*, 30-71.

Stott, D. H. Teaching reading: The psycholinguistic invasion. *Reading*, 1981, *15* (3), 19-25.

Strong, W. *Sentence combining: A composing book*. New York, NY: Random House, 1973.

Takahashi, B. L. Comprehension of written syntactic structures by good readers and slow readers. Unpublished Master's thesis, Rutgers University, 1975. (ERIC Document Reproduction Service No. ED 117 655).

Templeton, S. Spelling first, sound later: The relationship between orthography and higher order phonological knowledge in older students. *Research in the Teaching of English*, 1979, *13*, 255-264.

Tempo Center for Advanced Studies. *Evaluation of Title I ESEA-compensatory education* (GE 71 TMP-23). Santa Barbara, Calif.: General Electric, 1971.

Tenezakis, M. Linguistic subsystems and cognitive operations. *Child Development*, 1975, *46*, 430-436.

Thackray, D. V. A study of the relationship between some specific evidence of reading readiness and reading progress in the infant school. *British Journal of Educational Psychology*, 1965, *35*, 252-254.

Thackray, D. V. *Readiness to read with i.t.a. and t.o.* London, England: Geoffrey Chapman, 1971.

Thorndike, E. L. Reading as reasoning: A study of mistakes in paragraph reading. *Journal of Educational Psychology*, 1917, *8*, 323-332.

Thorndike, R. L. Reading as reasoning. *Reading Research Quarterly*, 1974, *9*, 135-147.

Tinker, M. A. Recent studies of eye-movements in reading. *Psychological Bulletin*, 1958, *55*, 215-231.

Tinker, M. A. *Bases for effective reading*. Minneapolis, Minn.: University of Minnesota Press, 1965.

Tomlinson-Keasey, C., Eisert, D. C., Kahle, D. C., Hardy-Brown, K., & Keasey, B. The structure of concrete operational thought. *Child Development*, 1979, *50*, 1153-1163.

Tovey, D. R. Children's perceptions of reading. *The Reading Teacher*, 1976, *29*, 536-540.

Treiman, R. A. *Children's ability to segment speech into syllables and phonemes as related to their reading ability*. New Haven, Conn.: Yale University, Department of Psychology, 1976.

Treiman, R., & Baron, J. Segmental analysis ability: Development and relation to reading ability. In G. E. MacKinnon & T. G. Waller (Eds.), *Reading re-*

search: Advances in theory and practice (Vol. 3). New York, NY: Academic Press, 1981.

Turnbull, K. Children's thinking: When is a letter a number? *Curriculum and Research Bulletin* (Victoria, Australia), 1970, pp. 126-131.

Tzeng, O. J. L., & Hung, D. Reading in the non-alphabetic writing system. In O. Tzeng & H. Singer (Eds.), *Perception of print: Reading research in experimental psychology*. Hillsdale, NJ: Erlbaum, 1981. Also in J. F. Kavanagh & R. L. Venezky (Eds.), *Orthography, reading, and dyslexia*. Baltimore, Md.: University Park Press, 1980.

Tzeng, O. J. L., Hung, D. L., & Wang, W. S.-Y. Speech recoding in reading Chinese characters. *Journal of Experimental Psychology: Human Perception and Performance*, 1977, *3*, 621-630.

Tzeng, O. J. L., & Singer, H. *Perception of print: Reading research in experimental psychology*. Hillsdale, NJ: Erlbaum, 1981.

Valtin, R. Increasing awareness of linguistic awareness in research on beginning reading and dyslexia. Paper presented at the University of Victoria/IRA Research Seminar on Linguistic Awareness and Learning to Read, Victoria, B.C., 1979.

Valtin, R. Deficiencies in research on reading deficiencies. In J. F. Kavanagh & R. L. Venezky (Eds.), *Orthography, reading, and dyslexia*. Baltimore, Md.: University Park Press, 1980.

Valtin, R. Zur "Machbarkeit" der Ergebnisse der Legasthenieforschung. In R. Valtin, U. Jung, & G. Scheerer-Neumann (Eds.), *Legasthenie in Wissenschaft und Unterricht*. Darmstadt, Germany: Wissenschaftliche Buchgesellschaft, 1981.

Valtin, R. Probleme der Erfassung sozial-kognitiver Fähigkeiten-analysiert am Beispiel der Perspektivenübernahme und der verbalen Kommunikation. In D. Geulen (Ed.), *Perspektivenübernahme und Soziales Handeln*. Frankfurt, Germany: Suhrkamp, 1982. (a)

Valtin, R. The disabled readers: What are their weaknesses in language and in the reading process? In K. Tuunainen & A. Chiaroni (Eds.), *Full participation. Proceedings of the second European conference on reading. Joensuu, 1981.* Joensuu, Finland: University of Joensuu, 1982. (b)

Valtin, R., Dovifat, S., & Thomalla, M. An interview study of children's understanding of communication rules. Research Rept., Berlin Free University, Germany, 1982. (ERIC Document Reproduction Service, No. ED 210 722).

Van Dongen, D. *Preventie van leesmoeilijkheden, Deelrapport 3, Opzet van het longitudinale onderzoek*. Nijmegen, The Netherlands: Instituut voor Onderwijskunde, 1979.

Van Dongen, D., Bosch, R., & Mommers, M. *Preventie van leesmoeilijkheden, Deelrapport 5. Eerste analyses op de data van het longitudinale onderzoek. Analyses op het niveau van de afzonderlijke meetinstrumenten voor het eerste leerjaar of voor de periode daaraan voorafgaand*. Nijmegen, The Netherlands: Instituut voor Onderwijskunde, 1981.

Van Dongen, D., & Van Leent, H. *Preventie van leesmoeilijkheden, Deelrapport 4. Opzet van het eerste deel van het exploratieve onderzoek*. Nijmegen, The Netherlands: Instituut voor Onderwijskunde, 1981.

Van Dongen, D., & Wolfhage, I. Predictability of reading ability. Unpublished manuscript. Instituut voor Onderwijskunde, Nijmegen, The Netherlands, 1982.

Van Leent, H. Het exploratieve onderzoek naar de ontwikkeling van de lees-vaardigheid in het eerste leerjaar van de lagere school. Paper presented at the Onderwijs Research Dagen, Tilburg, The Netherlands, 1982.

Van Leent, H. Auditieve analyse en leren lezen. *Pedagogische Studiën*, 1983, *60*, 13-27.

Van Metre, P. D. Syntactic characteristics of selected bilingual children. In M. P. Douglass (Ed.), *Claremont reading conference 38th yearbook.* Claremont, Calif.: Claremont Graduate School, 1974.

Vellutino, F. R. Alternate conceptualizations of dyslexia: Evidence in support of a verbal-deficit hypothesis. *Harvard Educational Review*, 1977, *47*, 334-354.

Vernon, M. D. *Backwardness in reading.* Cambridge, England: Cambridge University Press, 1957.

Vernon, M. D. *Reading and its difficulties.* Cambridge, England: Cambridge University Press, 1971.

Vogel, S. A. Syntactic abilities in normal and dyslexic children. *Journal of Learning Disabilities*, 1974, *7*, 103-109.

Vygotsky, L. S. *Thought and language.* Cambridge, Mass.: MIT Press, 1962. (Originally published in Russian, 1934.)

Wallach, M. A., & Wallach, L. *Teaching all children to read.* Chicago, Ill.: University of Chicago Press, 1976.

Waller, G. *Think first, read later. Piagetian prerequisites for reading.* Newark, Del.: International Reading Association, 1977.

Wang, W. S.-Y. Language structure and optimal orthography. In O. J. L. Tzeng & H. Singer (Eds.), *Perception of print: Reading research in experimental psychology.* Hillsdale, NJ: Erlbaum, 1981.

Waterhouse, L. H. The implications of theories of language and thought for reading. In F. B. Murray (Ed.), *Language awareness and reading.* Newark, Del.: International Reading Association, 1980.

Watson, A. J. Multiple seriation and learning to read. *Australian Journal of Education*, 1979, *23*, 171-180.

Waugh, N. C., & Norman, D. H. Primary memory. *Psychological Review*, 1965, *80*, 1-52.

Weaver, P. A. Improving reading comprehension: Effects of sentence organization instruction. *Reading Research Quarterly*, 1979, *15*, 129-146.

Weaver, P., & Shonkoff, F. *Research within reach: A research-guided response to the concerns of reading educators.* Newark, Del.: International Reading Association, 1979.

Weber, R. M. A linguistic analysis of first-grade reading errors. *Reading Research Quarterly*, 1970, *5*, 427-451. (a)

Weber, R. M. First-graders' use of grammatical context in reading. In H. Levin & J. Williams (Eds.), *Basic studies on reading.* New York, NY: Basic Books, 1970. (b)

Weed, K., & Ryan, E. B. Alphabetic seriation as a reading readiness indicator. Unpublished manuscript, University of Notre Dame, Ind., 1982.

Weinstein, R., & Rabinovitch, S. Sentence structure and retention in good and poor readers. *Journal of Educational Psychology*, 1971, *62*, 25-30.

Weintraub, S., & Denny, T. P. What do beginning first-graders say about reading? *Childhood Education*, 1965, *41*, 326-327.

White, C. V., Pascarella, E. T., & Pflaum, S. W. Effects of training in sentence construction on the comprehension of learning disabled children. *Journal of Educational Psychology*, 1981, *73*, 697-704.

White, S. H. Evidence for a hierarchical arrangement of learning processes. *Advances in Child Development and Behavior*, 1965, *2*, 187-220.

White, S. H. Some general outlines of the matrix of developmental changes between five and seven years. *Bulletin of the Orton Society*, 1970, *20*, 41-51.

Whiting, H. T. A. *Concepts in skill learning*. London, England: Lepus, 1975.

Whiting, H. T. A., & den Brinker, B. Image of the act. In J. P. Das, R. F. Mulcahy, & A. E. Wall (Eds.), *Theory and research in learning disabilities*. New York, NY: Plenum Press, 1982.

Wickelgren, W. A. Distinctive features and short-term memory for English vowels. *Journal of the Acoustical Society of America*, 1965, *38*, 583-588. (a)

Wickelgren, W. A. Short-term memory for phonemically similar lists. *American Journal of Psychology*, 1965, *78*, 567-574. (b)

Wickelgren, W. A. Distinctive features and errors in short-term memory for English consonants. *Journal of the Acoustical Society of America*, 1966, *39*, 388-398.

Wilkins, J. *An essay towards a real character and a philosophical language*. London, England: S. Gellibrand, 1668.

Will-Beuermann, H., Hinnrichs, J., & Valtin, R. *Lehrerhandbuch zur Bunten Fibel*. Hannover, Germany: Schroedel Verlag, 1980.

Williams, A. C., & Flexman, R. E. Evaluation of the school link as an aid to primary flight instruction. *University of Illinois Bulletin*, 1949, *46* (No. 71).

Williams, J. P. Linguistic awareness and reading. Paper presented at the University of Victoria/IRA Research Seminar on Linguistic Awareness and Learning to Read, Victoria, B.C., 1979.

Williams, J. P. Teaching decoding with an emphasis on phoneme analysis and phoneme blending. *Journal of Educational Psychology*, 1980, *72*, 1-15.

Willows, D. M., & Ryan, E. B. The differential utilization of contextual cues by skilled and less skilled readers in the intermediate grades. *Journal of Educational Psychology*, 1981, *73*, 607-615.

Willows, D. M., & Ryan, E. B. The role of a linguistic factor in the development of reading skill. Paper presented at the annual meeting of the American Educational Research Association, Montreal, Quebec, 1983.

Wilson, L. City kids: A multilingual experience-based program in Australia. *The Reading Teacher*, 1979, *32*, 674-676.

Wiseman, S. Curriculum evaluation, *Mimeograph*, 21, 1966.

Wohlwill, J. F. *The study of behavioral development*. New York, NY: Academic Press, 1973.

Wolfle, L. M. Strategies of path analysis. *American Educational Research Journal*, 1980, *17*, 182-209.

Woodcock, R. W., Clark, C. R., & Davies, C. O. *The Peabody rebus reading program*. Circle Pines, Minn.: American Guidance Service, 1969.

Zei, B. Psychological reality of phonemes. *Journal of Child Language*, 1979, *6*, 375-381.

Zhurova, L. Y. The development of analysis of words into their sounds by preschool children. In C. A. Ferguson & D. I. Slobin (Eds.), *Studies of child language development.* New York, NY: Holt, Rinehart, & Winston, 1973.

Zifcak, M. Phonological awareness and reading acquisition. *Contemporary Educational Psychology*, 1981, *6*, 117-126.

Author Index

Subject Index

Springer Series in Language and Communication

Continued from page ii